Electrical and Electronics Engineering: Principles, Technologies and Applications

Electrical and Electronics Engineering: Principles, Technologies and Applications

Edited by John Fenmore

CLANRYE INTERNATIONAL
www.clanryeinternational.com

Clanrye International,
750 Third Avenue, 9th Floor,
New York, NY 10017, USA

ISBN: 978-1-63240-666-8

Cataloging-in-Publication Data

Electrical and electronics engineering : principles, technologies and applications / edited by John Fenmore.
   p. cm.
Includes bibliographical references and index.
ISBN 978-1-63240-666-8
1. Electrical engineering. 2. Electronics. 3. Electronics--Design. I. Fenmore, John.
TK145 .E44 2018
621.3--dc23

For information on all Clanrye International publications
visit our website at www.clanryeinternational.com

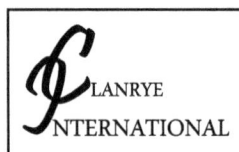

CLANRYE
INTERNATIONAL

Contents

Preface

The branch of engineering which focuses on the practical use of electricity, and studies the designing and maintenance of electrical devices is known as electrical engineering. It has a number of subdisciplines like instrumentation, electronics, telecommunication, signal processing, etc. This book outlines the processes and applications of electrical and electronics engineering in detail. Coherent flow of topics, student-friendly language and extensive use of examples make this book an invaluable source of knowledge. It aims to serve as a resource guide for students and experts alike and contribute to the growth of the discipline.

All of the data presented henceforth, was collaborated in the wake of recent advancements in the field. The aim of this book is to present the diversified developments from across the globe in a comprehensible manner. The opinions expressed in each chapter belong solely to the contributing authors. Their interpretations of the topics are the integral part of this book, which I have carefully compiled for a better understanding of the readers.

At the end, I would like to thank all those who dedicated their time and efforts for the successful completion of this book. I also wish to convey my gratitude towards my friends and family who supported me at every step.

Editor

An efficient fractional-pixel motion compensation based on Cubic convolution interpolation

Lung-Jen Wang[*], **Chia-Tzu Shu**

Dept. of Computer Science and Information Engineering, National Pingtung University, Pingtung, Taiwan, R. O. C.

Email address:
ljwang@mail.nptu.edu.tw (Lung-Jen Wang)

Abstract: The fractional-pixel motion compensation is used in the H.264/AVC algorithm, in order to improve the coding efficiency of fractional-pixel displacement, an efficient cubic convolution interpolation (CCI) with four coefficients is proposed. In this paper, the detailed derivation of the CCI filter and using CCI with fractional-pixel displacement are presented. It is shown by computer simulation that the presented method substantially reduces the computation complexity and also increases the precision of the motion compensation.

Keywords: H.264/AVC, Motion Compensation, Fractional-Pixel Displacement, Cubic Convolution Interpolation

1. Introduction

A video can be viewed as a time-ordered sequence of images-frames [1]. In general, the volume of uncompressed video data is so large that the use of video compression is almost mandatory. In High Definition TV (HDTV), if uncompressed, the bitrate could easily exceed 1Gbps. Therefore, video compression allows the video to be transmitted over the Internet in real time. Also it reduces the requirements for video storage.

The H.264/AVC algorithm is one of the latest international standard for the video compression technique, which was jointly implemented by ITU-T video coding experts group (VCEG) and ISO/IEC motion picture experts group (MPEG) [2][3]. In order to improve the precision for motion compensation prediction in the H.264/ AVC algorithm [4]-[8], the fractional-pixel displacement is calculated, which is used the interpolation process to estimate the fractional-pixel (1/2- and 1/4-) positions between the existing positions. In the H.264/ AVC standard, there are two processing steps in the fractional-pixel displacement. First, a fixed 6-tap Wiener filter is calculated for the 1/2-pixel displacement, and then the bilinear interpolation is used for the 1/4-pixel displacement. The disadvantage of the H.264/ AVC standard is that the computations required for 1/2-pixel displacement with 6-tap Wiener filter are substantially increased. The 6-tap Wiener filter requires not only huge computational cost, but also its accuracy is not guaranteed in the fractional-pixel

displacement [7].

Interpolation is the process of estimating the intermediate values of a continuous event from discrete samples. It is used extensively in image processing to magnify or reduce images and to correct spatial distortions. Because of the amount of image data, an efficient interpolation algorithm is essential. Rigorously speaking, the process of decreasing the data rate is called decimation and increasing the data samples is termed interpolation [9]-[11][16]-[18]. It is well known that several decimation and interpolation functions such as linear interpolation [10], cubic convolution interpolation [11], cubic B-spline interpolation [16], linear spline interpolation [17], cubic spline interpolation [9][18], etc. can be used in the image processing.

The authors proposed a cubic convolution interpolation (CCI) with four coefficients in [12], in order to reduce the computation complexity of the fractional-pixel displacement in the H.264/AVC standard. This paper proposes more detailed descriptions for the CCI and the fractional-pixel motion compensation prediction for the H.264/ AVC algorithm. That is, the detailed derivation of CCI and combination with motion compensation are presented in this paper. Finally, experimental results show that the proposed CCI method is used to speed up the 6-tap Wiener filter in the H.264/AVC standard and still obtain a superior performance for motion compensation. The primary advantage of the CCI method is that it increases the precisions and also substantially reduces the computation complexity for the fractional-pixel displacement.

This paper is organized as follows. Section 2 describes the background of this work for the interpolation function, H.264/AVC algorithm and interpolation in fractional-pixel displacement. Then the proposed CCI filter with four coefficients (4-tap CCI) is presented in Section 3. In this section, the cubic convolution function and 4-tap CCI are discussed in detail. In Section 4, the proposed CCI combined with motion compensation is illustrated for the H.264/AVC algorithm. The motion compensation with CCI and the CCI interpolation computation are also described in detail. Finally, experimental results and conclusions are discussed in Sections 5 and 6, respectively.

2. Background of this Work

2.1. Interpolation Functions

An interpolation function is a special type of approximating function. A fundamental property of interpolation functions is that they must coincide with the sampled data at the interpolation nodes, or sample points. In other words, if f is a sampled function, and if $\hat{f}$ is the corresponding interpolation function, then $\hat{f}(x_k) = f(x_k)$ whenever x_k is an interpolation node [11]. Thus, an interpolation function can be expressed in the following form.

$$\hat{f}(x) = \sum_{k=1}^{N} c_k R_k(x - x_k) \qquad (1)$$

where $c_k = f(x_k)$ are the coefficients to be determined from the input data, $R_k(x)$ is the chosen interpolation basis function, x and x_k represent continuous and discrete value, respectively, and N is the number of given data points. There are several interpolation functions, shown in Fig. 1, such as linear function, cubic B-spline function, sinc function, and cubic convolution function [10][11].

The linear function provides the first-order linear sample interpolation with triangle-shaped interpolation waveforms. This linear function may be considered to be the result of convolving a square function with itself. It requires only one addition and one shift in each interpolation. The convolution of the linear function with itself yields a cubic B-spline function. The cubic B-spline function is a particularly attractive candidate for image interpolation because of its properties of continuity and smoothness at the sample points [10]. The sinc convolution function uses many sample points for each interpolation point. It provides an exact reconstruction, but it cannot be physically generated by an incoherent optical filtering system [10]. It is possible to approximate the sinc function by truncating its tails. That is to say if we only consider the sinc function with a four sample interval. The problem is that the slope discontinuity at the ends of the sinc waveform will lead to amplitude ripples in a reconstructed function. In order to eliminate amplitude ripples in a reconstructed function, the cubic convolution interpolation is developed to force the slope of the ends of interpolation to be zero [11].

2.2. The H.264/AVC Algorithm

The H.264/AVC standard is the latest algorithm of video compression technique [2][3]. The block diagram of the H.264/AVC encoder is shown in Fig. 2. In this figure, the input video frame is partitioned into some blocks. For each block of input video frame, the H.264/AVC encoder applies either intra-frame prediction or inter-frame prediction to process the video signal. In the intra-frame prediction, each block is converted by the DCT transform into the DCT coefficients. Next the quantization (Q) and the entropy coding (ENC) are used for the DCT coefficients. In the inter-frame prediction, which is also called as the motion compensated prediction or temporal prediction, the motion estimation uses the current frame block from reference frames to predict the content of the current frame block [15]. Furthermore, the de-quantization (Q^{-1}) and the inverse DCT (IDCT) transform are used to obtain the reconstructed video frame.

2.3. Interpolation in Fractional-Pixel Displacement

In the H.264/AVC algorithm, the displacement vectors with fractional-pixel resolution are applied to perform the motion compensated prediction. In order to estimate and compensate the fractional-pixel displacement, the two-step interpolation process is used in [7]. In H.264/AVC, the block diagram of the two-step interpolation process is shown in Fig. 3. In the first step, the sampling rate of image I is increased by a factor of 2 and filtered by the 6-tap Wiener filter with the six coefficients: [1, -5, 20, 20, -5, 1]/32 is used to interpolate the 1/2-pixel positions in this step and an image I^{w} is generated after the first step. In the second step, the sampling rate of the resulting image I^{w} is also increased by a factor of 2 and filtered by a simple 2-tap bilinear interpolation filter with the

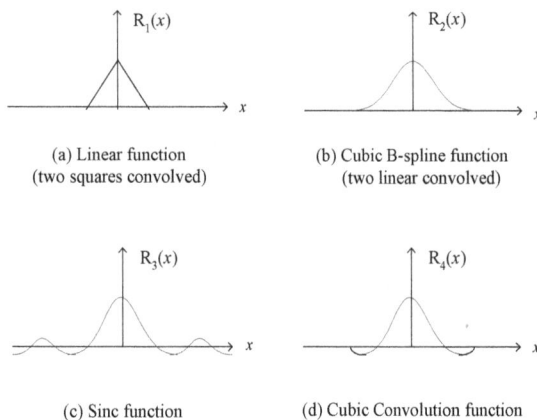

(a) Linear function
(two squares convolved)

(b) Cubic B-spline function
(two linear convolved)

(c) Sinc function

(d) Cubic Convolution function

Figure 1. Several interpolation functions.

two coefficients: [1, 1]/2 is used to interpolate the 1/4-pixel positions and an image I' is generated after the second step.

In Fig. 4, the interpolation relationship between the full-pixel positions (black), the 1/2-pixel positions (gray) and the 1/4-pixel positions (white) are shown. In this figure, at first, the 1/2-pixel positions aa, bb, b, q, cc, dd, and ee, ff, h, l, gg, hh are calculated, using a horizontal or vertical 6-tap Wiener filter, respectively. For example, aa = (A1 - 5A2 + 20A3 + 20A4 - 5A5 + A6) / 32, and ee = (A1 − 5B1 + 20C1 + 20D1 − 5E1 + F1) / 32, respectively. Using the same Wiener filter applied at 1/2-pixel positions aa, bb, b, q, cc, and dd, the 1/2-pixel position j is obtained as: j = (aa − 5bb + 20b + 20q − 5cc + dd) / 32. In the second step, the remaining 1/4-pixel positions are computed using the bilinear interpolation filter based on the calculated 1/2-pixel positions and existing full-pixel positions [8]. For example, a − (C3 + b) / 2 and d = (C3 + h) / 2 in horizontal and vertical interpolations, respectively.

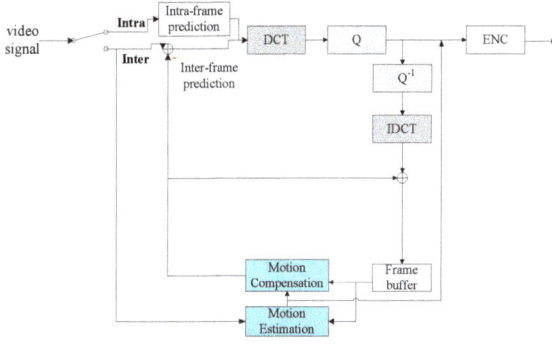

Figure 2. *Block diagram of the H.264/AVC encoder.*

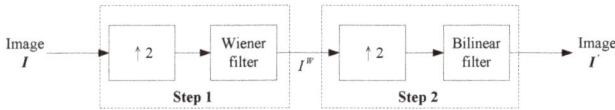

Figure 3. *Two-step interpolation process used in H.264/AVC.*

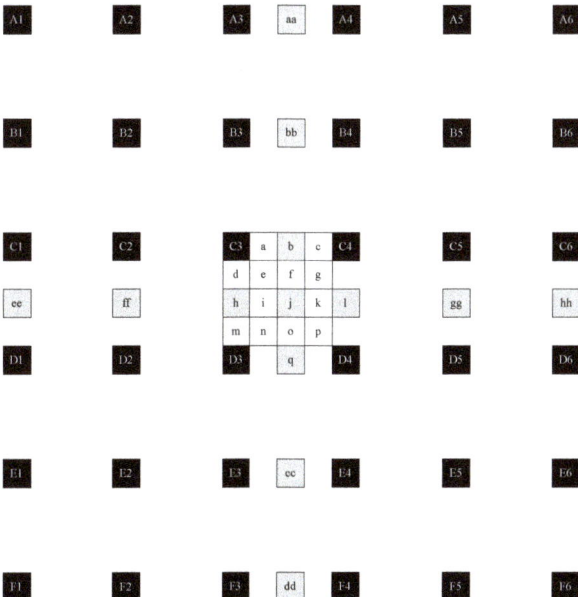

Figure 4. *Interpolation relationship.*

3. Cubic Convolution Interpolation

3.1. The Cubic Convolution Function

The cubic convolution function was given in [11], which is composed of piecewise cubic polynomials defined on the subintervals (-2, -1), (-1, 0), (0, 1), and (1, 2). Outside the interval (-2, 2), the function is zero. The function kernel must be symmetric. As a consequence of this condition, the cubic convolution function is defined by

$$R(t) = \begin{cases} A_1|t|^3 + B_1|t|^2 + C_1|t| + D_1 & , 0 \le |t| < 1 \\ A_2|t|^3 + B_2|t|^2 + C_2|t| + D_2 & , 1 \le |t| < 2 \\ 0 & , 2 \le |t| \end{cases} \quad (2)$$

where $R(0) = 1$, $R(\pm 1) = R(\pm 2) = 0$ and $R(t)$ and $R'(t)$ are continuous at $t = 0, 1, 2$ and $R'(\pm 2) = 0$. In (2), $A_1, B_1, C_1, D_1, A_2, B_2, C_2$ and D_2 are eight unknown coefficients. To solve these coefficients, Keys in [11] supposes $A_2 = a$, then the remaining seven unknown coefficients can be determined in terms of a. That is, the cubic-convolution function in (2) can be expressed in terms of a as

$$R(t) = \begin{cases} (a+2)|t|^3 - (a+3)|t|^2 + 1 & , 0 \le |t| < 1 \\ a|t|^3 - 5a|t|^2 + 8a|t| - 4a & , 1 \le |t| < 2 \\ 0 & , 2 \le |t| \end{cases} \quad (3)$$

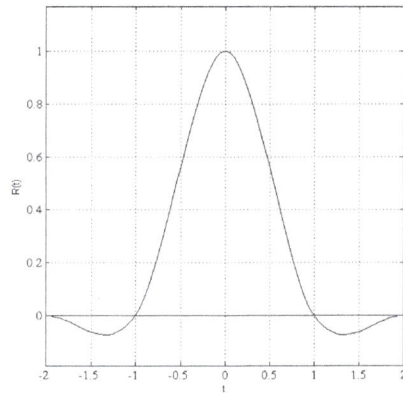

Figure 5. *The CCI function.*

Furthermore, Keys in [11] also uses $A_2 = a = -1/2$, finally, the cubic convolution function, shown in Fig. 5, is given by

$$R(t) = \begin{cases} (3/2)\ |t|^3 - (5/2)\ |t|^2 + 1, & 0 \le |t| < 1 \\ -(1/2)\ |t|^3 + (5/2)\ |t|^2 - 4\ |t| + 2, & 1 \le |t| < 2 \\ 0, & 2 \le |t| \end{cases} \quad (4)$$

For more details on this discussion, see Keys [11].

3.2. 4-Tap Cubic Convolution Interpolation

Let τ be a fixed, positive integer. Also, let $X(t)$ be the full-pixel samples in the current frame, where $0 \leq t \leq n-1$, and $X_0, \cdots, X_{n-1}$ be the n existing full-pixel samples in the reference frame. The shift function of the cubic convolution function can be defined as $R_k(t) = R(t - k\tau)$ for $0 \leq k \leq n-1$. Then the fractional-pixel displacement $\hat{X}(t)$ in the reference frame by the cubic convolution interpolation (CCI) as

$$\hat{X}(t) = \sum_{k=0}^{n-1} X_k R_k(t) = \sum_{k=0}^{n-1} X_k R(t - k\tau) \tag{5}$$

where $R(t)$ is the cubic convolution function in (4). The fractional-pixel function $\hat{X}(t)$ in (5) is the CCI function of the full-pixel samples $X_0, \cdots, X_{n-1}$ in the reference frame. One can be find the function $\hat{X}(t)$ in (5) that minimize the prediction error of $\hat{X}(t)$ to $X(t)$ is defined by

$$e(t) = \sum_{t=0}^{n-1} \left| X(t) - \hat{X}(t) \right| \tag{6}$$

Thus, the sequence of n fractional-pixel values $\hat{X}(t)$ in (5) can be obtained by the CCI function and used for the 1/2-pixel and 1/4-pixel displacements, respectively, in the H.264/AVC motion compensation. Furthermore, the fractional-pixel displacement $\hat{X}(t_a)$ between the two adjacent existing full-pixel samples X_k and X_{k+1} in the reference frame is illustrated in Fig. 6 and given by the sum of the CCI function,

$$\hat{X}(t_a) = X_{k-1} R_{k-1}(t_a) + X_k R_k(t_a) + X_{k+1} R_{k+1}(t_a) \\ + X_{k+2} R_{k+2}(t_a) \quad \text{for } t_k < t_a < t_{k+1} \tag{7}$$

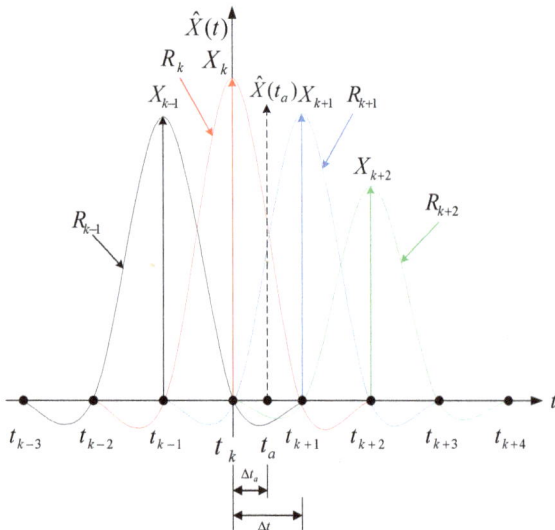

Figure 6. The fractional-pixel interpolation between samples.

In (7) and Fig. 6, let $\Delta t = \left| t_{k+1} - t_k \right|$ and $\Delta t_a = t_a - t_k$, then the displacement from t_a to t_{k-1}, t_{k+1} and t_{k+2} are $t_a - t_{k-1} = \Delta t_a + \Delta t$, $t_a - t_{k+1} = \Delta t_a - \Delta t$, and $t_a - t_{k+2} = \Delta t_a - 2\Delta t$, respectively.

Table 1. List of CCI coefficients for fractional-pixel displacement

fractional-pixel displacement	CCI coefficients
1/2-pixel displacement	[-1, 9, 9, -1]/16
1/4-pixel displacement	[-9, 111, 29, -3]/128
3/4-pixel displacement	[-3, 29, 111, -9]/128

In order to calculate the fractional-pixel displacement, if we set $\Delta t = 1$ and $\Delta t_a = 1/2$, then the 1/2-pixel displacement can be given by

$$\hat{X}(t_a) = \frac{1}{16}\left(-X_{k-1} + 9X_k + 9X_{k+1} - X_{k+2}\right) \tag{8}$$

Moreover, if we set $\Delta t = 1$ and $\Delta t_a = 1/4$, then the 1/4-pixel displacement can be given by

$$\hat{X}(t_a) = \frac{1}{128}\left(-9X_{k-1} + 111X_k + 29X_{k+1} - 3X_{k+2}\right) \tag{9}$$

That is, if we set $\Delta t = 1$ and $\Delta t_a = 3/4$, then the 3/4-pixel displacement can be given by

$$\hat{X}(t_a) = \frac{1}{128}\left(-3X_{k-1} + 29X_k + 111X_{k+1} - 9X_{k+2}\right) \tag{10}$$

In (8)-(10), the CCI filter with four coefficients (4-tap CCI) for the fractional-pixel (1/2-, 1/4- and 3/4-) displacements are summarized in Table 1.

4. Using CCI in Motion Compensation

4.1. Motion Compensation with CCI

In the fractional-pixel motion compensation, there are two processing steps in the H.264/AVC algorithm, shown in Fig. 3. In the first processing step, a fixed 6-tap Wiener filter is used for the 1/2-pixel displacement. The second processing step is to use the bilinear interpolation for the 1/4-pixel displacement. In this paper, the CCI filter with four coefficients (4-tap CCI) is proposed to reduce the computational complexity of the Wiener filter with six coefficients in the 1/2-pixel displacement. In Fig. 7, the first interpolation step is filtered by the 4-tap CCI process for the 1/2-pixel displacement, and the second interpolation step is then filtered by the 2-tap bilinear process for the 1/4-pixel displacement. This is also our proposed method and labeled as CCI+Bilinear. In Fig. 8, two processing steps are all filtered by the 4-tap CCI interpolation for both 1/2- and 1/4-pixel displacements. This method is also labeled as CCI.

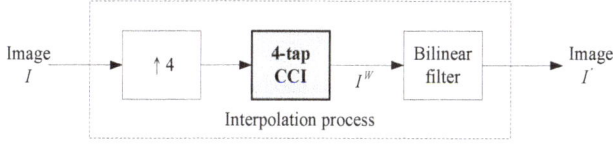

Figure 7. *The proposed CCI+Bilinear process.*

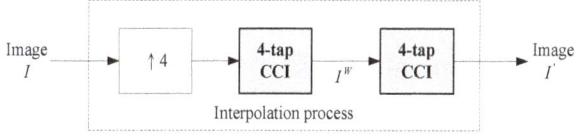

Figure 8. *The CCI process.*

4.2. The CCI Interpolation Computation

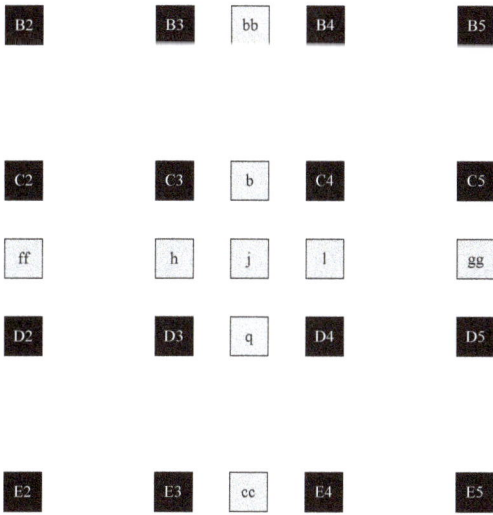

Figure 9. *1/2-pixel CCI operations: full-pixel positions (black); 1/2-pixel positions (gray).*

In Fig. 9, the 4-tap CCI interpolation is used to interpolate the 1/2-pixel positions and the 1/2-pixel CCI operations are illustrated: full-pixel positions (black); 1/2-pixel positions (gray). For example, b = (-C2 + 9C3 + 9C4 - C5) / 16. In Figs. 10 and 11, the 4-tap CCI interpolation is also applied to interpolate the 1/4-pixel and 3/4-pixel positions in horizontal and vertical interpolations, respectively. Fig. 10 shows the 1/4-pixel and 3/4-pixel CCI operations in horizontal interpolations: full-pixel (black); 1/2-pixel (gray); 1/4-pixel (white); 3/4-pixel (blue). For example, a = (-9C2 + 111C3 + 29C4 – 3C5) / 128, and c = (-3C2 + 29C3 + 111C4 – 9C5) / 128, respectively. In addition, Fig. 11 shows the 1/4-pixel and 3/4-pixel CCI operations in vertical interpolations: full-pixel (black); 1/2-pixel (gray); 1/4-pixel (white); 3/4-pixel (blue). For example, d = (-9B3 + 111C3 + 29D3 – 3E3) / 128, and m = (-3B3 + 29C3 + 111D3 – 9E3) / 128, respectively.

5. Experimental Results

The proposed CCI+Bilinear and CCI methods, which are described in Section 4 and shown in Figs. 7 and 8, respectively, are implemented in Microsoft visual C++ program and compared with the H.264/AVC standard method (labeled as

Standard), which Wiener filter and bilinear filter are used for the fractional-pixel displacement. These three algorithms are based on the JM 18.0 [13]. For the simulation common conditions of the H.264/AVC algorithm, the most important settings are summarized in Table 2.

The objective performance is measured by the peak signal to noise ratio (PSNR) and given by

$$Y\ PSNR = 10\ log_{10}\left(\frac{255^2}{MSE_Y}\right)\ (dB),\qquad(11)$$

where MSE_Y is the mean-square error between the original Y image and reconstructed Y image for the YUV format. Some standard QCIF (Bridge(close), Crew, HarBour, Highway), CIF (Bridge(close), Bridge(far), Foreman) and 4CIF (City, Crew, HarBour) frame sequences are selected and shown in Fig. 12. Performance comparisons are carried out on the above these frame sequences. All the experimental results (bit rates and PSNR performance) are computed from the reconstructed Y images.

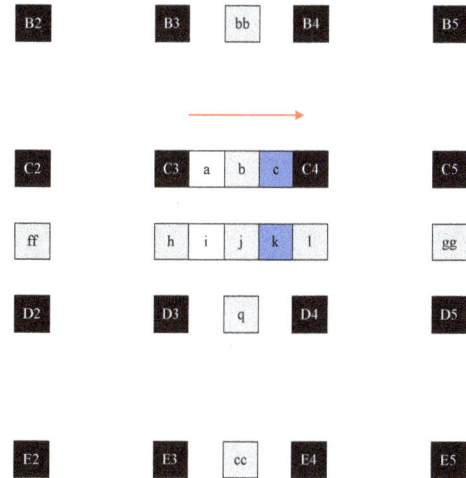

Figure 10. *1/4-pixel and 3/4-pixel CCI operations in horizontal interpolation: full-pixel (black); 1/2-pixel (gray); 1/4-pixel (white); 3/4-pixel (blue).*

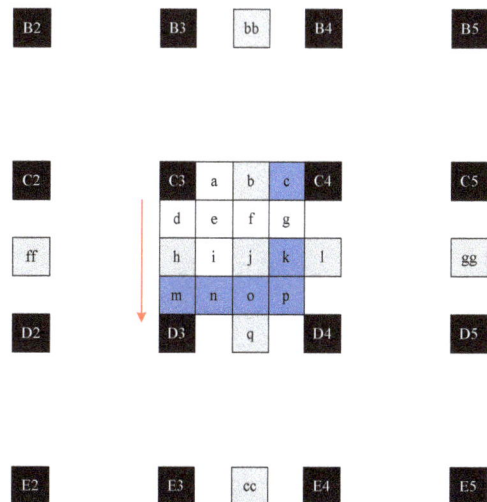

Figure 11. *1/4-pixel and 3/4-pixel CCI operations in vertical interpolation: full-pixel (black); 1/2-pixel (gray); 1/4-pixel (white); 3/4-pixel (blue).*

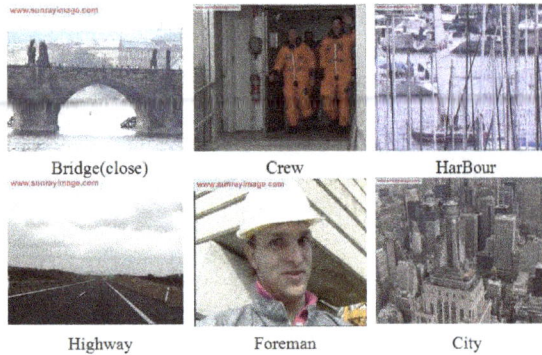

Figure 12. Some standard video test sequences.

Table 2. Most important H.264/AVC coder settings [14]

Parameter	Settings
Profile:	High
Number of reference images:	4
Number of B coded frames:	2
Sequence type:	I-B-B-P-B-B-P
Hierarchical Coding:	off
Weighted prediction:	on
Rate-distortion optimization:	on
Search range:	32 for QCIF and CIF
	64 for 720p and 1080p
YUV format	4:2:0
Quantization parameter (I/P/B):	22/23/24, 27/28/29,
	32/33/34, 37/38/39

For the computational complexity of the fractional-pixel displacement in the Standard (H.264/AVC), CCI+Bilinear and CCI methods, the number of addition(+), multiplication(*), and shift(>>) are estimated and compared in Table 3 and Table 4. Obviously, in Table 3, the estimated operation number of the proposed CCI+Bilinear method is less than those of both Standard and CCI methods. That is, the proposed CCI with bilinear (CCI+Bilinear) method is more fast and efficient than those of the H.264/AVC standard (Standard) and the CCI methods. In other words, for the number of operations, CCI+Bilinear < Standard < CCI.

In Table 4, in terms of the number of addition(+), multiplication(*) and shift(>>) operations, the CCI+Bilinear method is less than the H.264/AVC standard (Standard)

method by 152064, 76032 and 76032 for QCIF sequence, by 608256, 304128 and 304128 for CIF sequence, and by 2433024, 1216512 and 1216512 for 4CIF sequence, respectively. Thus, the proposed CCI+Bilinear method achieves a superior performance in the fractional-pixel displacement than the H.264/AVC standard method.

Table 3. Number of operations for three methods.

Methods	Operations	QCIF	CIF	4CIF
Standard	+	684288	2737152	10948608
(H.264/AVC)	*	228096	912384	3649536
	>>	684288	2737152	10948608
CCI+Bilinear	+	532224	2128896	8515584
	*	152064	608256	2433024
	>>	608256	2433024	9732096
CCI	+	1140480	4561920	18247680
	*	1368576	5474304	21897216
	>>	2433024	9732096	38928384

Some experimental results of the rate-distortion curves for the Standard (H.264/AVC), CCI+Binear and CCI methods are compared and shown in Figs. 13, 14 and 15. As shown in these three figures, the proposed CCI+Bilinear and CCI methods obtain the better PSNR values of reconstructed Y image than the Standard (H.264/AVC) method. That is, for the same bit rates, the PSNR values of the Y images, obtained by the CCI+Bilinear and CCI methods are higher than the Standard (H.264/AVC) method by 0.05 dB and 0.02 dB for QCIF sequence, by 0.04 dB and 0.02 dB for CIF sequence, and by 0.14 dB and 0.06 dB for 4CIF sequence, respectively.

Table 4. Comparison of operations for Standard and CCI+Bilinear.

Methods	Operations	QCIF	CIF	4CIF
Standard	+	684288	2737152	10948608
(H.264/AVC)	*	228096	912384	3649536
	>>	684288	2737152	10948608
CCI+Bilinear	+	532224	2128896	8515584
	*	152064	608256	2433024
	>>	608256	2433024	9732096
"Standard"	+	152064	608256	2433024
minus	*	76032	304128	1216512
"CCI+Bilinear"	>>	76032	304128	1216512

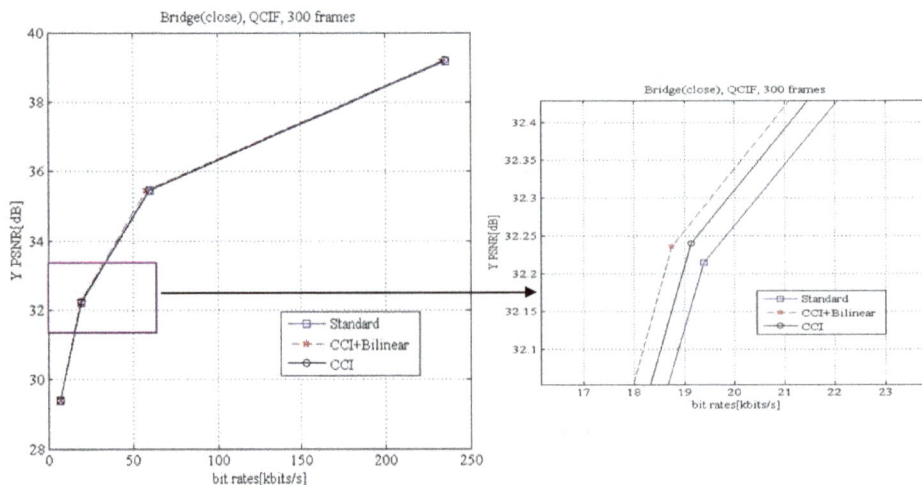

Figure 13. Rate-distortion comparison for QCIF sequence Bridge(close).

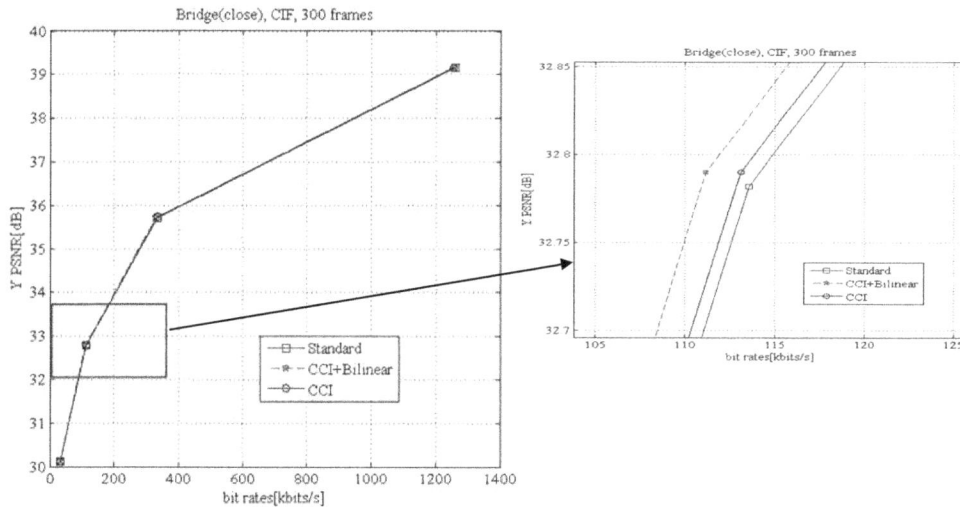

Figure 14. Rate-distortion comparison for CIF sequence Bridge(close).

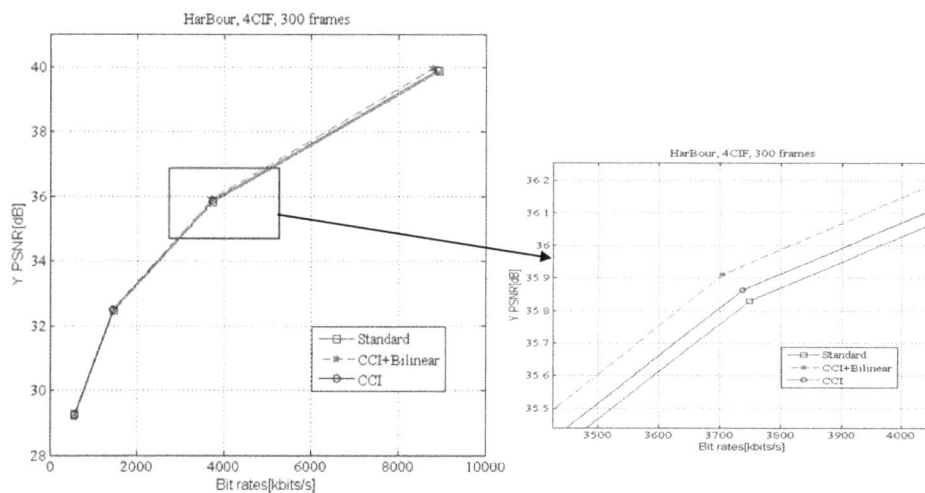

Figure 15. Rate-distortion comparison for 4CIF sequence HarBour.

6. Conclusions

The H.264/AVC algorithm is the international standard for video compression. In the H.264/AVC standard, the fractional-pixel displacement is used in the motion compensated prediction. There are two processing steps in the fractional-pixel motion compensation. In the first step, a fixed 6-tap Wiener filter is used for the 1/2-pixel displacement. The second step is to use the bilinear interpolation for the 1/4-pixel displacement. In this paper, the 4-tap CCI filter is proposed to reduce the computational complexity of the 1/4-pixel displacement and also increase accurately the motion compensated prediction. In addition, the derivation of CCI filter and CCI combined with motion compensation are discussed in detail. Finally, some computer simulations show that the proposed CCI algorithm with the bilinear interpolation can be used to speed up the H.264/AVC standard and also obtains a superior performance for motion compensation.

Acknowledgements

This work was supported by the National Science Council, R.O.C., under Grant NSC 102-2221-E-251-005.

References

[1] Z. N. Li and M. S. Drew, Fundamentals of Multimedia. Pearson Prentice Hall, 2004.

[2] T. Wiegand, G. J. Sullivan, G. Bjontegaard, and A. Luthra, "Overview of the H.264/AVC video coding standard," IEEE Trans. on Circuits and Systems for Video Technology, vol. 13, no. 7, pp. 560-576, July 2003.

[3] T. Wiegand and G. J. Sullivan, "The H.264/AVC video coding standard [Standards in a Nutshell]," IEEE Signal Processing Magazine, vol.24, no.2, pp.148-153, March 2007.

[4] T. Wedi, "Adaptive interpolation filter for motion compensated hybrid video coding," in Proc. Picture Coding Symposium (PCS), Seoul, Korea, Jan. 2001.

[5] T. Wedi, "Adaptive interpolation filter for motion compensated prediction," in Proc. IEEE International Conference on Image Processing (ICIP), Rochester, NY, pp. 502-509, Sept. 2002.

[6] T. Wedi and H. G. Musmann, "Motion- and aliasing-compensated prediction for hybrid video coding," IEEE Trans. on Circuits and Systems for Video Technology, vol. 3, no. 7, pp. 577-597, Jul. 2003

[7] T. Wedi, "Adaptive interpolation filters and high-resolution displacements for video coding," IEEE Trans. on Circuits and Systems for Video Technology, vol. 16, no. 4, pp. 484-491, Apr. 2006.

[8] Y. Vatis and J. Ostermann, "Adaptive interpolation filter for H.264/AVC," IEEE Trans. on Circuits and Systems for Video Technology, vol. 19, no. 2, pp. 179-192, Feb. 2009.

[9] T. K. Truong, L. J. Wang, I. S. Reed, and W. S. Hsieh, "Image data compressing using cubic convolution spline interpolation," IEEE Trans. on Image Processing, vol.9, no.11, pp.1988-1995, Nov. 2000.

[10] W. K. Pratt, Digital Image Processing, second edition. John Wiley & Sons, Inc., New York, 1991.

[11] R. G. Keys, "Cubic convolution interpolation for digital image processing," IEEE Trans. on Acoustic, Speech, and Signal Processing, vol. 29, no.6, pp.1153-1160, Dec. 1981.

[12] L. J. Wang and C. T. Shu, "A fast efficient fractional-pixel displacement for H.264/AVC motion compensation," in Proc.

of the 28th IEEE International Conference on Advanced Information Networking and Applications (AINA-2014), pp.25-30, Victoria, Canada, May 13-16, 2014.

[13] H.264/AVC Reference Software Version JM18.0, available online at: http://iphome.hhi.de/suehring/tml/download/old_jm/

[14] T. K. Tan, G. Sullivan, and T. Wedi, Recommended simulation common conditions for coding efficiency experiments, ITU-T Q.6/SC16, Doc. VCEG-AE10, Jan. 2007.

[15] Y. Ye, G. Motta, and M. Karczewicz, "Enhanced adaptive interpolation filters for video coding," in Proc. Data Compression Conference (DCC), pp. 435-444, March 2010.

[16] M. Unser, A. Aldroubi, and M. Eden, "B-spline signal processing: Part II-Efficient design and applications," IEEE Trans. Signal Processing, vol. 41, pp. 834-848, Feb. 1993.

[17] I. S. Reed and A. Yu, Optimal Spline Interpolation for Image Compression, United States Patent, No. 5822456, Oct. 13, 1998.

[18] L. J. Wang, W. S. Hsieh, T. K. Truong, I. S. Reed, T. C. Cheng, "A fast efficient computation of cubic-spline interpolation in image codec," IEEE Trans. on Signal Processing, vol.49, no.6, pp.1189-1197, June 2001.

2

Embedded system for speech recognition and image processing

Zhengxi Wei, Jinming Liang

School of Computer Science, Sichuan University of Science & Engineering, Zigong Sichuan 643000, PR China

Email address:
413789256@qq.com (Zhengxi Wei), ljm@suse.edu.cn (Jinming Liang)

Abstract: In recent years, the products of voice terminal and image retrieval show the intelligentized trend, but the mature commodities are rare in the market. This paper presents an embedded design method of intelligent voice terminal based on pattern recognition. The design adopts Samsung S3C2410 ARM as target board, Philips Uda1341TS as audio codec, embedded Linux OS as software platform, and speech recognition is implemented through small-vocabulary voice training. To improve the recognized effect, we use the image retrieval technology as an auxiliary tool, which helps speech recognition module create or more accurately find a personal voice-training library. By means of image recognition, the experimental results prove that the effect of speech recognition achieves an average increase of 10 percentages.

Keywords: Speech Recognition, Embedded Development, Image Retrieval, DTW Algorithm, ARM Development

1. Introduction

In recent years, multimedia-terminal [1] products to meet people's personalized requirements show the trend of intelligent, such as support for the conversion of text and voice, speech recognition on the basis of voice communication, and the recognition of digital image coming from the camera photography. These intelligent functions are implemented under the new development environment and background based on the theory of pattern recognition, VLSI hardware platform, and customized software operation system.

Operating system platform of intelligent terminal usually provides API functions, Software Development Kit, as well as integrated graphical development and testing tools. The developer not needing to understand the complex underlying hardware structure can develop the new business by its existing experience in this domain. As a result, the technical threshold of business development is greatly reduced.

At present, multimedia retrieval [2] is one of hot technologies in small intelligent terminal. It can be divided into three categories: (1) single cross-media integration index. In this method, when a medium is able to reflect the multimedia scene very well, the mark on this medium is also used to other media in the similar scene. (2) Multi-media integration. In this form, no one medium can well reflect the contents of the multimedia scene, we have to choose two or more medium based on multimedia content and give the judgments. As a result, these judgments are integrated together to form an interpretation to a multimedia scene. (3) dual-media feature integration. Various media feature may be combined together in accordance with the multimedia timing relationships, and we can use the feature combination to analyze the various multimedia scene.

Based on the research and the analysis of speech recognition [3] and image retrieval, this paper presents an overall design method on intelligent voice terminal. Our design mainly adopts Samsung S3C2410 ARM as the core in target development board, Philips Uda1341TS as audio codecs, embedded Linux OS as software platform, and speech recognition is implemented based on small-vocabulary voice training. To improve the effect of speech recognition, we use image recognition and retrieval technology as an auxiliary tool, which helps speech recognition module create or more accurately find out a personal voice-training library. Through capturing users' image information with a camera, we can design and achieve the intelligent voice terminal with image retrieval function.

2. Speech Recognition

The speech recognition is an advanced technology to make the machine convert the voice signal into the corresponding text or command through the identification and understanding of the process. Speech recognition has been widely used in the field of scientific research, and as to the daily life, it is more broad space for development.

2.1. Basic Principle

The speech recognition process can be attributed to pattern recognition and matching. Speech features can be extracted from the original speech signal, which should have been pre-processed and analysis-calculation, and finally speech recognition template is constructed. During the voice recognition, voice template stored in the system is to be compared to the characteristics of the input voice signal, according to certain algorithms and strategies, to identify the optimal template for matching the inputting voice, and finally to output recognition results.

In short, the speech recognition process [4] generally involves the following several key modules: signal pre-processing, speech feature extraction, matching training-library template, and outputting the matching results, as shown in figure 1.

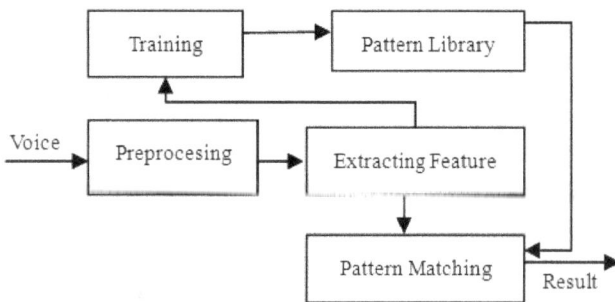

Figure 1. Speech recognition model

Signal pre-processing module includes sampling voice signal, removing differences of individual noise, excluding noise impact caused by the equipment and the environment, and involves the selection and endpoint detection of speech recognition unit. Speech feature-extraction module is used to extract the acoustic parameters that reflect the essential characteristics of voice, such as voice frequency, amplitude, and so on. The matching module is the core of the entire speech recognition system. It calculates the similarity (e.g. the voice speed and the likelihood probability) between the input characteristics and inventory models according to certain criteria such as word formation rules, grammar rules, semantic rules, and determines the semantic information of the inputting voice. Whereas outputting results module returns the final recognition results for its caller.

2.2. Speech Recognition Algorithm

Speech recognition algorithm is vital to the recognition effect. Good algorithms enable the signal processing to bring stable and excellent performance in the practice. DTW (Dynamic Time Warping) algorithm [5] applies the time sequence of the voice feature vector compare to every template in the reference template library. The most similar template will be acted as the recognition result. Embedded

Linux OS can run in embedded ARM board and Linux-based applications can get the external voice signal. DTW speech recognition program running the embedded platform determines the syntax and semantics of the reading content. The underlying hardware of ARM board is controlled by the Linux start-up and by device drivers to complete.

DTW algorithm is based on the dynamic planning idea, departs a complex global optimization problem into many local optimization problems to deal with, and tries to automatically find a path between two feature vectors whose total distortion amount can be minimum as possible, thereby avoiding introducing time-length error. Its mathematical principle is shown in figure 2.

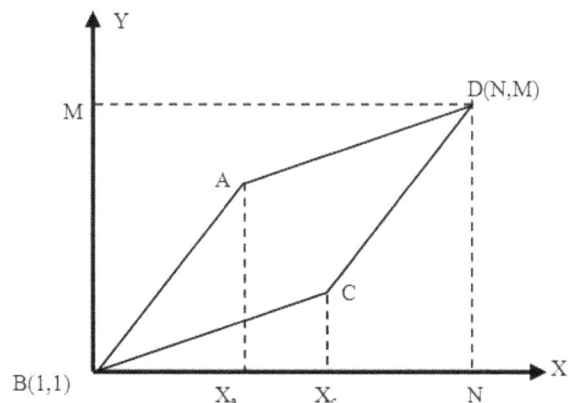

Figure 2. DTW Matching-path constraint graph

Assume there are M frames vector in the reference template and N frames vector in the test voice template, the dynamic-time-plan idea is to create a time revised function, as follows (1).

$$m = \omega (n) \tag{1}$$

The timeline will have test-vector n nonlinearly mapped to the reference-template-timeline m, making the cumulative-distance amount minimal between the test vector and the template vector of each frame, as the same time, the distance of matching-path between two vectors minimum, thus ensuring between the test template and reference templates with the maximum acoustic similar characteristics. Typically, the revised equation (1) is restricted to a parallelogram (assume ABCD) within the grid coordinates of the starting point is (1,1), the end point coordinates (N, M). The adjacent sides slope are 2 and 1/2. That means, simply needing calculate the matching distance of every frame that corresponds to various points in the parallelogram ABCD.

3. Image Processing

Image recognition theory contributes to image processing including image retrieval technology. It is the theoretical basis of the latter.

3.1. Image Recognition

An image recognition system [6] can be divided into three

main parts: (1) image preprocessing; (2) image segmentation and extraction feature; (3) the judgment or classification. The block diagram is seen in figure 3.

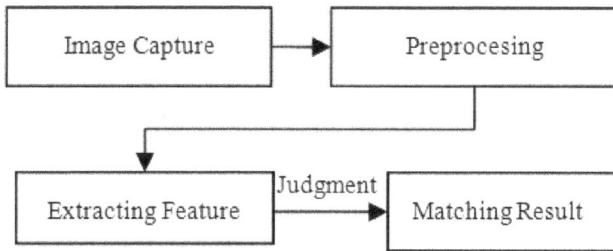

Figure 3. *Image recognition model*

Any kind of image recognition method first through a variety of sensors converts a variety of physical variables to values or set of symbols that the computer can receive. Traditionally, the space of this value or symbol is called the pattern space. In order to extract effective identified information from these numbers or symbols, it must be the following processing, removing noise, excluding irrelevant signals, calculating feature (such as the shape of the object, perimeter, area, etc.) as well as the necessary transformation (such as Fourier transformation).

Then by feature selection and extraction, the pattern feature-space is established. The subsequent pattern classification or pattern matching is based on the feature space. Finally, the system will output the object type or a model number that means this object in model database is the most correlative to the object to be searched.

3.2. Image Retrieval

Images can be manually labeled character information based on contents in the text annotations. Web Crawler can collect pictures from the web environment or extract some image marked similar text information in an HTML page, and establish originally keywords. Then it performs a pre-processes to these image, which includes de-nosing, setting standard size, and so on. The image is stored to the memory in development board and its feature-index will be further perfected after processing of relevant algorithms. Moreover, this index can later be retrieved and compared to the search keywords. In this way, it can determine whether they are the retrieved objects.

The image information will be abstracted to a generic string main through caliphemir algorithm library. Such as color histogram and other information can be extracted through the adjustment of parameters. Open source tools package (Java caliphemir) extracts the features such as color histogram and layout, convert them to the corresponding string from the image. The correspondence between extracted strings and images is established through the inverted algorithm used to file-search, co-exist in the index file. Different picture information can be stored in different fields, together constitute one document for the query. The feature string of image acts as a search keyword, and a picture of the maximum likelihood is found by querying the index file. Finally, a group

of image in the picture library are found and extracted from their path information so that these retrieved pictures could be displayed on terminal screen to users by the web-explorer.

4. Embedded Development Platform

4.1. Hardware Platform

The development board with Samsung S3C2410 microprocessor is selected as the hardware platform. CPU frequency is up to 203MHz in the board. Start-up codes, OS kernel and users' application programs are together stored in a FLASH whose capacity is 64MB. Application programs run in 64MB SDRAM, which can also be used as the room of various data and the stack. A camera capturing videos is connected to a USB interface in the board. The captured video is processed according to the image-matching rules. Subsequently, the result will be transmitted to speech recognition module. ARM Board of the system is shown in figure 4.

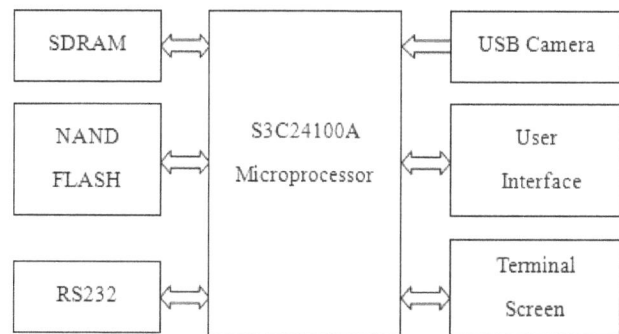

Figure 4. *ARM board*

The system uses UDA1341TS audio codec made by PHILIPS Company, which can achieve mutual conversion between stereo analog signal and digital signal. For digital signal, the chip also provides a DSP (digital signal processing) function. Input and output are composed of a microphone, a speaker and LCD (liquid crystal display). Voice analog signal inputting from microphone is first preprocessed, including A/D converter, AGC etc. A/D sampling frequency is set at 6 kHz, which is right the sampling frequency of voice signal, and Flash chip is used as the storage.

4.2. Embedded Operating System

The embedded Linux2.6.12 is a kind of miniature operating system, which is designed for the demand of the embedded OS. It has some advantages, such as small code amount, fast running speed, strong stability, and so on. This OS cuts out the normal Linux and becomes much smaller in size. It can even be solidified in a memory chip with a few KB or MB. The kernel of Linux2.6.12 can be customized by development engineers in terms of the actual demand. So it is regarded as the ideal software platform to develop embedded application programs. Speech recognition application adopts the DTW algorithm to implement speech processing and matching.

The implementation details are as follows. DTW algorithm is firstly to compare time series of speech feature vector with the reference template library to find the highest similarity template as the recognition result. Its core code based on DTW in the embedded Linux platform is encoded by C language as follows. In the program, f denotes the amount of test template, dist[i] represents the shortest distance to a feature vector.

```
for(f=0;f<2;f++){ //read parameters of test template
 wave_read(t,filename_test[f],f);
 for(d=1;d<=M;d++){
 // read parameters of reference template
 wave_read(r,filename_ref[d],d);
  for (i=0;i<n;i++) {
   for (j=0;j<n;j++) {
    float sum1=0•0;
     for(g=0;g<M;g++)
      sum1+=sqrt( t[i][g]- r[j][g]));
   D1[i][j]=sum1; //machting distance matrix D1[n][n] } }
   D2[0][0]=D1[0][0];
  for(i=1;i<n;i++)
   for(j=0;j<n;j++){
    D3=D2[i-1][j];
    if(j>0) D4=D2[i-1][j-1];
    else D4=REALMAX;
    if(j>1) D5=D2[i-1][j-2];
    elseD5=REALMAX;
    D2[i][j]=D1[i][j]+zmin(D3,D4,D5);
    }// zmin( ) calculate the minimum among D3,D4,D5
   strdata[d].data=D2[i-1][j-1];
   aa=strdata[1].data; g=0;
   for(k=1;k<=10;k++){
   if(aa>strdata[k].data){
   aa=strdata[k].data; g=k; }
   }strdata[g].data=aa;
   printf(" %s and %s is min and value""dist[%d] is %e\n\n",
       filename_tdata[g-1].chstr[g-1], g, strdata[g].data);}
}
```

In summary, the development platform includes the target board with the S3C24100A microprocessor and the embedded Linux2.6.12. The former constitutes hardware system architecture and development environment. In addition, the latter, as embedded OS provides powerful support for the development of search retrieval software.

4.3. Image Retrieval Module

Image Retrieval module is main to create image index and search users' picture. The image information will be abstracted to a generic string main through Caliph & Emir algorithm [7] library. Such as color histogram and other information can be extracted through the adjustment of parameters. Open source tools package (Java Caliphemir) extracts the features such as color and layout, converts them to the corresponding string from the image.

The correspondence between extracted strings and images is established through the inverted algorithm used to file-search, co-exist in the index file. Different picture information can be written in different fields, together constitutes one document for the query. The feature string of images acts as a search keyword, and a picture of the maximum likelihood is found by querying the index file. The most similar image in the picture library is obtained from its path information.

Finally, Image Retrieval module will establish a connection with the speech library. The information of matching-image or a new-user will be sent to the speech recognition module in order to accurately find or create a personal voice-training library, and complete speech recognition.

5. Experimental Results

We have firstly used a simple prototype system with sine-lifted cepstral-coefficients. With context-independent templates and no optimization, an error rate of about 2% is obtained. This setup is used in the comparative experiments on database size, and template selection. It should be noted that the error rate is readily reduced to below 1% when using context-dependent templates.

After that, we have more completed tests to our hardware and software system is as following. Five testing persons whose ages and sex are different select several sentences randomly from a 100-sentence library. Everyone says a few words in turn by a microphone. After voice signals inputting from the microphone are preprocessed, converted and encoded, recognition results are obtained by the DTW algorithm and templates, and voice signals will be output from the speaker. Listeners hearing the output voice give a score and the scoring criteria are as follows: 1-worst, 2-bad, 3-general, 4-good, and 5-excellent. The score reference is the pronunciation of an average person, and its value is set as 5. Mean opinion score (MOS [8]) of subjective perception experiment is 4.0, shown in figure 5(a).

Mean Opinion Score

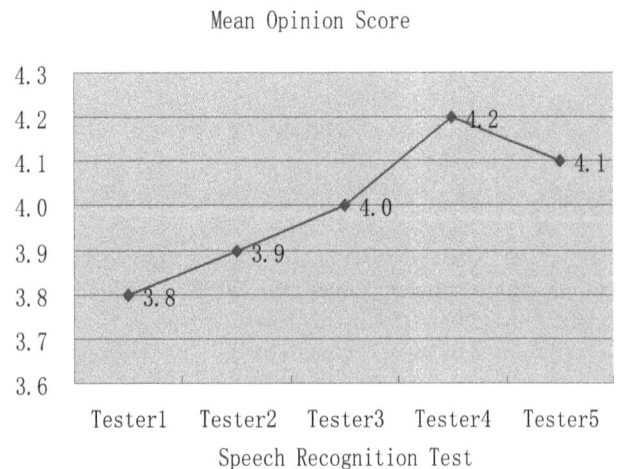

(a)

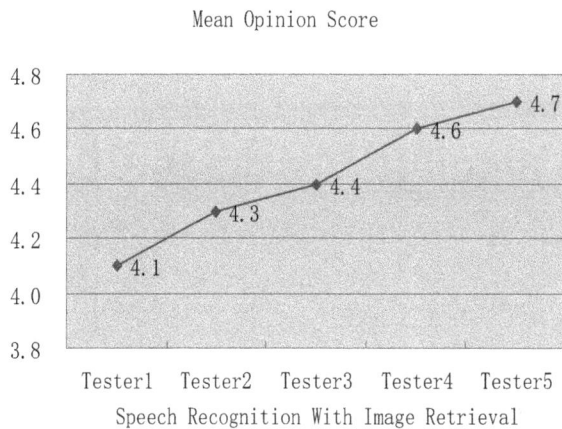

Mean Opinion Score

Speech Recognition With Image Retrieval

(b)

Figure 5. MOS result

For further testing the auxiliary effect of image recognition, the voice terminal under the same conditions loads the image recognition module in the second test. Camera captures different people's face image to the development board, for each one to establish image index, so that the next step could be automatically identify the speaking person. Subsequently, application programs are able to create or change the voice-training file of that user and support for the speech recognition. The experimental results are shown in Figure 3(b). From the experimental results, Mean Opinion Score is 4.4. The effect of speech recognition achieves an average increase of 10 percentages when the intelligent terminal is assisted through image retrieval module.

6. Conclusions

Combined with image-recognition technology, the paper presents an embedded-design method about developing speech recognition and image retrieval for small intelligent terminal. The experimental results show that the speech recognition accuracy is improved 10%, and the design idea for developing intelligent voice terminals can be referenced.

After a few modifications, the intelligent terminal can be used for much equipment, such as used in mobile phones, automatic answering machines and other portable devices; can also be used for the intelligent building systems, instrumentation-control, smart toys and other occasions, with good application prospects.

Acknowledgements

The research was supported by Artificial Intelligence Key Laboratory of Sichuan Province (No. 2013RYY04) and the Sichuan Provincial Education Department's Key Project (No.14ZA0210).

Our work was also supported by university Key Laboratory of Sichuan Province (No. 2013WYY09) and Fund Project of Sichuan Provincial Academician (Experts) Workstation (No.2014YSGZZ02).

References

[1] Shen Y I. Portable personal multimedia terminal: U.S. Patent D689, 856[P]. 2013-9-17.

[2] Rasiwasia N, Costa Pereira J, Coviello E, et al. A new approach to cross-modal multimedia retrieval[C]//Proceedings of the international conference on Multimedia. ACM, 2010: 251-260.

[3] Rabiner L R, Schafer R W. Digital Speech Processing [J]. The Froehlich/Kent Encyclopedia of Telecommunications, 2011, 6: 237-258.

[4] Hinton G, Deng L, Yu D, et al. Deep neural networks for acoustic modeling in speech recognition: The shared views of four research groups [J]. Signal Processing Magazine, IEEE, 2012, 29(6): 82-97.

[5] Muscillo R, Schmid M, Conforto S, et al. Early recognition of upper limb motor tasks through accelerometers: real-time implementation of a DTW-based algorithm [J]. Computers in biology and medicine, 2011, 41(3): 164-172.

[6] Zhu B B, Yan J, Li Q, et al. Attacks and design of image recognition CAPTCHAs[C]//Proceedings of the 17th ACM conference on Computer and communications security. ACM, 2010: 187-200.

[7] Lux M, Klieber W, Granitzer M. Caliph & Emir: semantics in multimedia retrieval and annotation[C]//Proceedings of the 19th International CODATA Conference. 2004: 64-75.

[8] Viswanathan M, Viswanathan M. Measuring speech quality for text-to-speech systems: development and assessment of a modified mean opinion score (MOS) scale [J]. Computer Speech & Language, 2005, 19(1): 55-83.

3

Fast Type Recognition of Missive Insulator Leakage Current Data Using Spark

Song Yaqi

Department of Computer Science, North China Electric Power University, Baoding, China

Email address:
bdsyq@163.com

Abstract: Performance is the key issue in power big data applications. One of main challenges is how to exploit these technologies in building power big data processing platform and facilitating science discoveries such as those in electric power systems. This paper explores how Spark and Cloud computing can accelerate performance of missive insulator leak current data pattern recognition. We have designed and implemented the Parallel KNN(k-NearestNeighbor) algorithm using Spark and then deployed onto the Aliyun E-MapReduce cloud computing platform. The results from experiments shows the performance and scalability can be enhanced through these advanced technologies.

Keywords: Insulator Leakage Current, Electric Power Big Data, Spark

1. Introduction

Electric power equipment monitoring are developing from single-parameter monitoring to all-round and multi-parameter monitoring. The monitoring data showed exponential growth. Remote monitoring center of power equipments is faced with heavy tasks, such as data collection, processing, storage and analysis when large amounts of monitoring data flock in.

The traditional single-machine environment which using a single task is only applicable to small amount of data and will be difficult to finish data processing tasks on time or even unable to deal with when facing large volume of data [1].

[2] stores and manages massive neuroimaging data by integrating database management systems (DBMS) with Grid computing. [3] propose a dynamic programming algorithm for pattern-based time series classification on GPUs. Given the industrial real-time demand, [4] propose a parallelized method to model the Elman network, which shifts the computational intensive tasks of network training on GPU. [5] study a large scale EEG (electroencephalogram) distributed data storage method on Hadoop [6]. Cloudera used Hadoop to store and manage around 1.5 trillion points of time-series data in 15TB of PMU archive files at TVA [7]. Although the

Hadoop MapReduce [8] can effectively deal with large amounts of data, it will frequently access disks and the task need to take up several minutes of even hours. In view of complex iterative computing tasks in electric power equipment condition evaluation, the MapReduce can not finish analysis and pattern recognition tasks for large amounts of alarm monitoring data in a short limited time and performance is difficult to meet the requirements [9].

Apache Spark is a fast and general engine for large-scale data processing [10]. It has an advanced DAG execution engine that supports cyclic data flow and in-memory computing. In some computing tasks, Spark run programs up to 100x faster than Hadoop MapReduce in memory, or 10x faster on disk. Spark offers over 80 high-level operators that make it easy to build parallel apps. And one can use it interactively from the Scala, Python and R shells. Spark has been widely used in seismic data analysis [11], data analysis in smart grid [12], GATK DNA analysis [13], data reduction [14], etc.

This paper studied fast pattern recognition of electric power equipment monitoring data using Spark. Spark-based K-Nearest Neighbor algorithm (KNN) is designed and implemented in Aliyun E-MapReduce platform and used for insulator leakage current data type identification. The results from experiments show that Spark-KNN runs up to 2.97x faster than MapReduce-based one.

2. Distributed Storage of Monitoring Data in RDD

Resilient Distributed Dataset (RDD) is the core concept in Spark framework. It can hold any type of data. Spark stores data in RDD on different partitions. It helps with rearranging the computations and optimizing the data processing. It also has fault tolerance because an RDD know how to recreate and recompute the datasets. RDDs are immutable and can be modified with a transformation with the result of a new RDD returned, whereas the original RDD remains the same. RDD provides a rich set of operations to manipulate data, including map, flatMap, filter, join, group By, reduce By Key, etc which facilitate distributed data processing.

The waveform or extracted features of power equipment monitoring data is stored and organized as RDDs. RDD can be understood as a large array, but the array is distributed on the cluster. A RDD logically is composed of multiple partitions which are corresponding to physical block in data node memory. The process of executing analysis includes a series of transformations and actions for RDDs. Monitoring data is stored in the RDD as shown in Figure 1.

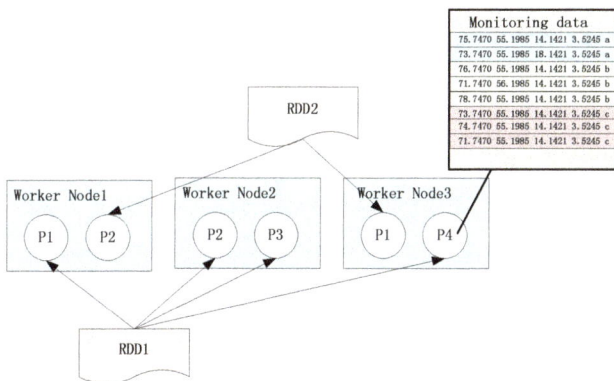

Figure 1. *Distributed storage of monitoring data in RDD.*

In Figure 1, RDD1 contains four partition (P1, P2, P3, P4) and are distributed stored in three nodes (Worker Node1, Worker Node2 and Worker Node3). RDD2 contains two partitions (P1 and P2) and are stored in 2 nodes (Worker Node3 and Worker Node1).

3. The Spark-KNN Algorithm for Fast Pattern Recognition

In pattern recognition, the k-Nearest Neighbors algorithm (k-NN) is a non-parametric method used for classification and regression [15]. In the classification phase, k is a user-defined constant, and an unlabeled vector (a query or test point) is classified by assigning the label which is most frequent among the k training samples nearest to that query point. A commonly used distance metric for continuous variables is Euclidean distance. The k-NN algorithm is among the simplest of all machine learning algorithms.

This paper studied parallel KNN algorithm using Spark,

called Spark-KNN. The input and output can be from local file system, HDFS, or OSS, etc. Spark-KNN algorithm is described in Table 1.

Table 1. *Spark-KNN algorithm.*

Spark-KNN algorithm
Input:
TrainSet file(cvs format);TestSet file(cvs format);
ResultSet file path; Parameter k;
Output:
ResultSet file(cvs format);
Procedure:
1: Initialize SparkContext environment parameters;
2: Load TrainSet to RDD using SparkContext.textFile( );
Do format conversion using RDD.map( );
map(line => {vardatas = line.split(" ") (datas(0), datas(1), datas(2))})
3: Executing RDD.collect( ) to get a scala Array for TrainSet in Driver node named TrainSet_Array;
4: Broadcast the TrainSet_Array to every node using SparkContext.broadcast( );
5: Broadcast the parameter k to every node using SparkContext.broadcast( );
6: Load TestSet to RDD using SparkContext.textFile( );
Do format conversion using RDD.map( ), as described at step 2;
7: Get the result set using RDD.map( ):
7.1: Parsing a test sample tuple, extracting the characteristics;
7.2: distance_set= trainDatas.foreach(trainData =>(characteristics，distance，category)});
7.3: Sorting the distance_set according to the ascending order of distance;
7.4 Get the first k points and their categories;
8: ResultSet.saveAsTextFile(ResultSet file path);

The flowchart for Spark-KNN is shown in Figure 2.

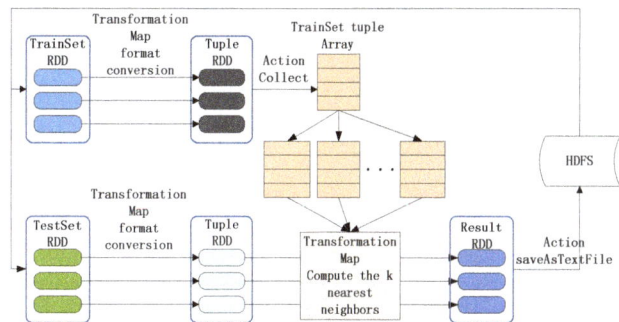

Figure 2. *Data processing flow in Spark-KNN.*

In Figure 2, Input data are loaded from HDFS using textFile function of SparkContext class. Then input data are organized as RDDs. Format conversion operations are conducted by the map transformation. A new RDD will be produced after map transformation. Colletc is a kind of Action and return all the elements of the dataset as an array at the driver program. This is usually useful after a filter or other operation that returns a sufficiently small subset of the data. Broadcast is a kind of Action and allows the programmer to keep a read-only variable cached on each machine rather than shipping a copy of it with tasks. They can be used, for example, to give every node a copy of a large input dataset in an efficient manner. Spark also attempts to distribute broadcast variables using efficient broadcast algorithms to reduce communication cost. 'saveAsTextFile' is a kind of Action and can write the elements of the dataset as a text file (or set of text files) in a given directory in the local filesystem, HDFS or any other Hadoop-supported file system. Spark will call 'toString' on each element to convert it to a line of text in the file.

4. Experimental Evaluation in the Cloud

4.1. The Experiment Environment

We have conducted the experiments based on the Cloudcomputing platform. We deploy our prototype in the AliyunE-Map Reduce platform by renting virtual computer nodes (ECS.S3. Large type) from Aliyun ECS. The configuration of each machine is described in Table 2.

Table 2. Configuration of a virtual machine fromAliyun ECS.

Item	Configuration
CPU	Intel Xeon CPU,4 cores
Memory	8GB memory
Storage	80GB SSD cloud disk
Network bandwidth	8MB
Operating system	Federa Core 8 (2.6.21.7-2.ec2.v1.2.fc8xenLinux Kernel)
Platform major version	EMR 1.0.0
Software	hive 1.0.1; ganglia 3.7.2; Spark 1.4.1; yarn 2.6.0; pig 0.14.0

We deploy the Spark cluster using Spark on YARN mode. The ganglia (3.7.2) is a scalable distributed monitoring system for high-performance computing systems such as clusters and Grids. We use It to monitor the CPU and memory utilization so as to make adjustment and optimization of parallel tasks configuration parameters, such as 'number-exector', etc. The following 4 parameters are very important as shown in Table 3.

Table 3. Parameter Configuration of Spark job.

Parameter	Description	Default value	Our value
--executor-memory	Memory per executor	1GB	2GB
--driver-memory	Memory for driver	512MB	1GB
--num-executors	Number of executors to launch	2	4
--executor-cores	Number of cores per executor	1 in YARN mode	4

4.2. Experimental Data

This paper focus on the insulator leakage current data pattern recognition intransmission line monitoring system. We select the widely used four features (maximum leakage current, the 50-Hz, 150-Hz and 250-Hz amplitudes after Fourier transform) to form a 4-dimension vector for pattern recognition using Spark-KNN. Some samples selected from training set are shown in table 4.

Table 4. Samples in training set.

Type	maximum (mA)	50Hzamplitude (mA)	150Hzamplitude (mA)	250Hzamplitude (mA)
Stage A	14.8936	12.4707	0.1082	0.1016
Stage A	18.0136	14.7075	0.8962	0.1175
Stage A	59.1919	44.0040	11.5511	2.7286
Stage B	87.6251	63.2405	15.7299	3.7481
Stage B	92.7320	68.5759	17.1612	4.2756
Stage C	138.9287	102.3116	20.6031	5.8952
Stage E	20781	1603	348	161

Insulator leakage current types [16] are described in table 5.

Table 5. Category of iced insulator leakage current samples.

Type	Type description
Stage A	Faint discharges, a subtle audible sound, no visible signal
Stage B	Some visible point discharges, a continuous sound
Stage C	Liner weak local arcs
Stage D	Intermittent, stronger local arcs
Stage E	Flash over

Table 6. Train set.

Train Set ID	Sample size (piece)
T1	50
T2	500
T3	1000

Experimental data is from artificial experiments and real-measured insulator leakage current. We make several replications to simulate large scale concurrent alarm data. The experiment simulates 6 million monitor points, and by setting up the fault rate(0—100%), simulates the different size of alarm data due to bad weather. In a short period of time, the alarm data size needed to deal with is in the range of zero to 6 million pieces of data. The data set includes raining set and test set as shown in Table 6 and Table 7.

Table 7. Testset.

Test Set ID	failure rate	Sample size (kilo pieces)
C1	10%	60
C2	30%	180
C3	50%	300
C4	80%	480
C5	100%	600

4.3. Performance Evaluation

We use the data set in table 5 to test the performance of Spark-KNN. We run the KNN program respectively in single computer, Hadoop cluster and Spark cluster, which named KNN, MR-KNN and Spark-KNN respectively. The MR-KNN and Spark - KNN run at the same hardware environment as shown in table 1.

In a single computer, the run time of KNN changes along with the data size variation as shown in Figure 3.

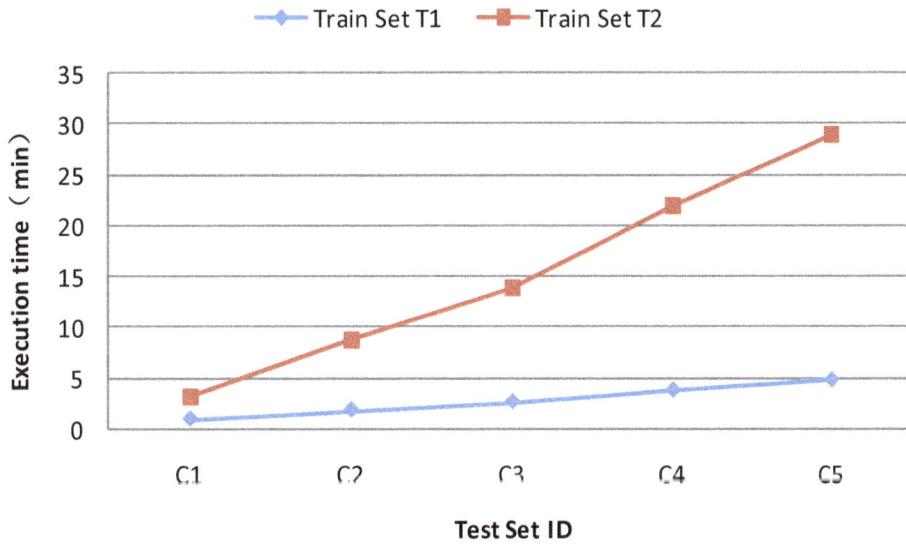

Figure 3. Execution time of KNN on a single computer.

As can be seen from the Figure 3, the execution time is close to half an hour when training set is T2 and test set is C5. While choosing training set T3, the execution time is so long that the computer is 'died'. As a result, T3 curve is not drawn in figure 3. The experimental results show that the single computer environment is not up to the fast pattern recognition task of large-scale alarm data.

We have measured the execution time of MR-KNN and Spark-KNN by varying the training set, as shown in Figure 4.

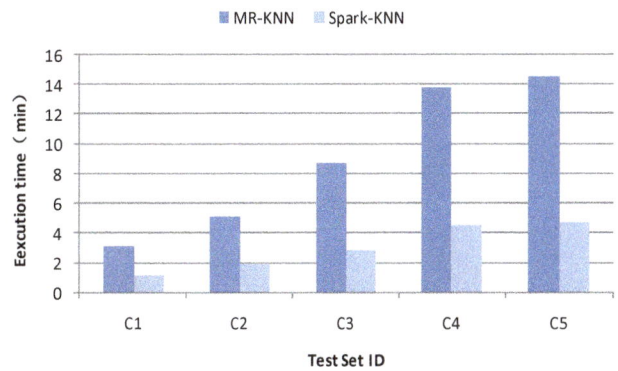

a) Training set T1

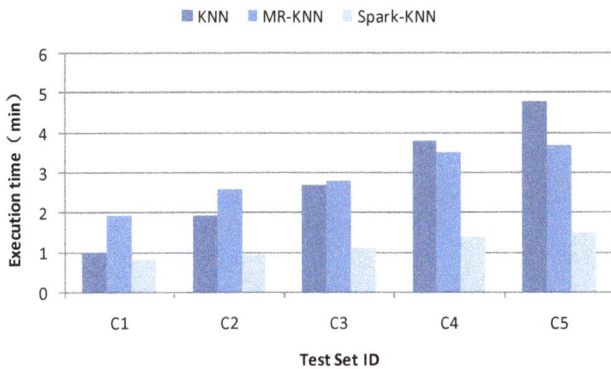

b) Training set T2

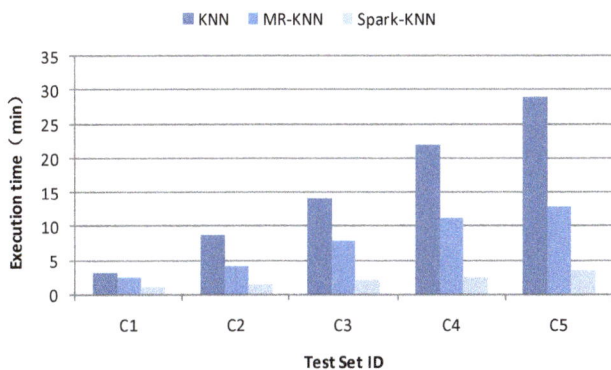

c) Training set T3

Figure 4. Execution time comparison between Spark-KNN and MR-KNN.

As shown in Figure 4, the Spark - KNN performance is superior to MR-KNN under various training set. Spark-KNN runs up to 2.97x faster than MR-KNN.

The execution time trend of Spark-KNN under different training sets is shown in Figure 5.

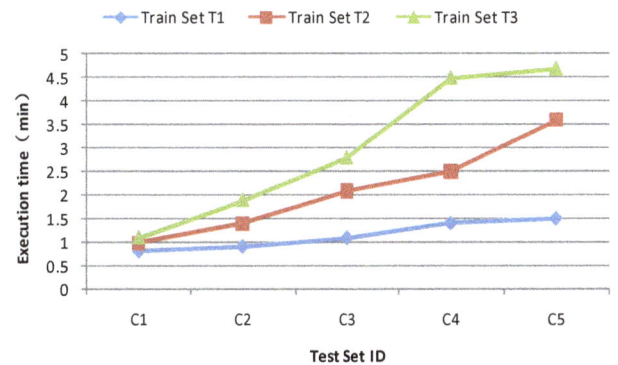

Figure 5. Execution time trend of Spark-KNN.

As can be seen from Figure 5, the programexecution time grows slowly as the growth of the test set size.

Speedup is a metric for improvement in performance between two systems processing the same problem. Speedup is calculated by formula (1). T_s denotes the execution time with a single CPU core. T_h denotes the execution time with h CPU cores.

$$S_{peedup} = \frac{T_s}{T_h} \qquad (1)$$

We calculate the speedup for Spark-KNN by varying core number of the cluster and using various training set and test set, as shown in Figure 6.

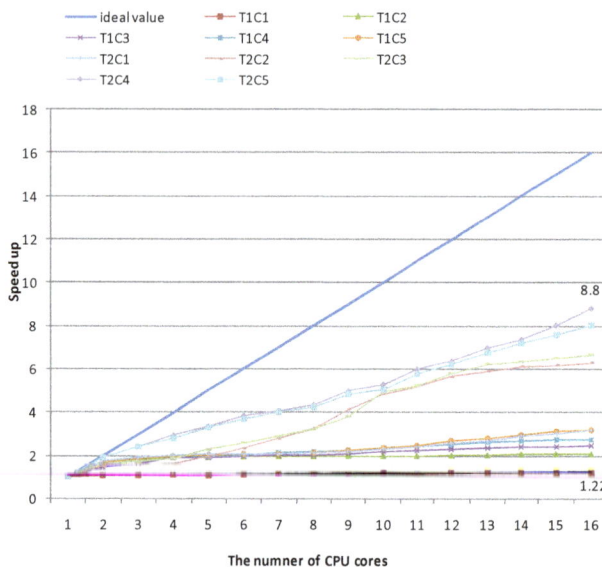

Figure 6. *Speedup of Spark-KNN.*

In Figure 6, the speedup increases with the growth of data scale. The maximum is 8.8 when using the dataset T2 and C4 with 16 CPU cores. The minimum is 1.22 when using the dataset T1 and C1. The speedup does not increase when using dataset T1 and C1 even if add more CPU cores, while increases almost linearly when using dataset T2C4 and T2C5.

5. Conclusion

In this paper, we have investigated how to apply Spark and Cloud computing to insulator leak current data pattern recognition use case in order to understand how well these advanced technologies can accelerate the performance in supporting data-intensive applications. We have adapted the KNN pattern recognition task to Spark program. The prototype was deployed on the Aliyun E-MapReduc cloud computing platform for experimental evaluation. We have used the speed up as the standard metric.

The results from the experiment show that the performance can be improved by using Spark. We have also compared execution time of the Spark and the Hadoop MapReduce. The result shows that Spark-KNN is much faster than MR-KNN and more suitable for real-time data processing for electric power equipment monitoring system.

Acknowledgements

This work was supported by the Fundamental Research Funds for the Central Universities (2016MS117, 2016MS116).

References

[1] Zhou, G., Zhu, Y., Wang, G., & Song, Y. "Real-time big data processing technology application in the field of state monitoring". Diangong Jishu Xuebao/transactions of China Electrotechnical Society, vol.29, pp. 432-437.

[2] Uri Hasson, Jeremy I Skipper, Michael J Wilde, Howard C Nusbaum, and Steven L Small. Improving the analysis, storage and sharing of neuroimaging data using relational databases and distributed computing. NeuroImage, 39(2):693–706, 2008.

[3] Kai-Wei Chang, Deka, B Hwu, W.-M.W, etc. Efficient Pattern-based Time Series Classification on GPU[C]. 2012 IEEE 12th International Conference on Data Mining (ICDM 2012). Los Alamitos, CA, USA, 2012: 131-40.

[4] Zhao Jun, Zhu Xiaoliang, Wang Wei, etc. Extended Kalman filter-based Elman networks for industrial time series prediction with GPU acceleration [J]. Neurocomputing, 2013, 118: 215-224.

[5] Haimonti Dutta, Alex Kamil, Manoj Pooleery, Simha Sethumadhavan, and John Demme. Distributed Storage of Large-Scale Multidimensional Electroencephalogram Data Using Hadoop and HBase [J]. Grid and Cloud Database Management.2011.9.

[6] White T. Hadoop: The definitive guide [M]. O'Reilly Media, Inc, 2012:260-261.

[7] Christophe Bisciglia. The smart grid: Hadoop at the Tennessee Valley Authority (TVA) [EB/OL]. 2009.6 [2013.2]. http://www.cloudera.com/blog/2009/06/smart-grid-hadoop-tennessee-valley-authority-tva/

[8] Dean J, Ghemawat S. MapReduce: simplified data processing on large clusters [J]. Communications of the ACM, 2008, 51(1): 107-113.

[9] Zaharia M, Chowdhury M, Das T, et al. Resilient distributed datasets: A fault-tolerant abstraction for in-memory cluster computing [A]. Proceedings of the 9th USENIX conference on Networked Systems Design and Implementation[C]. USENIX Association, 2012: 2-2

[10] Zaharia M, Chowdhury M, Das T, et al. Fast and interactive analytics over Hadoop data with Spark[J]. USENIX; login, 2012, 37(4): 45-51

[11] Yan Y, Huang L, Yi L. Is Apache Spark scalable to seismic data analytics and computations? [C]// IEEE International Conference on Big Data. IEEE, 2015.

[12] Shyam R, Bharathi Ganesh H. B, Sachin Kumar S, et al. Apache Spark a Big Data Analytics Platform for Smart Grid[J]. Procedia Technology, 2015, 21:171-178.

[13] Mushtaq H, Al-Ars Z. Cluster-based Apache Spark implementation of the GATK DNA analysis pipeline[C]// Bioinformatics and Biomedicine (BIBM), 2015 IEEE International Conference on. IEEE, 2015:1471-1477.

[14] Ram&#, Rez-Gallego S, Garc&#, et al. Distributed Entropy Minimization Discretizer for Big Data Analysis under Apache Spark [C]// IEEE Trustcom/bigdatase/ispa. IEEE Computer Society, 2015.

[15] Cover, T., Hart, P. Nearest neighbor pattern classification [J]. IEEETrans. Inf. Theory, 1967, 30(1): 21–27

[16] Suda T. Frequency characteristics of leakage current waveforms of an artificially polluted suspension insulator [J]. Dielectrics & Electrical Insulation IEEE Transactions on, 2001, 8(4): 705-709.

4

Optimum bit rate for image transmission over underwater acoustic channel

Hamada Esmaiel, Danchi Jiang

School of Engineering, University of Tasmania, Hobart, Australia

Email address:

Hamada.Esmaiel@utas.edu.au (H. Esmaiel), Danchi.Jiang@utas.edu.au (D. Jiang)

Abstract: In this paper, image transmission in underwater channels is considered. The images are encoded with forward error correction using an unequal error protection technique together with the Reed-Solomon codes and dynamic bit-rate allocation before transmitted. This paper proposes a novel rate allocation scheme for efficient image bit stream transmission in underwater acoustic channels with optimum bit rates. The optimality is achieved in the sense that the comprehensive peak signal–to–noise ratio of the image transmission is maximized under channel bit rate and bit error rate constraints. Based on a modified set partitioning in hierarchical trees (M-SPIHT) image coder, four different flocks of bit-streams based on their significance levels are generated. According to their significance levels, the blocks of the significant bits, the sign bits, the set bits and the refinement bits are transmitted with different protection levels, so as to reduce the total distortion of received image. In addition to the careful selection of each component and intuitive justification in the detailed system design, simulation results have also been included. It is demonstrated that the proposed scheme outperforms the equal error protection for image transmission in underwater channels, significantly improves the peak signals–to–noise ratio (PSNR) performance in comparison to existing coding schemes.

Keywords: Rate Allocation, Reed Solomon Coder, SPIHT Coding, Underwater Acoustic Channel, Unequal Error Protection (UEP)

1. Introduction

Wireless underwater acoustic (UWA) communication systems are sensitive to numerous of scientific and civilian tasks in the water of the ocean, such as observation and reconnaissance for the oceans, undersea rescue, and emergency response to undersea disasters, etc. Modern high data rate communications for image transmission are desirable for efficient underwater expeditions. As the attenuation is significant in cases where electromagnetic waves are used in the wireless communication system for the salt water, acoustic waves are the viable wave to carry the information in this case. However, existing acoustic communication technologies currently available can only support a limited data rate communication due to limited channel bandwidth available in addition to multipath spreading in the UWA channel [1-3]. In fact, as the special physical characteristic of the underwater channel, wave propagation is significantly affected by temperature, salinity of water, activity of hydrogen ions (pH) of water, depth of the water column or pressure and surface/bottom roughness [4].

The high error rates, significant path loss, and low channel capacity are typical features of underwater channels and that make the task of high speed data transmission even more challenging than that with terrestrial radio links [5].

One of challenges of image transmission in an UWA channel is a large amount of long burst errors. The underwater acoustic channel fading can causes significant long burst errors and sporadic failure, for that the image transmission is far more challenging than that in stationary and more predictable wired or wireless environment. With a conventional image coder, image transmission signals over UWA channel can be significantly dispersed and distorted. For that sophisticated coding techniques will be needed. However, that may increase the overhead in spectrum use. In [6] is known that effective transmission over UWA channels is dependent on two factors: (1) efficient compression techniques and (2) well-organized bandwidth. In this paper, we attempt on both aspects where efficient data compression will be achieved by using effective wavelet based progressive image compression technique called set partitioning in hierarchical trees (SPIHT) [7], aiming to decrease the mean-squared error (MSE) between encoded and decoded image. The bandwidth efficiency is achieved using unequal error protection optimization technique formulated by taking the particular channel feature into account at the transmitter [8, 9].

In this paper, a system will be designed to minimize the transmission error of image bit-streams under a total broadcasting rate constraint. Unequal error protection (UEP) [10] will be proposed for high quality image transmission. UEP uses four different bit stream groups outputed from a modified SPIHT [7] source code. The expected respective contributions of each group at the overall peak signal–to–noise ratio (PSNR) in the decoded image will be estimated based on the corresponding level of protection in each group determined. The reason that SPIHT is selected to use for underwater acoustic (UWA) channel because it is inherently designed for image encoding, and the associated redundancy can be managed conveniently for communications and its capability to adapt to actual channel constraints [11]. A Reed Solomon (R–S) channel coder will be adopted in this paper to enhance error correction capability. Authors in [11] have used a fixed UEP technique where more important data is protected with more redundancy and the less important data is protected by limited redundancy. The redundancy amount can be fixed for each packet. In this paper, we try to develop an adaptive UEP technique based on the SPIHT bit stream sensitivity, the noise level of the underwater acoustic channel, and its effects on the overall channel distortion.

This paper is organized as follows: The set partitioning in hierarchical trees SPIHT with unequal error protection algorithm, together with underwater channel will be introduced in Section 2 as preliminaries. The proposed coding algorithm for underwater acoustic channel to reduce the channel distortion will be developed in Section 3. An intuitive explanation behind the various aspects of the proposed UWA communication system will be given in Section 4. The proposed system will be simulated under a set of typical underwater channel parameters and the illustrative outcomes are given in Section 5. Finally, the conclusion and summary of the major techniques and contributions of this paper will be included in Section 6.

2. SPIHT Coder with UEP and Underwater Acoustic Channel

Sophisticated image compression techniques based on wavelet algorithms have been reported for the underwater transmission. Please see [5, 10, 11] for some examples. In these works, SPIHT coder has been demonstrated of advantages in many aspects including advanced bit error ratio, continuous-tone with a high compression ratio, lossy and lossless compression, and progressive processing. SPIHT was originally introduced by Said and Pearlman [7]. It

improves PSNR for certain compression ratios for a broad assortment of images compression coders [12]. As such, it has been widely used in image application, and set as an essential performance benchmark for the evaluation of new image coding techniques. These features make the SPIHT compression technique a reasonable choice for realistic underwater transmissions. In this line, a simple source compression scheme incorporated with SPIHT source coding scheme has been investigated earlier in [5], for underwater transmissions. The encoded bitstream was divided into consecutive chunks before transmitting over underwater links. The expected attenuation of UWA channel is one of the major channel challenges. In addition, restrictions due to distance, depth, frequency, temperature and acidity to salinity of the UWA channel make the communication channel to be dynamic and fast time varying, which significantly affects communication system behavior. In order to make further improvements, it is necessary to apply appropriate underwater channel models to lay a foundation for the discussions it in this article, which will be detailed in this section.

2.1. SPIHT Coded Output Bit Stream Classification

Set partitioning in hierarchical trees (SPIHT) is an algorithm established based on the hierarchical partitioning of the wavelet coefficients set and organize it into a spatial orientation trees, where all sub-bands coefficients are included in the corresponding branches to identify their particular position in the image [7, 13, 14]. All insignificant coefficients are sequentially encoded together as an insignificant symbol set. An appropriately defined partitioning rule can be utilized to further divide this insignificant set to progresively extract the significant coefficients iteratively. SPHIT coder is considered the best available coder due to progressive rate control and transmission, as well as coding process simplicity, etc. [7]. In particular, after image decomposition with 9/7 tap wavelets by the method reported by Antonini et al [15], the general SPHIT coding algorithm encodes image by splitting the decomposed image into considerable sections on the basis of the significance classification function as:

$$S_n(\Gamma) = \begin{cases} 1, & if \quad \max_{(i,j) \in \Gamma} \{|Y(i,j)| \geq 2^n\} \\ 0, & otherwise \end{cases}, \quad (1)$$

where $Y(i, j)$ represent wavelet coefficient at the pixel (i, j) for n bit plane. The algorithm works on the basis of two tiers of loops, the sorting loop and the refinement loop. To implement SPHIT coder three related groups [16] are obtained: first index of insignificant pixels , the index of significant pixels, and the index of insignificant sets. A quantization process is applied to sort the wavelet coefficients of these different groups using a different quantization level. On the basis that each subset is divided into different groups ordered according to their significance levels, the most important wavelets coefficients in the image are encoded with a bit-stream to realize the best encoding quality than the previous schemes [17]. While the algorithm initially generates a list of insignificant pixels (LIP), then evaluates its element's significance versus updated threshold value and encodes their relevance by 0 or 1. For the significant coefficients, the signs of the coefficients are encoded together with the corresponding pixel. Then the encoded pixels are sent to a list of significant pixels (LSP) group. After checking the list of insignificant sets (LIS), the threshold for the significant set coefficients will be updated. The particular tree/set significant bits are classified into children or grandchildren based on the updated significance thresholds. Otherwise, for the insignificant set it is represented as a zero-tree in the encoded bits in the output bit streams. At the end of each iteration, the classification process determines the refinement bits. Its output from the SPIHT algorithm at the updated bit significance pixels is moved to the LSP at each threshold update, leading to the refinement of significant pixels with rate reducing error. The threshold is incremented with two steps until the output bit-stream ends.

In this paper, the sorting pass will be performed on separate four lists rather than three: LIS, LIP, LSP, and the roster for refinement path. The LSP and LIP consist the nodes indicate the single pixels while the descendant nodes are represented by the LIS. The highest coefficients can be represented by the number of bits in the following formula:

$$n_{max} = \log_2\left(\max_{\{i,j\}}\{|Y(i,j)|\}\right), \tag{2}$$

which is similar to that in [7]. While screening image pixels, classified LIP of pixels can be obtained based on the significance degree using Equation (1). The resultant $S_n(\Gamma)$ is then forwarded to the coded bits. All those significance pixels will be sent to the LSP group with their sign bits sent to the output stream. Similarly, for sets in LIS they will be tested and sent to LSP or LIP according to their significance levels. For refinement loops, the nth generality significant bit of the coefficients in the LSP is used. By iteratively going through all n that is decreased by one, the desired rate can be obtained until every LSP node is processed.

2.2. Underwater Acoustic Channel

The UWA channel can be highly distorted. One of the most challenging issues in it is the significant propagation delay, in addition to the high attenuation level with low propagation velocity. All of this makes the UWA channel to be one of the most challenging environments for wireless communication [18]. Many models have been developed to predict the UWA channel attenuation, which is dependent on several factors of UWA channel. For example, the carrier frequency, communication distance, acoustic sensor depth, temperature and acidity to salinity of the UWA channel [19].

2.2.1. Propagation Loss

Before The communication distance is inversely proportional to the power of acoustic signal in UWA channel. UWA channel is influenced by physical channel

characteristics such as boric acid, magnesium sulphate, geometric spreading, and particle motion. UWA channel propagation loss is affected by three parameters characterizing of the geometrical spreading, attenuation and the abnormality of spreading. Other factors cannot be easily modeled. Accordingly, the expected attenuation of the propagated signal, in dB, with respect to the transferred range l and a carrier frequency signal f can be approximated as [4, 8, 20, 21]:

$$10 \cdot \log A(l, f) = 10 \cdot k \cdot \log l + 10 \cdot l \cdot \log \alpha \tag{3}$$

where k is the geometrical UWA channel spreading factor. Its value is usually between 1 and 2. In this paper, we consider value $k = 1.5$ for illustrative purpose. α is the absorption coefficient in dB/km.

To characterize UWA multipath effects, we consider the p-th path as a generic case. Its effect with a delay t_p and a mean magnitude gain α_p can be estimated. The UWA channel multipath magnitude and tab delay values are a function in the length of the communication path l_p. The UWA channel path gain is given by $\alpha_p = \Gamma_p / \sqrt{A(l_p, f_c)}$. Assuming that the path gain is remains the same value over each path, $\Gamma_p = \left(1/\sqrt{2}\right)^{r_p}$. r_p is an integer associated with the number of reflected paths. i. e., $r_p = 0,1,3,5,7$, for the main path, first path, second path, … , respectively. The propagation loss in this path can be obtained by Equation (3), $A(l_p, f_c)$, calculated by using the following Equation:

$$A(l_p, f_c) = l_p^k [\alpha(f_c)]^{l_p}, \tag{4}$$

The path delay t_p is given by $t_p = l_p/c$ (the speed of sound in the water is $c = 1500$ m/s) [22]. Path length l_p can be obtained based on geometry plane model as in [22].

2.2.2. Ambient Noise Model

The UWA channel ambient noise can have a continuously changing power spectral density (p.s.d.) [4]. The significant factors on the level of the UWA channel ambient noises are the shipping, turbulence, wind waves and thermal noise. Each factor has an ambient noise p.s.d. in dB re µPa per Hz. It can be approximated using the formulae (5), (6), (7), (8) shown below, respectively [4]:

$$10\log N_s(f) = 40 + 20(s - 0.5) + 26\log f - 60\log(f + 0.03), \tag{5}$$

$$10\log N_t(f) = 17 - 30\log f, \tag{6}$$

$$10\log N_w(f) = 50 + 7.5\sqrt{\omega} + 20\log f - 40\log(f + 0.4), \tag{7}$$

$$10\log N_{th}(f) = -15 + 20\log f. \tag{8}$$

Ambient noises are functions of the carrier frequency. The overall effect of the ambient noise can be formulated as [4]:

$$N(f) = N_t(f) + N_s(f) + N_w + N_{th}(f). \qquad (9)$$

UWA channel performance is dependent on the composite ambient noise and channel multipath attenuation.

2.2.3. Signal–to–Noise Ratio

UWA channel signal–to–noise ratio (SNR) is a function of attenuation $A(l, f)$ as well as the ambient noise $N(f)$ [23]. The SNR of the UWA channel at the receiver can be calculated as:

$$SNR(l, f) = \frac{P}{A(l, f)N(f)\Delta f}. \qquad (10)$$

Notice that SNR in Equation (10) is calculated in μPa per Hz. The expected SNR is a function in communion range l, the signal carrier frequency f, transmitted signal power P and expected receiver noise bandwidth Δf. As such, the expected UWA channel SNR can be maximized by using an optimal carrier frequency. The SNR for signal transmission can be predicted using the knowledge of the transmission range and the carrier frequency.

2.2.4. Doppler Shift in UWA Channel Multipath

The Doppler shift is one of the most challenging factors in the fast time varying channel. A Doppler shift in the UWA channel is different from wireless channels. It is not only a relevant to the frequency shift, but also to the low propagation speed. In low propagation speed a small receiver movement produces an impact Doppler shift [24]. The phase of the Doppler shift φ_d depends on the receiver motion velocity v and ratio of the modulated symbol rate $S_R = 1/T$ to carrier frequency f_c [25]. The UWA channel Doppler shift is given as:

$$\varphi = -2\pi f_c T \frac{\Delta}{1-\Delta} = -2\pi \frac{f_c}{S_R} \frac{v}{c-v} \qquad (11)$$

3. Coding Operation over Underwater Acoustic Channel

Advanced underwater acoustic with error correction coding is proposed to long range underwater acoustic communication [26]. In that paper, R–S block channel code is proposed with adaptive unequal error protection. It has also been reported in [27] that UEP enhance significantly the quality of the transmission operation as well as minimize the transmitted rate in highly distorted channels. Albanese et al. succeessfully adopted the UEP to transmit packets over networks in their priority encoding transmission (PET). Albanese technique is based on optimized rate as addressed in [28, 29]. In this paper, the UEP technique will be applied from another perspective. A modified SPIHT (M-SPIHT) source encoder will be proposed to generate four various types of bit-streams based on encoded bits significance in decoded image quality. M-SPIHT output bit-streams are obtained based on the contribution of each bit-steam in the

PSNR of the received image. Then the protection level of each bit-stream will be adjusted adaptively by using R–S channel code.

3.1. Problem Formulation

M-SPIHT is proposed to classify bits to four different types of groups based on their sensitivity levels and their contributions in the reconstructed image PSNR.

Bit error sensitivity (BES) analysis can be carried out assuming the image is encoded using an M-SPIHT coder. For analysis purposes, assume the encoded image is corrupted progressively bit by bit orderliness from beginning until the output bit-stream ends. For oblique bits, the decoded image MSE is calculated. The oblique bit is repaired first, then the same procedure is applied to the next bit. The proposed M-SPHI produces four major types of bit-steams with different levels of sensitivity. They can be given in order as: (1) the significance bit (2) the sign bits; (3) set bits to determine if setting significantly, (4) refinement bits. In Figure 1, it can be observed that the bit's groups are ordered by their significances from highest to lowest for a 256*256*8 Lena image and it's bits can be ordered as: significance bits > sign bits > set bits > refinement bits. The four different bit-streams have different significant levels in the received image PSNR. Channel coder overhead information is added for each transmitted packet adaptively by an optimization algorithm. The optimization algorithm is based on the length of the packet, the expected UWA channel SNR and the expected MSE is decreasing in each packet when a packet is received correctly. Depending on the rate allocation vectors and R–S overhead length the bit-stream is transmitted.

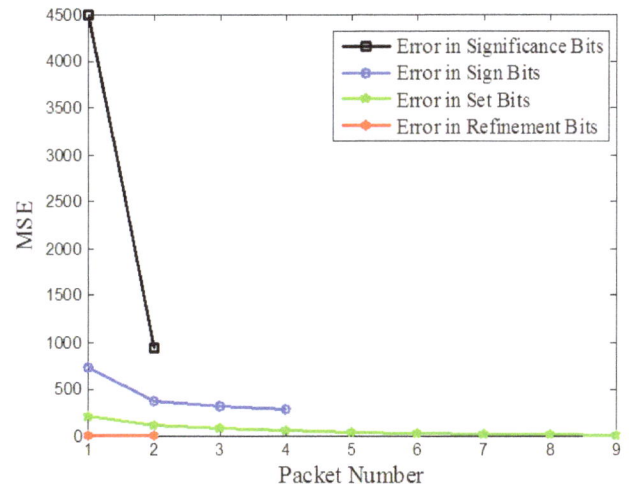

Figure 1. M-SPIHT expected packet sensitivity to error.

The forward error correction (FEC) will be using a R–S coding for data transmitted over UWA channel. Assuming the M-SPHIT encode image and its output splits into a series of packets, if $\Delta D_i \geq 0$ is the predict MSE reduction in the decoded i^{th} packet, the total deformation in total received packets can be formulated as:

$$D(l) = D_o - \sum_{i=1}^{l} P_i \Delta D_i, \qquad (12)$$

Here, D_o is the MSE expected for transmitting packet without any overhead information. P_i is the probability that i^{th} packet correctly received over the channel, and l is the total transmitted packet number. P_i is the probability of correctly received packet over the UWA channel. It is a function in the transmission data rate and it can be written as:

$$P_i = \prod_{j=1}^{i} Q_j(r_j), \qquad (13)$$

where $Q_j(r_j)$ is the probability of receiving the jth layer correctly at receiver side when it is set at a rate of r_j. Applying Equation (13) to Equation (12) yields:

$$D(l) = D_o - \sum_{i=1}^{l} \left(\prod_{j=1}^{i} Q_j(r_j) \right) \Delta D_i. \qquad (14)$$

Optimal allocation rate r_j is the amount of total bits minimizing the received packet distortion based on the Equation (14) at a constant transmission rate. The minimal distortion problem can be formulated generally as follows:

$$\min_{r} D(r) \quad Subject \quad to \quad \sum_{j=1}^{l} r_j \leq R, \qquad (15)$$

where R is the total transmission rate in the equal error protection system (EEP) and it must not exceed the channel capacity.

3.2. UWA Channel Reed Solomon Coder

Reed Solomon is a type of forward error correcting codes. It is a systematic linear block code [9, 30]. Each code block is organized into m-bit data-words units with a specific length. Its length is usually between 3 and 8 bits. For example, with 8 bit code length all probable data-words are suitable for coding. Each systematically generated code-word is composed of the original version of the encoded data-word in addition to the "parity" symbols to be subjoined it. Symbol error probability (SEP) of R–S over communication channel is P_E. For hard input R–S decoder and assuming the independence of the errors to the channel SEP and P_E is a function in symbol-error correcting capability p_e the maximum number of correctable errors t can be written as follows [31]:

$$P_E \approx \frac{1}{2^m - 1} \sum_{s=t+1}^{2^m - 1} s \binom{2^m - 1}{s} p_e^s (1 - p_e)^{2^m - 1 - s}, \qquad (16)$$

where m is the number of bits/symbol. The maximum number of correctable errors t for a (N, K) R–S code can be calculated as $t = 0.5(n - k)$.

Shallow water UWA noise is neither Gaussian nor white distributed [32-34]. In [35] the authors reported a model of probability density function (pdf) for the underwater acoustic noise issue based on real measurements at sea. An analytical expression is given for the UWA channel SEP in case of using BPSK transmission. For binary signaling, the related likelihood functions are obtained and an SEP expression is obtained. The SEP of the binary UWA noise channel can be estimated as a Student's t pdf and it is expressed by [35]:

$$P_e = k \sqrt{2 E_b / N_o} \int_0^\infty \left[(4 E_b / N_o)(x + 1)^2 + 5 \right]^{-2.75} dx \qquad (17)$$

Here, x is a vector of amplitude noise M level. E_b / N_o is the energy per bit to noise power (normalized SNR) and k is an integer $k = 1, 2, \ldots, M$.

The noise average power spectral density is $N_o = \sigma^2 / B$, where B is the bandwidth and $B = 1/2T_b$. When the amplitude of the pulses is unitary, the noise variance σ^2 could be related to the SNR per bit E_b / N_o, and it will be:

$$\sigma^2 = \frac{1}{2 E_b / N_o}. \qquad (18)$$

3.3. Optimization Technique

The output transmitted bit-stream can be split into different coding blocks. Each of these blocks includes k source packets. Each block is encoded by (N, K) R–S channel coder. In case of (N, K) R–S channel coder (N–K) is the maximum overhead information required to protect each block. Then, these packets are transmitted over UWA channel. Let P_e be the symbol error probability of the channel. Optimal rate allocation vector r for Equation (15) can be obtained by using the Lagrange multipliers Equation as:

$$J(r, \lambda) = D(r) + \lambda \sum_{i=1}^{l} r_i. \qquad (19)$$

The correct receiving conditional probability at transmission rate r_j of the jth source packet is $1 - P_E(r_j)$. Then, Equation (13) becomes:

$$P_i = \prod_{j=1}^{i} Q_j(r_j) = \prod_{j=1}^{i} (1 - P_E(r_j)), \qquad (20)$$

Apply the Equation (14) and Equation (20) into equation (19) one obtains:

$$J(r, \lambda) = D_o + \sum_{i=1}^{l} \left[\left(- \prod_{j=1}^{i} Q_j(r_j) \right) \Delta D_i + \lambda r_i \right]. \qquad (21)$$

This function in Equation (21) can be optimized to determine the appropriate r_j. It is dependent on the set of distortion increments ΔD_i, as well as the probability that the jth layer source packet is received correctly. There are many methods available to solve this optimization problem. It can be solved based on an alternating variables iterative method [36]. $J = (r_1, \ldots\ldots, r_l)$ in Equation (21) is the objective optimization function. It is used to minimize one variable each time. The objective optimization function is keeping other variables constant until the iterative process reach to the convergence. Let $r(0)$ is the initial rate allocation vector and $r(t) = (r_1(t), \ldots\ldots, r_l(t))$ is obtained values at $t = 1, 2, \ldots$ as follows: for $x \in \{r_1, \ldots\ldots, r_l\}$ to be optimize at t step this done in a round–robin way. In the next step, for $x = r_i$, following rate optimization is performed:

$$f(r) = r_i^{(t)} = \underset{D_i}{\arg\min} J\left(r_1^{(t)}, \ldots\ldots, r_l^{(t)}\right) - \underset{r_i}{\arg\min} \sum_{i=1}^{l}\left[\left(-\prod_{j=1}^{v}Q_j(r_j)\right)\Delta D_i + \lambda r_i\right], \quad (22)$$

$$D(r_1, r_2, r_3, \ldots\ldots, r_l) = D_o - \sum_{i=1}^{l}\left(\prod_{j=1}^{i}Q_j(r_j)\right)\Delta D_i, \quad (23)$$

$$R(r_1, r_2, r_3, \ldots\ldots, r_l) = \sum_{i=1}^{l}r_i, \quad (24)$$

where r is the overall transmission rate in the unequal error protection system (UEP) for each transmitted packet with total transmission rate $\sum_{j=1}^{l}r_j \le R$.

R is the maximal total transmission rate in the equal error protection (EEP) system.

The Lagrange multipliers method can also be used to find the point of minimum overall distortion D, across all transmission rates R_n in the unequal error protection system

(UEP) with such maximal rate constraint. It leads to:

$$\left.\frac{\partial}{\partial \vec{r}}\left(D + \lambda(R_n - R)\right)\right|_{\vec{r}=\vec{r}^*} = 0, \quad (25)$$

$$\left.\frac{\partial}{\partial \vec{r}}\left\{D_o - \sum_{i=1}^{l}\left(\prod_{j=1}^{i}Q_j(r_j)\right)\Delta D_i + \lambda\left(\sum_{j=1}^{l}r_j - R\right)\right\}\right|_{\vec{r}_k=\vec{r}_k^*} = 0. \quad (26)$$

The gradient-descent type algorithms as a standard non-linear optimization procedure can also be used to minimization optimization problem for a fixed λ [36].

4. Proposed UWA Communication System

Based on the considerations listed in Sections 1, 2, and 3, we propose a communication system scheme for UWA channels as shown in Figure 2. The major components are: (1) the M-SPIHT coder used to generate a four types of bitstream regards to its significance order; (2) data outputs which is split into a series of stream packets; (3) the predictable decrements in ΔD_i that can be approximately estimated based on the expected change in MSE; (4) the optimization algorithm to generate the optimum bit rate distribution of packets, which is adaptive to the optimization algorithm inputs, including packet length, expected decrease in distortion ΔD_i and channel SNR; (5) R–S encoder used the output rate allocation vectors to generate the transmitted bit-stream; (6) the receiving bit-stream to be decoded by using R–S decoder the M-SPIHT decoder.

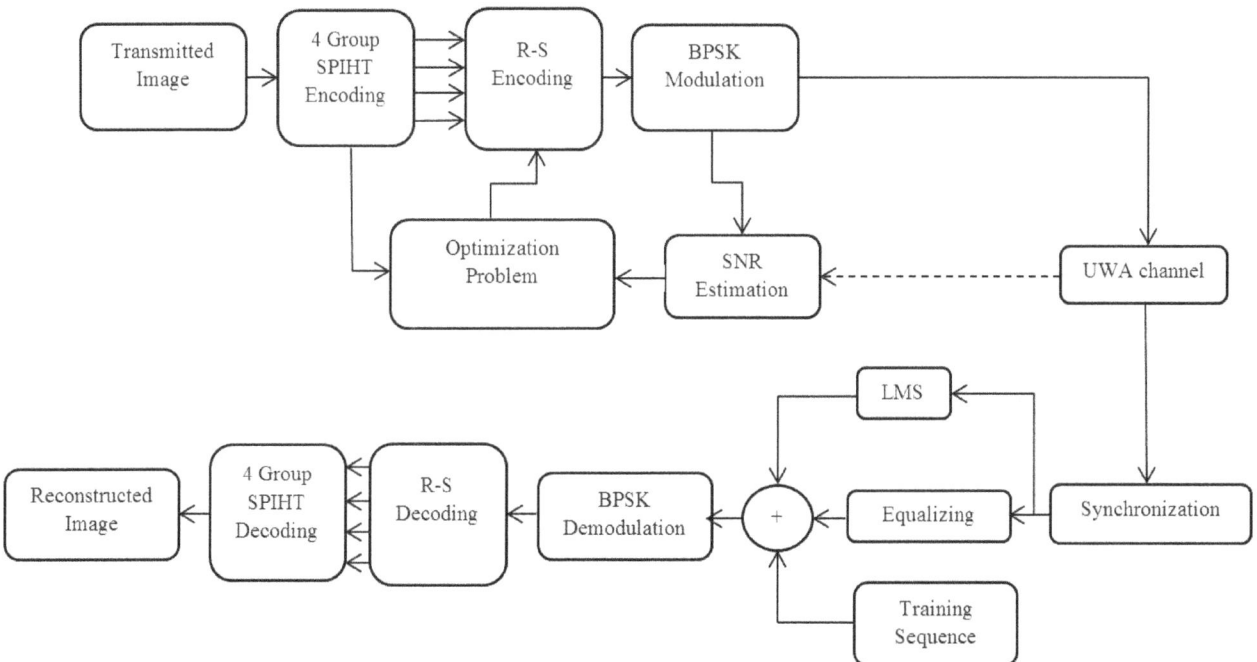

Figure 2. System configuration.

An SNR estimation block is included in the diagram in Figure 2. In this paper, SNR is estimated using equation (10) which depends on the distance between the transmitter and receiver. The carrier frequency, signal transmission power and receiver noise bandwidth are all known at the transmitter. Another important parameter in the equation (10) is the ambient noise (considered as feedback signal) and it depends on physical channel characteristics. Equation (9) can be used to approximate the ambient noise. This block can also be integrated with other information source such as the bit-error rate estimation from network management protocols to facilitate direct feedback from the real time communication status to the optimal bit protection level allocation.

At the beginning, the SPIHT coder output bit-stream modification has been introduced. The proposed UWA communication system for image transmission is based on the sensitivity of each block of M-SPIHT output and its contribution in reconstructing image PSNR.

The BES can be evaluated with corrupted M-SPIHT encoder outputs bit–by–bit, starting from the first bit to the last one. For each corrupted bit, the MSE of original and the reconstructed image is calculated. For each iteration the last corrupted bit is corrected first. Then, the next bit is corrected and transmitted.

5. Simulation Results

5.1. Simulation Setup

In this section, the simulation results using proposed M-SPHIT coder algorithm for image transmission over the UWA channel will be presented. Communication system described in Section 4 is implemented. 256*256*8 Lena image is used as the source image encoded using M-SPHIT with 0.5 bpp compression rates. The resultant BES is depicted in Figure 1. As shown in the Figure 1, the output is ordered starting from the highest significant bits (a first group of bits). We consider a link of 5 Km in shallow water and take into account different wind speeds, while water depth is 40m, the relative velocity is set to be 8 m/s. The attenuation-noise (AN) factor is obtained as 8.6 KHz as the optimal carrier frequency for our suggested underwater channel physical parameter according to Equation (10). Table 1 shows the calculation results of channel relative delays and magnitudes for each path. The wind causes significant waves, scatters the surface reflection, and gives these reflections a Doppler shift. For that decision-feedback equalizers (DFEs) have been used. DFEs in conjunction with the training data sequence are used to equalize the received data received from these types of channels [37] with this behavior. A minimum mean square error (MMSE) criterion is used to estimate the coefficients of equalizer as well as carrier recovery parameters.

UWA channel is a time variant channel. Least mean squares (LMS) approach can be used to estimate the time delay and equalize such variant channel based on the DFEs equalizer. After demodulation R–S decoder is used to recover the decoded image stream, finally decoders of 4 group SPIHT bit strems recover the transmitted image.

Table 1. Underwater acoustic channel profile for simulations.

Parameter	value				
	1st path	2nd path	3rd path	4th path	5th path
Amplitude	0.206	0.145	0.073	0.036	0.018
Relative delay time (ms)	0	0.7	2.24	4.64	9

The Equation (11) is applied to detect the Doppler shift effect and its amplitude in each path. Data symbols are modulated using the BPSK with symbol duration $T = 0.4$ *ms*. After demodulation, the R–S decoder is used to recover the decoded image stream.

The M-SPHIT encoder output is divided into packets. Each packet has a 2000 bit, which is further divided into 25 blocks with each block to be 80 bits. Each block has 10 symbols with each encoded using one byte. In the simulation results, the encoded Lena image bit-streams are split into 17 packets. M-SPHIT output divides these packets to four separated groups based on its significance. In particular, (1) two of the encoded packets are for the significant bits and they will have the highest entropy, (2) four packets are used to represent the sign bits in the encoded image, (3) nine of the encoded packets are used to represent the set bits, (4) the last two encoded packets are used to represent the refinement bits in the image. At first, the outcome of UEP method are compared with that of the EEP method. For the EEP case, the number of R–S overhead symbols is selected to be six in an encoded packet row. That makes the total number of symbols per packet sent to the modulation to be 400.

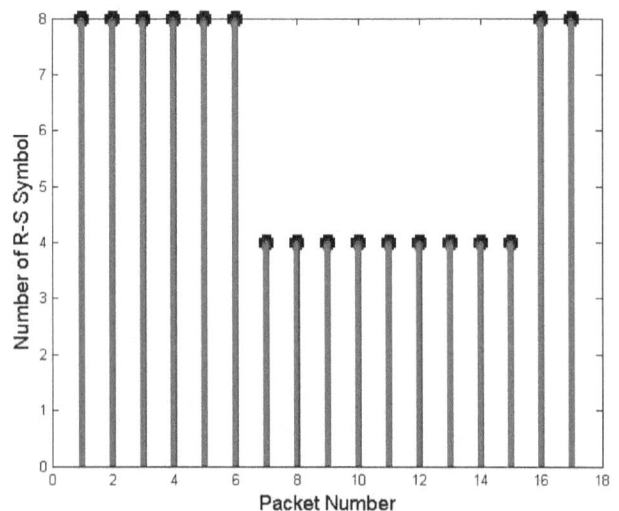

Figure 3. Transmitted "Lena" packets R–S protection levels as determined using the UEP algorithm.

In the case of UEP case, the total number of symbols in all encoded packets in Equation (22) are selected to be no more than that of EEP with R–S symbols equal to 6 symbols, which should be less than channel capacity. For each row of the packet and R–S symbols is determined for each encoded packet based on the optimization algorithm to minimize the total distortion. As shown in Figure 3, the sign and significance bit-stream blocks are protected by an eight-symbols, considering the low UWA channel signal-to-noise ratio in this case. The set packets groups are protected using four symbols and eight symbols are used to protect the refinement bit-steams.

5.2. Results

Image transmission over the underwater channel is simulated under different physical conditions to assess the system performance for both unequal and equal error protection with the R–S channel coder. The parameter of wind speed is selected at different values. Each experiment runs for 50 times to collect the average. As shown in Figure 5 and Figure 6, where several variations of images originated from the standard Lena image and dolphin swimming [8, 21], respectively are transmitted through using the proposed underwater communication system. The reconstructed images are obtained at different wind speed levels using the two protection algorithms, unequal and equal error protection, with channel rates distributions for the protection with unequal error protection algorithm case.

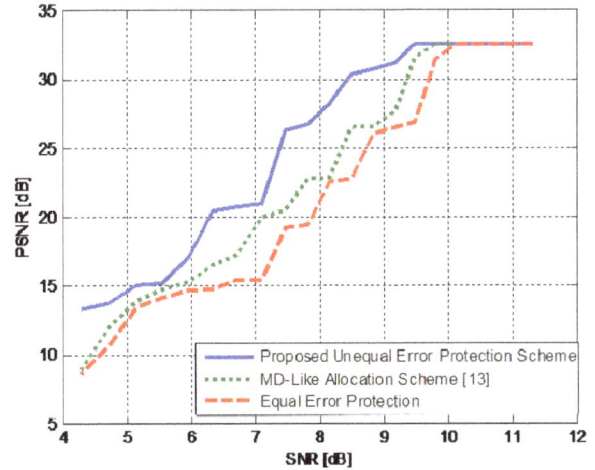

Figure 4. *The average PSNR of the received Lena image transmitted over the UWA channel for EEP, MD-Allocation [13] & proposed UEP channel coding, at channel Doppler shift = 29.6 Hz.*

Original Lena Image

Lena Image received using (UEP), wind=10 Knots, PSNR=29dB, total symbol transmitted=67487

Lena Image received using (EEP), wind=10 Knots, PSNR=26dB, total symbol transmitted=67487

symbol rates distributions of R-S coder per packet for UEP

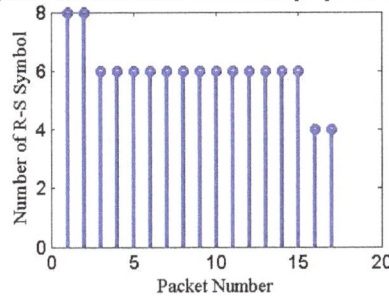

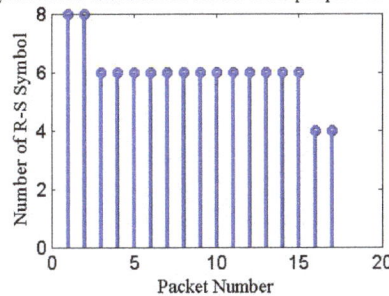

Lena Image received using (UEP), wind=40 Knots, PSNR=20.4dB, total symbol transmitted=67487

symbol rates distributions of R-S coder per packet for UEP

Lena Image received using (EEP), wind=40 Knots, PSNR=17.8dB, total symbol transmitted=67487

(a) (b) (c)

Figure 5. *Reconstructed "Lena" image transmitted over the UWA channel with equal and unequal error protection with channel rate distribution of unequal error protection.*

Original Dolphin Swimming Image

Dolphin Swimming received using (UEP), wind=10 Knots, PSNR=34.87dB, total symbol transmitted=65999

symbol rates distributions of R-S coder per packet for UEP

Dolphin Swimming received using (EEP), wind=10 Knots, PSNR=26dB, total symbol transmitted=67487

Dolphin Swimming received using (UEP), wind=25 Knots, PSNR=23.3dB, total symbol transmitted=67487

symbol rates distributions of R-S coder per packet for UEP

Dolphin Swimming received using (EEP), wind=25 Knots, PSNR=21.4dB, total symbol transmitted=67487

Dolphin Swimming received using (UEP), wind=40 Knots, PSNR=19.9dB, total symbol transmitted=66987

symbol rates distributions of R-S coder per packet for UEP

Dolphin Swimming received using (EEP), wind=40 Knots, PSNR=18.1dB, total symbol transmitted=67487

(a)   (b)   (c)

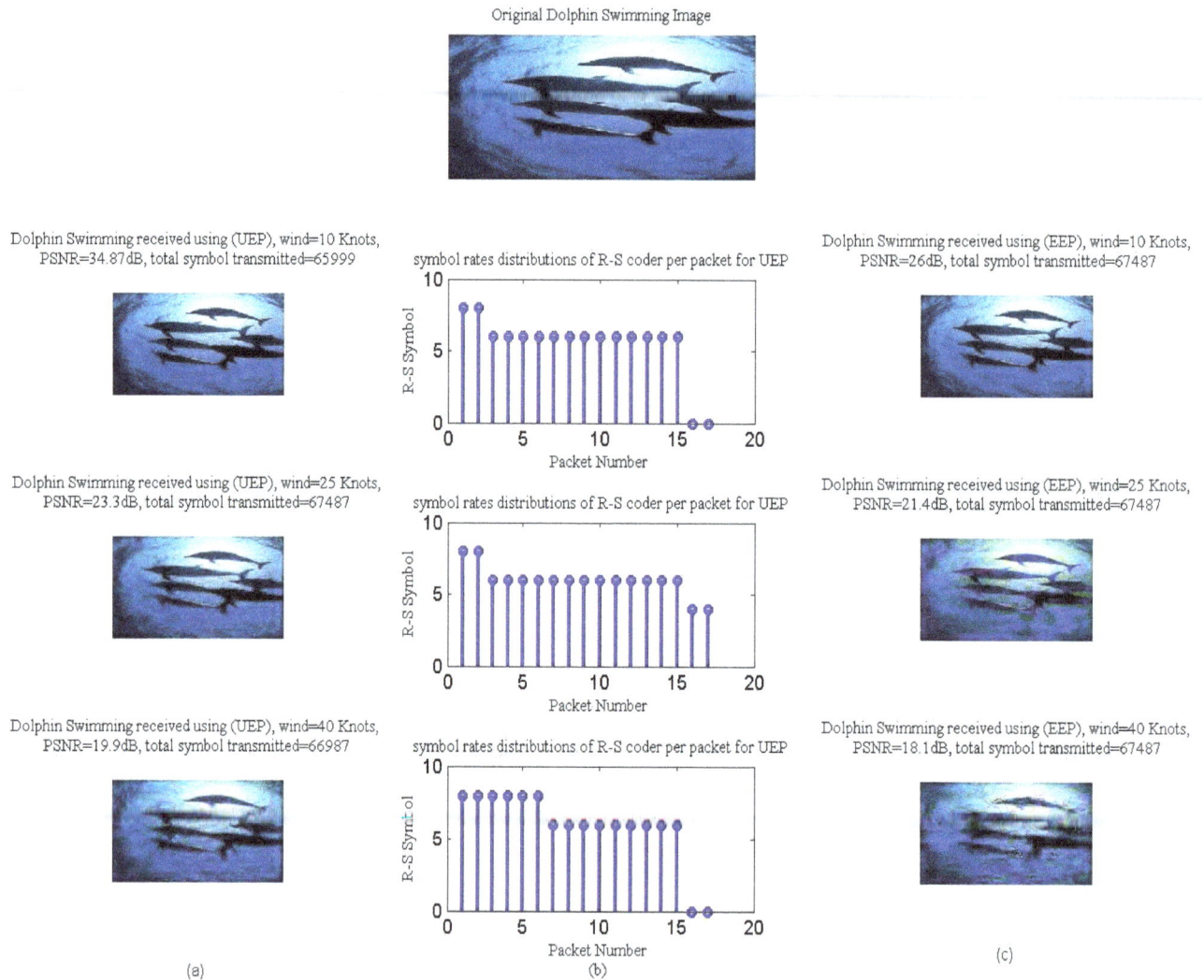

Figure 6. *Reconstructed "Dolphin Swimming" image transmitted over the UWA channel with equal and unequal error protection with channel rate distribution of unequal error protection.*

Table 2. *Proposed scheme comparison to others.*

Items	Proposed scheme	MD-Like allocation scheme, [11].	HQAM- approaches [21]
Complexity	Complex	Complex	Simple
CSI (channel state information)	Not required	Required	Not required
Feedback from receiver to transmitter	Not required	Required	Not required
Validity for different communication systems	Valid for any communication system	Valid for any communication system	Specific only for HQAM systems
Simulation computation time (seconds)	151.19477	135.6961	88.381125

In *UEP* case, the total number of transmitting symbols (R in (15)) in the simulation is 67487. The redundancy symbol number for each encoded group, including the first two packets are significance packet, followed by four sign packets, then nine set packets and two refinement packets in the final, is determined by solving the optimization problem defined in (15). R–S symbol redundancy is shown in Figure 5 (b). In that Figure, the significance packets have eight symbols protection level, the sign and set packets have six symbols protection level, and refinement packets are protected by four symbols. In the simulation of dolphin swimming image, R–S rate distribution changes under various channel wind

condition, as shown in Figure 6 (b). In particular, at 10 Knot wind the packets of significance type are protected by eight symbols, six symbols are used to protect the sign block and set type of the packets, and the refinement packets are completely unprotected. As the result, the total number of transmitting symbols using UEP is 65999 symbols, which are 1488 symbols less than that obtained using traditional EEP technique.

Figure 4 shows the average PSNR of the received Lena image transmitted over the UWA channel by using the proposed scheme, MD-Like allocation shame [11], and equal forward error correction, respectively, with R–S coder all coded images transmitted at certain total bit budget.

Due to high UWA channel noise, curves in Figure 4 are not smooth. The unequal error protection encoding method improves the PSNR of the received image. From the simulation results, the proposed UEP method outperforms the conventional EEP system by 3.5 dB and the system of MD-Like allocation schemes [11] by 2.2 dB at underwater acoustic channel with Doppler shift equal 29.6 Hz at fixed total bit budget. The iterative procedure in the proposed system increases the time needed to encode an image, but results in higher PSNR for reconstructing images in low SNR channel which make it more suitable for underwater acoustic channel. Also from the simulation results R–S code is shown to be suitable for an underwater acoustic channel.

In our work in [21], we proposed unequal error protection with only certain communication system Hierarchical Quadrature Amplitude Modulation (HQAM). Table 2 has a conclusion of a comparison between our proposed and other methods.

6. Conclusion

In this paper an efficient rate allocation image transmission scheme in underwater acoustic system is discussed. Based on a modified SPIHT coder scheme proposed, a novel image transmission system has been developed. SPIHT coder algorithm has been modified based on the order of significance of the coded bits as well as its contribution in the PSNR of the received image. In particular, four different types of bit-stream are generated by a modified SPIHT coder. The Reed Solomon channel coder is used together with the output of rate allocation vectors to generate the output bit-stream for transmission. While such technique can be considered for generic image transmission issues to achieve reasonable efficiency, an optimization problem is carefully formulated by taking the particular channel feature of underwater acoustic communication channel, such as Doppler shift and probability bit error distribution, into account. As the result, noticeable reduction of the overall distortion over underwater acoustic channel using our proposed technique is observed, together with the improvement in PSNR for the reconstructed image. Simulation experiments are included for illustrative purpose with different image types and different physical parameters such as wind speed. It is shown that the proposed system can effectively reduce the overall distortion in the reconstructed image and improve PSNR with approximately 1.5 dB in one of the most distorted channels.

References

[1] I. Iglesias, S. Aijun, J. Garcia-Frias, M. Badiey, and G. R. Arce, "Image transmission over the underwater acoustic channel via compressive sensing," in *45th Annual Conference on Information Sciences and Systems*, 2011, pp. 1-6.

[2] D. B. Kilfoyle and A. B. Baggeroer, "The state of the art in underwater acoustic telemetry," *IEEE Journal of Oceanic Engineering*, vol. 25, pp. 4-27, 2000.

[3] J. Rice, B. Creber, C. Fletcher, P. Baxley, K. Rogers, K. McDonald, *et al.*, "Evolution of seaweb underwater acoustic networking," in *Oceans*, 2000, pp. 2007-2017.

[4] M. Stojanovic, "Underwater acoustic communication," *Wiley Encyclopedia of Electrical and Electronics Engineering*, 1999.

[5] C. Murphy and H. Singh, "Wavelet compression with set partitioning for low bandwidth telemetry from AUVs," in *Proceedings of the Fifth ACM International Workshop on UnderWater Networks*, 2010, pp. 1-8.

[6] C. Pelekanakis, M. Stojanovic, and L. Freitag, "High rate acoustic link for underwater video transmission," in *OCEANS*, 2003, pp. 1091-1097.

[7] A. Said and W. A. Pearlman, "A new, fast, and efficient image codec based on set partitioning in hierarchical trees," *IEEE Transactions on Circuits and systems for video technology*, vol. 6, pp. 243-250, 1996.

[8] H. Esmaiel and D. Jiang, "SPIHT coded image transmission over underwater acoustic channel with unequal error protection using HQAM," in *IEEE International Conference on Information Science and Technology (ICIST)*, 2013, pp. 1365-1371.

[9] W. C. Cox, J. A. Simpson, C. P. Domizioli, J. F. Muth, and B. L. Hughes, "An underwater optical communication system implementing Reed-Solomon channel coding," in *OCEANS*, 2008, pp. 1-6.

[10] P. A. Chou, A. E. Mohr, A. Wang, and S. Mehrotra, "Error control for receiver-driven layered multicast of audio and video," *IEEE Transactions on Multimedia*, vol. 3, pp. 108-122, 2001.

[11] B. Tomasi, L. Toni, P. Casari, J. Preisig, and M. Zorzi, "A study on the SPIHT image coding technique for underwater acoustic communications," in *Proceedings of the Sixth ACM International Workshop on Underwater Networks*, 2011, p. 9.

[12] M. A. Kader, F. Ghani, and R. B. Ahmed, "Unequal Error Protection for SPIHT Coded Image Transmission over Erroneous Wireless Channels," *Asian Transactions on Fundamentals of Electronics, Communication & Multimedia*, vol. 1, pp. 1-6, 2011.

[13] U. S. Mohammed and H. Hamada, "An efficient rate allocation scheme with selective weighted function for optimum peak-to-average power ratio for transmission of image streams over OFDM channels," *International Journal of Video & Image Processing & Network Security*, vol. 9, 2009.

[14] U. S. Mohammed and H. Hamada, "Image transmission over OFDM channel with rate allocation scheme and minimum peak-toaverage power ratio," *Journal of Telecommunication*, vol. 2, p. 70:78, 2010.

[15] Y. Sun, R.-m. Li, and X.-l. Cao, "Image compression method of terrain based on Antonini wavelet transform," in *IEEE International Geoscience and Remote Sensing Symposium*, 2005, pp. 692-695.

[16] E. Christophe and W. A. Pearlman, "Three-dimensional SPIHT coding of volume images with random access and resolution scalability," *Journal on Image and Video Processing*, vol. 2008, pp. 1-13, 2008.

[17] Z. Lu, D. Y. Kim, and W. A. Pearlman, "Wavelet compression of ECG signals by the set partitioning in hierarchical trees algorithm," *IEEE Transactions on Biomedical Engineering*, vol. 47, pp. 849-856, 2000.

[18] J. Preisig, "Acoustic propagation considerations for underwater acoustic communications network development," *ACM SIGMOBILE Mobile Computing and Communications Review,* vol. 11, pp. 2 10, 2007.

[19] K. Prasanth, "Modelling and Simulation of an Underwater Acoustic Communication Channel," Master, Electronic Engineering University of applied sciences, Bremen, Germany, 2004.

[20] H. Esmaiel and D. Jiang, "Review Article: Multicarrier Communication for Underwater Acoustic Channel," *International Journal of Communications, Network and System Sciences,* vol. 6, pp. 361-376, 2013.

[21] H. Esmaiel and D. Jiang, "Image transmission over underwater acoustic environment using OFDM technique with HQAM mapper," in *IEEE International Conference on Information Science and Technology (ICIST),* 2013, pp. 1596-1601.

[22] Y. Chen, X. Xu, L. Zhang, and Z. Zou, "Design and Application of dynamic coding in shallow water acoustic communications," in *OCEANS,* 2012, pp. 1-6.

[23] A. Sehgal, I. Tumar, and J. Schönwälder, "AquaTools: An Underwater Acoustic Networking Simulation Toolkit," *Oceans,* pp. 1-6, 2010.

[24] A. Fish, A. Sayeed, S. Gurevich, R. Hadani, and O. Schwartz, "Delay-Doppler channel estimation with almost linear complexity: To Solomon Golomb for the occasion of his 80 birthday mazel tov," in *IEEE International Symposium on Information Theory Proceedings,* 2012, pp. 2386-2390.

[25] J. Trubuil and T. Chonavel, "Accurate Doppler estimation for underwater acoustic communications," in *OCEANS,* 2012, pp. 1-5.

[26] A. Katariya, A. Arya, and K. Minda, "Coded under Water Acoustic Communication (UWA) with Cryptography," in *International Conference on Computational Intelligence and Communication Networks,* 2010, pp. 493-497.

[27] S. Sandberg and N. Von Deetzen, "Design of bandwidth-efficient unequal error protection LDPC codes,"

IEEE Transactions on Communications, vol. 58, pp. 802-811, 2010.

[28] A. Albanese, J. Blomer, J. Edmonds, M. Luby, and M. Sudan, "Priority encoding transmission," *IEEE Transactions on Information Theory,* vol. 42, pp. 1737-1744, 1996.

[29] R. Puri and K. Ramchandran, "Multiple description source coding using forward error correction codes," in *Conference Record of the Thirty-Third Asilomar Conference on Signals, Systems, and Computers,* 1999, pp. 342-346.

[30] A. Goalic, J. Trubuil, C. Laot, and N. Beuzelin, "Underwater acoustic communication using Reed Solomon Block Turbo Codes channel coding to transmit images and speech," in *OCEANS,* 2010, pp. 1-6.

[31] J. P. Odenwalder, "Error control coding handbook," DTIC Document1976.

[32] M. Chitre, "A high-frequency warm shallow water acoustic communications channel model and measurements," *The Journal of the Acoustical Society of America,* vol. 122, p. 2580, 2007.

[33] M. Stojanovic, J. Catipovic, and J. G. Proakis, "Adaptive multichannel combining and equalization for underwater acoustic communications," *The Journal of the Acoustical Society of America,* vol. 94, p. 1621, 1993.

[34] M. A. Chitre, J. R. Potter, and S.-H. Ong, "Optimal and near-optimal signal detection in snapping shrimp dominated ambient noise," *IEEE Journal of Oceanic Engineering,* vol. 31, pp. 497-503, 2006.

[35] J. S. Panaro, F. R. Lopes, L. M. Barreira, and F. E. Souza, "Underwater Acoustic Noise Model for Shallow Water Communications," in *Simpósio Brasileiro de Telecomunicações,* 2012, pp. 1-4.

[36] C. T. Kelley, *Iterative methods for optimization* vol. 18: Society for Industrial and Applied Mathematics, 1987.

[37] M. Stojanovic, "Recent advances in high-speed underwater acoustic communications," *IEEE Journal of Oceanic Engineering,* vol. 21, pp. 125-136, 1996.

5

Design and implementation of a teleprotection system with digital and analog interfaces

Vahid Hamiyaty Vaghef, Maryam Shabro, Behnam Gholamrezazadeh Family

Communication Department, Niroo Research Institute, Tehran, Iran

Email address:
vvaghef@nri.ac.ir (V. H. Vaghef), mshabro@nri.ac.ir (M. Shabro), behnam_gh_f@yahoo.com (B. G. Family)

Abstract: Safe and un-interrupted power transmission requires a widespread use of control and protection equipment to avoid fault propagation in power network. Protection systems have been used to detect faults and block them to strike the high voltage lines. The teleprotection system continuously sends commands to remote substation in case of fault identification. Thus, the reliability and transmission time are vital issues to avoid intolerable errors. Teleprotection systems are interface between protection system and communication equipment. This paper proposes a digital teleprotection system to connect to digital communication network. The system is based on line interfaces and in accordance to ITU-T G.703 standard (64kbps and 2Mbps rates). Furthermore, the system benefits from the analog interface and direct analog voice band or optical fibers connection using eight independent commands. The system supports 8 channels simultaneously while its transmission time is 4ms at most.

Keywords: Protection, Relay Interface, Communication Interface, Type Test, Dependability and Security

1. Introduction

High voltage networks, transmission lines, bus bars, power transformers, etc. are mainly installed and operated in outdoor and are exposed to natural disasters. Therefore, they should be protected against hazard events and they should be separated from other parts of the network, to prevent the fault expansion. In order to protect the equipment upon an error occurrence and/or abnormal situations, a protective system should be used to detect and isolate the faulty parts from the whole network and prevent the fault from propagation [1]. Using the teleprotection system protection commands can be sent to remote substation through the telecommunication network in order to isolate the damaged sections from the entire network according to protection scheme. "Fig. 1" shows the architecture of a protection link between two substations.

Figure 1. Systems for a protection link

Teleprotections are used in conjunction with communication channels to establish a link between protection relays in substations of both line sides. In situation of breakdown in voltage lines or equipment, the protection equipment with an available and reliable communication media, provide the possibility to isolate damaged sections of the entire network by sending commands in the shortest possible time.

The performance of a teleprotection system is characterized by several parameters. Traditionally, the Propagation Delay, Dependability and Security have been considered the most important ones.

The teleprotection system works as an interface between the protection system and the telecommunication devices. Therefore, it includes some sections which prepare inputs and outputs based on these equipments. In general, teleprotection system is divided into three parts:

- Protection equipment interface
- Processing and making decisions
- Telecommunication equipment interface

The protection interface (or high voltage relay interface) and the communication interface provide an isolated full-duplex communication. The processing unit controls the states and makes decisions based on the receiver conditions.

The majority of current analog teleprotection systems (ATPS) transmits up to four commands and connects to analog telecommunication systems such as analog PLCs. While ATPS are mainly used in power utilities, digital communication networks are growing rapidly in many industrial applications such as electrical power industry. Therefore, a comprehensive protection and control of electrical power systems is achieved using digital communication networks.

Digital teleprotection systems (DTPS) are used in conjunction with digital communication equipments. These systems provide signal conditioning, data encoding/decoding and error detection capabilities. Due to their digital nature, DTPSs offer built-in high security and dependability along with fast response in comparison with analog ones.

In this article, the specification of commands in DTPS, the block diagram of different parts and the features of the designed DTPS are discussed. Besides, the type tests according to the IEC60834-1 standard for DTPS are presented.

2. Specification of Commands in the Designed System

One of the main advantages of DTPS with respect to ATPS is the communication mechanism. While in ATPS, the frequency and power level of commands is transferred, in DTPS each protection command is mapped into an n-bit code word. These code words are chosen to reduce the probability of receiving incorrect code due to the channel noise. In this study, a convolution coding is employed to generate 16-bit code words. This ensures that the hamming distance of a code word is higher than 8 bit and leads to the more flexible framing and to provide a much greater diversity of commands. Utilizing innovative techniques such as iterative frame sending, decision threshold and acceptance window to code words minimizes the probability of receiving unwanted or missed commands. Therefore, higher security and dependability is achieved in higher BER channels.

The *guard* code is sent in normal condition while in other cases, the command codes are sent. The effective parameters in commands transmit and receive is illustrated in "Fig. 2". When a change in the status of input command happened, the input filter checks the validity of it, and the transmitter sends a command code word instead of Guard code word. This code word will be received after some delay, because of transmission media and receiver execution delay in the receiver of the opposite side. These delays make the nominal transmission time that is one of the main parameter in teleprotection systems.

It should be noted that an analog communication interface translated the code words to voice band frequencies. Each command is mapped to single or dual tone frequencies. Transmission delay and robustness against noise and interference are taking into account using the appropriate algorithm.

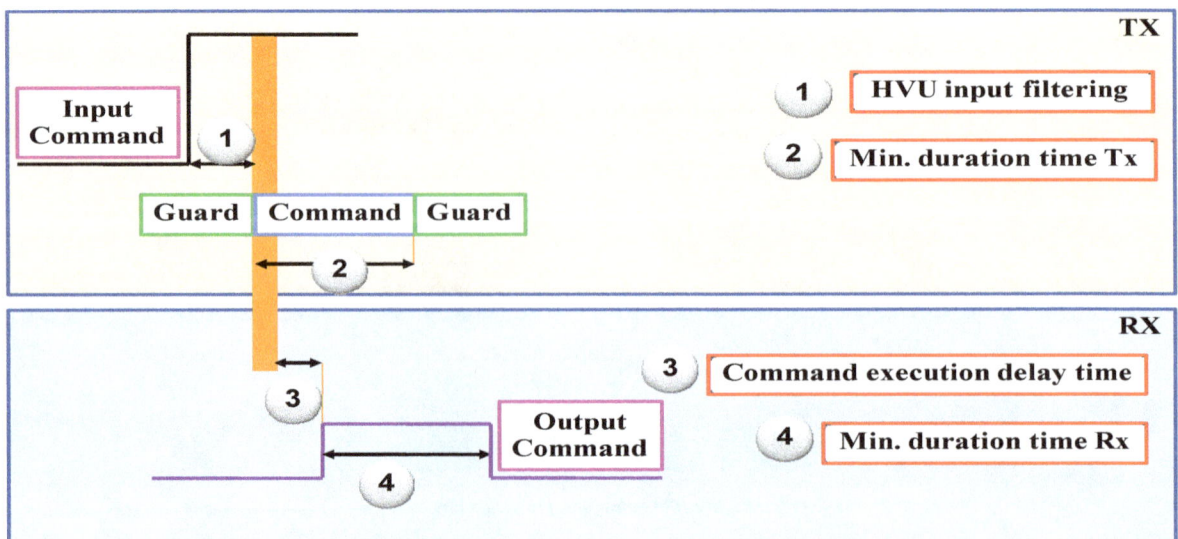

Figure 2. Parameters in commands transmit and receive.

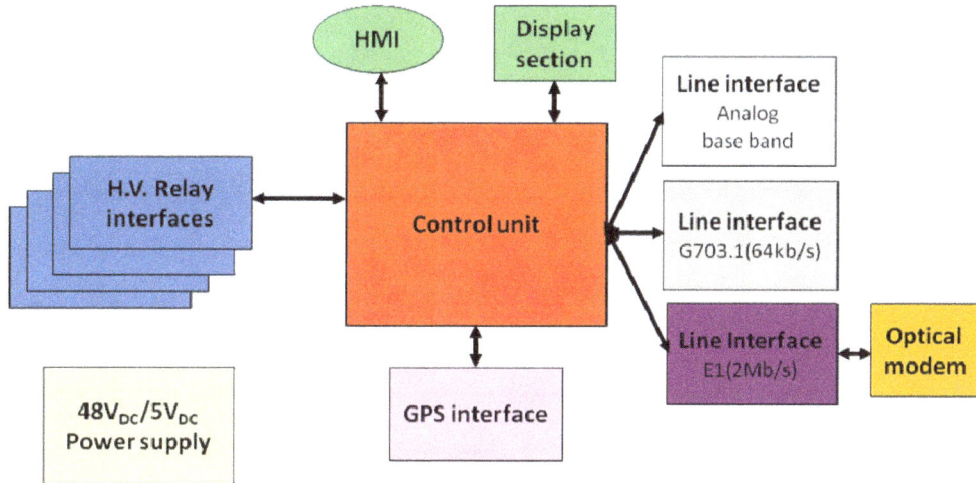

Figure 3. Generic blocks of the designed DTPS.

3. Block Diagram of the Designed System

The generic block diagram of the designed DTPS and their interactions is shown in "Fig.3". It has a central controller unit, communication/line interfaces, H.V. relay interfaces, interface with GPS receiver, display units, user interfaces (HMI), and power supply converter. The main characteristics of the designed and implemented DTPS are [2]:

- Possibility to send and receive up to eight commands as independent or simultaneous
- Isolated interfaces for connecting to a protection system
- The ability to connect telecommunication equipment with digital interface (E1 and 64kbps interfaces in accordance with ITU-T G.703 standard)
- The ability to connect communication equipment with the analog interface card in voice band frequency
- Possibility to device configuration through the RS232/USB/LAN interfaces
- Parameters adjustment by software
- Possibility to record the history of commands and events
- Possibility for synchronization with GPS
- An isolated DC power supply
- Modular parts with a 3U standard sub rack
- In compliance with IEC60834-1 standard

The main tasks of the units are described in below

3.1. Control Unit

Some important processing in this unit are: making decisions according to relay status in the transmitter section, choosing appropriate code words, processing the received commands in the receiver side, checking the validity of commands, check the alarm conditions, controlling all the tasks and parameters configuration, interface for GPS receiver and HMI software.

3.2. Line Interface Units

These units make a full duplex communication with telecommunication equipments including line drivers, framer,

and isolators. The block diagram of LIU cards is shown in "Fig.4" and "Fig.5".

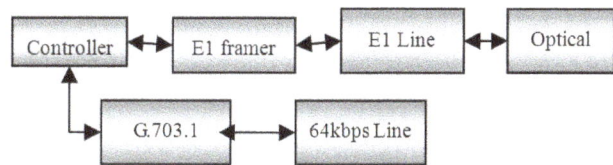

Figure 4. Blocks of Digital LIU cards

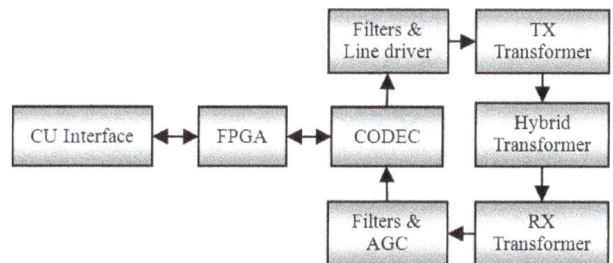

Figure 5. Blocks of Analog LIU card.

3.3. H.V. Relay Interface Units

These units receive the state of protection relays and transmit the necessary commands to protective relays. It includes level shifters, comparators, buffers, and isolators (opto-coupler and relay). The block diagram of RIU cards is shown in "Fig.6".

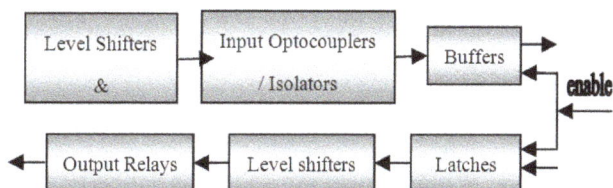

Figure 6. Blocks of H.V. RIU card.

3.4. Power Supply Unit

The power supply unit makes an isolated power for internal use in system; it includes line protection, EMI filter, line filter,

DC/DC converters. The block diagram of PSU card is shown in "Fig.7".

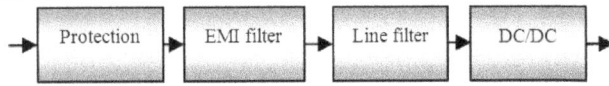

Figure 7. Blocks of PSU card.

4. Type Tests of the Designed System

The functional and performance tests are the most popular tests. Besides this issue, due to a variety of electronic equipments which are used in the industrial environments, the insulation and electromagnetic compatibility between them are very important. Also environmental tests will be done on the device. The IEC60834-1 standard is published for Teleprotection systems [3]. It defines the tests level, and their acceptance criteria. The tests are listed as below:

1) Power Supply tests, including:
   - Power supply variation
   - Interruptions
   - LF disturbance emission
   - Reverse polarity
2) TeleProtection System Performance tests, including:
   - Transmission time
   - Security
   - Dependability
   - Security with sudden signal interruption
   - Recover time
   - Jitter
   - Nominal impedance
3) Insulation Voltage Withstand tests, including:
   - Power frequency voltage withstand test and insulation resistance
   - Impulse voltage withstand test
4) Electro Magnetic Compatibility tests, including:
   - Damped oscillatory wave immunity test
   - Fast transient burst (EFT)
   - Electrostatic discharges (ESD)
   - Radiated electromagnetic field
   - RF disturbance emission
5) Environmental Conditions tests, including:
   - Temperature and humidity
     - Change of temperature test
     - Damp heat test
     - Dry heat test
     - Cold test

- Mechanical tests, including:
  - Shock
  - Vibration

All of the above mentioned type tests have been performed on the designed DTPS in reference laboratory. Despite the practical and theoretical considerations, a variety of problems have been faced during the EMC tests. After some modifications, the final test results showed the compatibility to the IEC60834-1 standard.

5. Discussion and Conclusion

In this paper the design and implementation of a digital teleprotection system is presented. The main features of the implemented DTPS have been described. Incorporating the E1 and 64kbps (according to G.703 standard) interfaces are main features of this system. Connecting to analog telecommunication through analog voice band, direct connecting to optical fibers and send/receive up to eight independent or simultaneous commands are the salient features of the system. Using a graphical user interface, system parameters are set. Therefore, using the presented system, different protection schemes (direct, permissive and blocking) in 400kV, 230kV and 63kV substations are available. The results of type tests on this product meet the IEC60834-1 requirements.

Acknowledgements

This work was supported by the Niroo Research Institute (NRI), Iran and PKG under Grant No. JCMPN02.

References

[1]   Cigre, Working Group B 5.19 of Cigre, "Protection relay coordination", October 2010.

[2]   NRI, Communication research group, final report of "Design and implementation of a industrial Teleprotection system for digital communication networks", document code: JCMPN02/E, 2012.

[3]   IEC60834-1, "Teleprotection equipment of power systems, performance and testing, Part 1: command systems", Second Edition, IEC, 1999.

[4]   M. Shabro, V. H. Vaghef, and B. Gh. Family, "Design and Implementation a Digital Teleprotection System", 7thSASTech 2013, Iran, Bandar-Abbas. 7-8 March, 2013.

6

Risk management methods applied to renewable and sustainable energy

Lee Cheuk Wing*, **Zhong Jin**

Department of Electrical and Electronic Engineering, The University of Hong Kong, Hong Kong, China

Email address:
cwlee@eee.hku.hk (Lee Cheuk Wing), jzhong@eee.hku.hk (Zhong Jin)

Abstract: Renewable energy policy has always been recognized as a major incentive to the growth of renewable energy and market. In particular, in the last decade, renewable energy sources are considerably increased due to the supportive renewable energy policy worldwide. Policymakers keep on updating and revising policies in response to market changes and advances in technologies. At the same time, policymakers have shifted their perspectives from cost and benefit to risk and return so as to align with investors' perspectives. As a result, risk management has to be kept accordance with the changing policy of renewable energy. The dynamic process is important to make certain that major risks are not unattended and managed. The intent of the research is to provide stakeholders in renewable energy projects, including policymakers, financiers, developers and risk management instrument providers, a thorough review of risk management of renewable energy policy and to better define those risks so that they can be adequately mitigated to attract future investment. Five major risks which include market, credit, operational, liquidity and political risks associated with renewable energy developments and markets have been identified. Particularly, renewable energy policy risk is investigated and commonly used risk management tools are reviewed and proposed to address the associated risks and uncertainties faced by financers, developers and investors. It is also intended to setup a place for stakeholders to start, either when they want to replicate current, or are trying to develop new, workable risk management measures for renewable energy policy.

Keywords: Renewable Energy, Renewable Energy Policy, Risk Management

1. Introduction

Over the last decade, the growth of renewable technologies is tremendous. By 2012, the renewable energy industry was investing \$244 billion annually [1]. Around the world, developed and developing countries are continuously seeking to boost renewable energy investment. The development of renewable energy is important to address concerns about climate change and energy diversification [2]. Renewable energy policy has been recognized as one of the main credits of the growth. In the absence of level playing ground, national, state and provincial policies have taken an important role in turning renewable energy resources to be more competitive [3]. Detailed design and proper implementation are always the keys to success. Consequently, policymakers continue to update and revise policies in response to changing environment. At the same time, policymakers have adopted risk and return perspectives in

supporting investments, rather than traditional cost and benefit perspectives. Simply relying on the evolution of renewable energy policy, but still using the same risk management paradigm, will potentially leave risks unmanaged. Appropriate risk management instruments are undoubtedly essential to financers, developers and investors. In this paper, major merits and deficiencies of each renewable energy policy are identified. Uncertainties due to the deficiencies are individually investigated and handled with suitable risk management instruments.

This paper considers the fundamental renewable energy policies to evaluate the five key risk factors which include market risk, credit risk, liquidity risk, operational risk and political risk. Section 2 provides an overview of renewable energy policy and a classification of risks. Section 3 investigates the deficiencies of renewable energy policies and recommends some of corresponding risk management methods. Finally, discussion and conclusion are presented in Section 4.

2. Renewable Energy Policy Analysis

2.1. Renewable Energy Policy Overview

Renewable energy policy is a vital element for development and deployment of renewable energy. Policies aimed at supporting renewable energy developments are often adopted to capture a wide range of benefits. Common objectives for renewable energy policy include the following [4]-[7]:
- Reducing reliance on non-renewable energy sources
- Reducing emission of greenhouse gases and other air pollutants as well as their impacts
- Reducing environmental impacts
- Enhancing the diversification of electricity generation mixes
- Enhancing renewable energy involvement
- Enhancing competitiveness of renewable energy sources

The above objectives are designed to generalize benefits of increasing the use of renewable energy. In addition, return and risk are always the primary concerns for financers and developers [8]. To align with their perspectives, the rationales of renewable energy policies are often set to either increase revenues or reduce uncertainties [9]. National and state policies for establishing an enabling environment for renewable energy developments can be classified into three categories which are regulatory policies, fiscal incentives, and public financing [3]. The policies can be further sub-categorized into as follows:
- Regulatory policies
- Feed-in tariffs
- Utility quota obligations
- Net metering
- Obligations and mandates
- Tradable renewable energy certificates (RECs)
- Fiscal incentives
- Capital subsidies, grants or rebates
- Tax incentives
- Energy production payments
- Public financing
- Public investments, loans, or financing
- Public competitive bidding

Feed-in tariff is a policy scheme created to expand the growth of renewable energy technologies. The policy guarantees a sale price for renewable energy resources and grid access. This provides investors, including small-scale and large-scale developers, with incentives by securing the future income streams on their investment. In practical, long-term contracts are often signed and tariff is set high enough to recover the cost and earn an appropriate profit [10]. As of 2013, feed-in tariff had imposed on 71 countries and 28 states/provinces [3]. Since feed-in tariff is usually known in advance, this effectively stabilizes the profit of a renewable energy project and hence reduces the market risk faced by renewable energy developers and investors [11].

Utility quota obligation and mandate are other means to promote renewable energy developments. The policies define the minimum shares of generations that are generated by renewables or specific renewable sources so as to make sure renewable energy developments align with the national target. The policies are effective only if penalties are adequately set and strictly enforced [12]. In addition, literatures revealed that the effectiveness of assigning a renewable energy target relies on both of the framework of overall supporting policies and the design and barriers of electricity market [13]. In 2013, 22 countries and 64 countries have implemented utility quota obligation and obligation and mandate respectively [3]. Since the policies only define the minimum shares of renewable energy generations, the policies neither enhance returns nor lower risk. Investors and developers are mainly exposed to market risk.

According to the database of Renewable Energy Policy Network for the 21st Century (REN21), net metering has been adopted in 32 countries. The policy aims to support distribution-level renewable energy developments, which permits customers to offset their electricity consumptions by feeding renewable energy generation back to the grid [14]. Studies have investigated how net metering is effective for rewarding the deployment of renewable energy technologies [15]-[17]. The achievement of the policy should not be underestimated, although its target beneficiaries are small-scale developers. For an instance, Germany was dominated in small-scale renewable energy developments in 2010, 2011 and 2012, reflecting its attractive net metering [1], [18], [19]. Unlike feed-in tariff, electricity price of net metering is usually unknown to investors. The income received from net metering can only be estimated and hence developers and investors face market risk.

REC is a transferable energy certificate that is represented as every megawatt-hour generated from renewable energy technologies. Once REC is created, investors are flexible to trade via voluntary market or compliance market to gain additional revenue to finance renewable energy projects [20]. The REC market mechanism has been widely promoted as the solution to drive renewable energy development and investment [21], [22]. According to the database of REN12, RECs have been applied in 26 countries in which the majority is in Europe lately. Similar to net metering, developers and investors face market risk due to the price uncertainty from the sale of RECs [23]. In addition, they also face liquidity risk depending on the type, size and regulation of exchange REC market as well as the activeness of market participants [24].

While high upfront costs of renewable energy developments are usually the most significant barrier for investors and developers, even in the occasion that the project is economically feasible in a long run [25], [26]. Several renewable energy policies are designed to address the high upfront cost issue, which include capital subsidies, grants, rebates, investment tax credits and loans. Capital subsidies, grants or rebates are direct cash incentives provided to renewable energy developers while investment tax credits are indirect non-cash incentives. These policies are one-time incentives and effectively reduce the upfront costs as well as the levelized cost of energy (LCOE) [27].

Instead of providing a one-time incentive, loan programs are revolving and can be used to support renewable energy developments again. Therefore, although the effectiveness of LCOE reduction is lower comparatively, it is treasured by policymakers. Energy production payment and production tax credit are other policies to increase earnings. The former is a direct cash incentive to one unit of renewable energy generation [28] while the latter is tax credit to one unit of renewable energy generation [29]. These two policies aim to reward developers based on projects' performance. Similar to capital subsidy, grant or rebates and investment tax credit, these policies effectively reduce the LCOE. In effect, government subsidies, grants and rebates are less efficient compared to tax incentives. The reason is that government subsidies, grants and rebates are often biased by the ideological positions of the responsible politicians and by the short-term economic benefits of undertaking the project, which ignore the social impact on the entire country and the actual risk-return trade-off of the project [30]. Under the policy frameworks, developers and investors are mainly exposed policy risk and market risk.

Public competitive bidding is a tendering system by which construction and operation contracts of specific quantities of renewable capacity are awarded [31]. Investors and developers are invited to enter into a bidding process. In general, the winner will be the project developer that satisfies the descriptions and requirements of tenders with the lowest bid. A long-term contact is often rewarded. A detail example is shown in [32]. To be successfully implemented, strict development requirements need to be imposed on bidders to avoid price dumping and shortfalls or delays of developments [33]. In a typical bidding scheme, price is the most important determinant. Therefore, market risk is most likely faced by developers and investors.

2.2. Types of Risk

Table 1. Risk Type and Sources of Risk

Risk Type	Sources of risk
Credit Risk	Default of renewable energy projects
	Non-performance of renewable energy projects
	Low capacity factor
Market Risk	Low connection rate
	Low dispatch priority
	Discontinuous electricity output
Operational Risk	Volatile electricity output
	Outdated operating paradigm of the grid
	Non-existence of secondary market
Liquidity Risk	Long payback period
Political Risk	Unstable renewable energy policy

The main categories of risks exposed to renewable energy developments and markets are market risk, credit risk, liquidity risk, operational risk and political risk. The main sources of each risk type are summarized in table 1.

Credit risk defined as the risk that a borrower will default by failing to repay principal and interest in a timely manner [34]. Due to default or non-performance of renewable energy project, developers may fail to make required payments.

Security agreement is one of the risk instruments to mitigate the credit risk [35]. The security requirement can be fulfilled in many ways, including a parent or affiliate guarantee, a stand-by letter of credit or a direct equity contribution. If default or non-performance occurs after the renewable energy project is developed, the developer releases this security and forfeits all rights to the project. Debt financer then becomes the project owner, with access to the power and any revenues generated by the project. Credit default swap (CDS) is another risk instruments to mitigate the risk [36]. CDS is a specific kind of counterparty financial agreement which provides credit risk protection. It functions like an insurance policy. In the event of default or downgrade, the buyer of the CDS receives a payoff from the seller. In general, the seller of the CDS receives payment from the buyer regularly to compensate for providing protection. Therefore, project developer can subscribe a CDS to protect themselves in case of default. To assess the protection needed, risk managers often quantify the credit risk by analyzing the probability of default, loss given default and exposure at default [37].

Market risk refers to the potential loses amount due to market movements. A capacity factor is the ratio of its actual output to its full capacity over a period of time. Typically, capacity factors of renewable energy are lower than traditional energy sources because of intermittent nature and idle capacity [38]. As a result, the amounts of renewable energy generation are often unpredictable and it causes substantial risk to investors. A low capacity factor adversely impacts on the stability of future income stream from renewable energy project. Low connection rate and low dispatch priority further reduce the competiveness of renewable energy projects and result unstable electricity sales [39]-[41]. Hedging instruments, such as derivatives and forwards, are commonly used to transfer market risk and lessen the impact on business [42]. Further measures to mitigate market risk include improving connection and dispatch policies [43], [44].

Broadly speaking, operational risk is the risks resulting from breakdowns in people, systems and internal processes. Operational concerns include personnel, equipment, testing, commissioning, operation and maintenance. History tells us that operational risk can create great impact to society and economic loss [45], [46]. As mentioned before, most of sources of renewable energy technologies are intermittent and volatile in nature that creates many problems in the operational aspect. Also, outdated operation paradigm of the grid has hindered the full potential of renewable energy projects and cause operational problem [47]. A good illustration of the problem is the situation for wind energy providers in China. Due to variability and lack of updated operation paradigm, an average of 20 to 50 percent of wind power is curtailed and subsequently not connected into the national electric grid in 2011 [48]. Catastrophe bond is one of risk management tools to transfer the operational risk to bond investors. Catastrophe bonds are risk-linked securities. It allows investors transferring a specified set of risks to the bond sponsor, such as natural disasters [50]. Renewable energy developer can utilize the bond to secure lower-cost protection from capital market. In the event of corresponding operational failure, risk is transferred to bondholders and hence developer is protected [51]. A typical and comprehensive operation risk management model can also mitigate the operational risks [49]:

- Evaluates and quantifies operational risks
- Implement appropriate risk management tools and frameworks
- Monitors the operational risks
- Investigate the causes of expected and unexpected loss events based on probability of occurrence
- Evaluate the trends, correlations and patterns of the operational risks
- Evaluate the potential losses from operational risk and impact on revenue and investment

Funding liquidity risk and asset liquidity risk are two main streams of liquidity risk concerning renewable energy development and market. Funding liquidity risk refers to the capability of a firm to access financing and capital sources to meet its liabilities while asset liquidity risk refers to the capability of a firm to trade and realize its asset on existing market at the fair value. Renewable energy investment generally requires long investment period. For an instance, the average payback periods for small (20-50kW), medium (100kW-500kW) and large (500kW-5MW) wind turbines with feed-in tariff are 12, 8 and 3 years respectively [52]. In the meantime, secondary market is seldom existed. Developer and investors are difficult to sell the asset. Therefore, liquidity risk is always a big concern in renewable energy development and investment. Debt financing and renewable energy pooled funds are means to solve the liquidity problem due to non-existence of secondary market [20], [53]. Further measures to mitigate liquidity risk include improving project revenue and improving renewable technologies. An effective liquidity risk management model can also mitigate the liquidity risks, which should have the following key factors [54]:

- A well-defined risk governance framework
- A sound liquidity management practice
- A prudent risk liquidity risk analysis, control and monitoring

3. Risk Management of Renewable Energy Policy

3.1. Renewable Energy Policy Risk

Political risk refers to the risk of investment loss in a given country caused by changes in policy or political structure. There are two main categories of political risk which are macro-level and micro-level political risks. Policy risk belongs to micro-level political risk and is defined as project specific risk. Renewable energy policy risk is the risk of investment loss in a given country caused by changes in renewable energy policy. Prospective policy risk and retroactive policy risk are two classifications of renewable energy policy risk [55]. Prospective policy risk considers the impact on the planning of new project caused by the overall uncertainty and instability of the regulatory framework, while retroactive policy risk considers the impact on the financial stability of existing projects due to policy changes. Of the two types of policy risks, the impact of retroactive changes is higher because the changes directly break down the assumptions and forecasts made by developers, financers and investors [56]. The three parties are three key parties involved in renewable energy project and they face renewable energy policy risk differently according to the timeline of project. Figure 1 shows a typical timeline of renewable energy project from the perspectives of developers, financiers and investors.

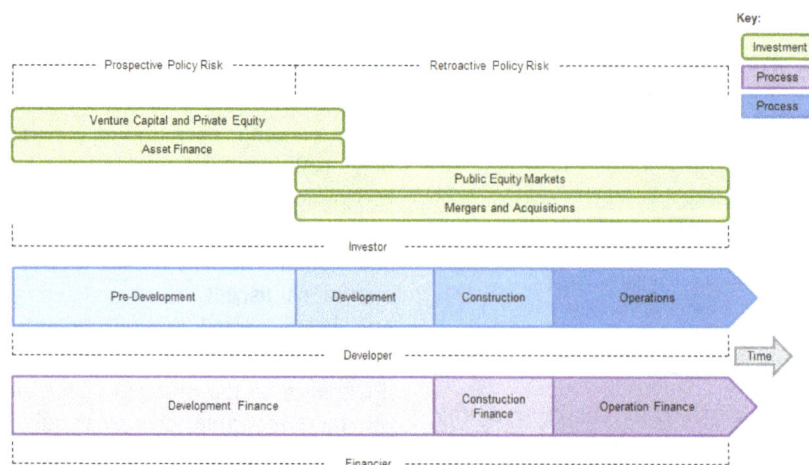

Figure 1. A typical renewable energy project timeline for developer, financier and investor

As shown in figure 1, developer's perspective can be divided into four distinct phases of activity. In the first phase of activity, developers consider possible market fundamentals that affect the renewable energy project's developing, constructing and operating environment. They identify market opportunities and focus on a set of renewable technologies or resources. Then developers will screen identified projects and only move forward the most promising project to the next phase. In this phase, technical analysis and financial analysis are usually performed to reveal major hurdles that deter the project execution. Renewable energy developers perform their own proprietary pro forma analysis to assess the project based on their risk tolerance and professional judgment. Renewable energy policy risk is limited because amount of money invested and time involved are not significant. In phase two, investment of capital and time required by the developer increases substantially. It is because all of the necessary documentations for the project have to be prepared and completed for the financing and construction of projects within this phase. A project development framework called SROPTTC is one of the decision making tools to access the risks connected to the renewable energy project [57]. Renewable energy policy risk is the highest, as considerable capital and time are involved. In phase three, developers start construction. The primary concern is that the developers have to deliver the service and operate the renewable energy project according to the requirements of the contract. Policy risk is high in this phase. Because amendment of renewable energy policy can significantly affect the assumptions and forecasts of the project and reversal of the project is difficult. However, many of the risks have been mitigated by the creation of asset. In the last phase, the timeline shifts from construction phase into operation phase. Renewable energy project has been commissioned and starts to operate. From the perspectives of developers, they have developed the renewable energy project capable of operating at the requirements of the contract. From then on, developers are responsible to operate and maintain the renewable energy project in accordance to the contact. Since the project approaches to the end of its planning horizon, the effect of policy amendment decreases. Therefore, policy risk decreases gradually.

The financier's perspective can be divided into three sections which are development finance, construction finance and operation finance. In the first phase, it represents the most speculative phase. In the event a deal is not completed, financiers face the risk of total investment loss. Since this phase is highly speculative, debt is often not available. Most of the capitals for development finance are come from the developers and other equity investors. Similarly, in the event policymakers amend renewable energy policy in ways that adversely impact the completion of deal, financiers face the risk of total investment loss. Hence, renewable energy policy risk is the highest in this phase. In the second phase, it represents the total capital cost of a renewable energy project. Because many of the risks have been mitigated by the creation of asset, equity and debt financing are usually provided at the construction phase. Policy risk is still high, but is dropped due to the backup of asset. In the last phase, it represents the operation finance of a renewable energy project. The high risk project has been transformed to a stable asset that is not exposed to development and construction risks anymore. The effect of policy amendment and policy risk decrease gradually.

There are four main categories of investments throughout the process: venture capital and private equity, asset finance, public equity markets and mergers and acquisitions by referring to figure 1 [1]. Venture capital and private equity and are renewable energy investment at the early stage. The investments are long-term and illiquid strategy [58]. The main difference between them is that private equity investors invest in mature companies while venture capital investors invest in startup companies. Hence, the expected return and risk from venture capital is generally higher than private equity. The most widely used valuation methodologies include price of recent investment, earnings multiple, net assets, discounted cash flows, discounted earnings and industry valuation benchmarks [59]. Depending on the valuation methodologies, the impact of prospective policy risk can be significant, particularly discounted cash flows and earnings. It is because the changes directly break down investors' assumptions and forecasts. Similar to venture capital and private equity, asset finance is an investment at the early stage. Internal company balance sheets, loans and equity capital are the main sources of funding. Investors generally have excessive information and knowledge to make investment decisions. Therefore, the expected return and risk are low compared to other categories of investments. Net assets, discounted cash flows and discounted earnings are commonly used valuation methods. The impact of prospective policy risk is significant for the same reason as venture capital and private equity investment. Unlike the above mentioned categories, public equity markets and mergers and acquisitions are investment at the late stage. Investment of public equity markets is publicly traded. Investors have flexible investment time horizon, high liquidity and accessible market information. The common valuation methodologies are discounted cash flows, discounted earnings, dividend discount model and earnings multiple [60]. Since these methodologies are all based on accurate assumptions and forecasts, retroactive policy changes could dampen the investment return [61]. The impact and retroactive policy risk is varied depending on business diversification. In general, the more the diversification of business is, the lessor the impact and policy risk are. Hence, a risk-adverse investor is more preferable to invest in a multi business company. Mergers and acquisitions are both strategic investment with the buying, selling, dividing and combining of entities with the aim to create synergy [62]. A merger is that two or more firms join forces for mutual benefit while an acquisition is that one firm takes control of another firm by purchasing the majority of its assets or shares. The five common categories of mergers are conglomerate merger, horizontal merger, market extension merger, product extension merger and vertical merger. On the other hand, the five common categories of acquisitions are value creating acquisition, consolidating acquisition, accelerating acquisition, resource acquiring acquisition and speculating acquisition [63]. Merger and acquisition investment are highly complicated procedures from per-deal planning, deal completion, post-deal integration and extraction of value. Although investors generally have enough knowledge and information to evaluate target firm, the valuation of synergy is difficult to be predicted and determined. The common valuation methods are discounted cash flows, discounted earnings, earnings multiple and net

assets [64]. The impact of retroactive policy risk is significant, especially with the use of discounted cash flows and earnings. It is because the changes could destroy the value of synergy.

3.2. Risk Management

Table 2. Primary Merits and Deficiencies of Renewable Energy Policy

Renewable Energy Policy	Merits	Deficiencies
Feed-in tariff	Stable revenue streams Guaranteed profitability Guaranteed grid access Performance based incentive	Overpriced/underpriced feed-in tariff Inappropriate contract duration Delay in payment Delay in grid access Lack of degression rate Revised existing/future feed-in tariff
Utility quota obligation Obligation and mandate	Cost reduction due to competition Less government expenditures Centralized way to achieve national target Market based policy	Unstable electricity price Inappropriate obligation Inappropriate penalty Revised obligation Excessive focus on low cost renewables
Net metering	LCOE reduction Capable of driving small-scale projects Performance based incentive	Overpriced/underpriced net metering Unfair charges/fees imposed by utilities Impact on profitability of utilities
RECs	Additional revenue/LCOE reduction Market based incentive	Inappropriate market rules Issues of market risk Issues of liquidity risk
Capital subsidy, grant or rebates Investment tax credit Loans	LCOE reduction Upfront investment cost reduction Easier access to project financing	No guarantee of project performance Over/under reward Revised policies Interest rate risk (loans)
Energy production payment Production tax credit	Additional revenue/LCOE reduction Performance based incentive	Renewal uncertainty Revised policies Ineffective to debt financing and cash flow (production tax credit)
Public competitive bidding	Cost reduction due to competition Centralized way to achieve national target	Cost uncertainty Price dumping Shortfalls or delay in development

Table 2 summaries the primary merits and deficiencies of renewable energy policy. The uncertainties of feed-in tariff policy are mainly come from inappropriate tariff, inappropriate contract duration, improper implementation, improper design and unstable policy [12], [65]-[69]. Excessive price or duration leads to ineffective use of public funds and redundant renewable energy projects while underpriced or short term feed-in tariff is insufficient to attract investment and leads to deficit investment. Excessive feed-in tariff rates can also put upward pressure on electricity prices, especially if large-scale of high cost renewable energy technologies are included [70]. In addition, excessive feed-in tariffs can create heavy burden on the public budget [71]. An under developed transmission network and long delay in grid connection also affect the capacity factors and profits of renewable energy projects, which could ultimately cause projects default [72]. Yet, risks are manageable in this situation with adequate due diligence. Since the terms of feed-in tariff are usually made clear to investors in advance to develop renewable energy projects, investors are able to reach final decision based on the existing feed-in tariff. Instead, revise in feed-in tariff can create a bigger impact. In most of the time, governments reserve the final right to amend the feed-in tariffs. As a result, investors never know

the exact duration of policy. An unexpected increment in feed-in tariff leads to an unnecessary competition and create an unfavorable business environment for existing investors while a sudden decrement can harm the growth of renewable energy and create an unfavorable business environment for new entrants [73]-[75]. Portfolio management is one of the tools to diversify the risks and impact [76]. Considering portfolio theory, if assets are not perfectly and positively correlated to each other, risk can be reduced via portfolio management due to diversification effect [77]. Therefore, a portfolio of generation resources, renewable energy projects or any combination is able to mitigate the risks. Another approach is scenario analysis which is a tool to ascertain probable future outcomes with the consideration of probable alternative events that can take place in the future. Investors can use the probable outcomes to minimize the uncertainty and choose the optimal solution based on their perspectives of risk and return. For instance, a scenario-based approach is applied to search the optimal decision by taking into account different feed-in tariff schemes and risk and return perspectives [78]. In addition, Monte Carlo simulations and mean-variance analysis are also applied to quantify the risk-return profiles of renewable energy projects [79]. Another mean to diversify the policy risk is to purchase political risk insurance products which are well designed to compensate the impact of policy change [55]. Concerning delay in payment and grid access, although portfolio management and scenario analysis are able to reduce the risks, supportive policy is still the best way to address the problems [10].

The primary risks connected with obligation/mandate and utility quota obligation are unstable electricity prices, inappropriate implementation, excessive development on low cost renewable energy technologies and revised obligation [12], [80]-[83]. Policies support the concept of free market and leave price unregulated, which increase market risk faced by investors [84]. Market-based approach and game-theoretic approach to model generation expansion planning in deregulated market can be used to handle the price uncertainties [85], [86]. Real option approach has also been applied to address the market risk in different renewable energy developments, such as hydropower power plant [87] solar power plant [88] and wind power plant [89]. Other problems are inappropriate obligation and penalty. The problems create dilemma for the project developers. On the one hand, project developer may face financial and practical issues, such as insufficient funding and profit, to achieve the targets. On the other hand, project developer may be fined for non-compliance with the obligation. Therefore, a balance between obligation and penalty should be well arranged by policymakers. An additional uncertainty associated is an unexpected obligation amendment. If policymakers suddenly revise the obligation and set a higher target, some of the renewable energy project developers could be compelled to develop an economically infeasible project in order to fulfill the obligation and avoid penalty. On the contrary, if policymakers suddenly lower a target, some of renewable energy generations could be redundant and eliminated due to excessive competition. Staging real option approach is a mean to tackle the problem. Broadly speaking, real option approach models the flexibility in response to changes in business environments, which includes the capability of deferring, abandon or adjusting the project so as to react with the evolution of uncertainty [90]. Rather than building a renewable energy project at a single stage, developer divides the project into different stages. At each of the stage, developer preserves the flexibility to abandon or to expand the project. This segmentation improves both the learning and risk reduction effect [91]. In addition, diversification is also a key to reduce risk exposure. A portfolio-based approach is one of the methods to diversity the energy mix by resources so as to lessen the impact due to policy uncertainty [92].

Referring to table 2, there are three main deficiencies of net metering, which include inappropriate net metering price, inappropriate implementation and negative impact of profitability of utilities [93]-[95]. If net metering is underpriced or unfair conditions, such as high minimal connection fee and high standby charge, are imposed by utilities, renewable energy developments are deterred. On the contrary, if the buyback prices are overpriced and the

required amounts of buyback are too high, profitability of utilities will be threatened. Therefore, a proper implementation is essential to balance the benefits to both investors and utilities. Similar to feed-in tariff, net metering is usually known in advance. Hence, uncertainties due to overprice and underprice can be handled with sufficient risk assessments. To tackle the risk of unfair conditions, policymakers may impose regulations to protect small-scale investors. A study has shown that net metering lessened the impact on utility company in the case that the more efficient unit fails and has to be substituted by a less efficient one [96]. As a result, the issues of charging additional fees on net metering which punishes customers for choosing a more energy efficient appliance should be addressed. For example, a safe harbor provisions has been imposed in Minnesota so as to eliminate the disincentive conditions [97]. Concerning the impact of profitability of utility companies, business diversification strategy is a way to reduce the associated policy risk. Utility companies often have superiority in economies of scale and knowledge. As a result, it is easy for them to expand their business to retail level to lessen the impact. Literature shows that concentric diversification strategy is one of the business strategies [98]. For an instance, generation and transmission companies can extend their services to net metering equipment, installation and maintenance to hedge the risks.

As mentioned before, the sale of RECs advances the revenue and is able to finance renewable energy projects [20]. However, the overall contribution is uncertain and fluctuates significantly due to different REC markets, energy policies and other possible reasons. The impact can be as high as 50 percent or as low as 1 percent of total revenue from a renewable energy project [99]. Besides, demand uncertainty, supply uncertainty and price uncertainty induce additional market risk and liquidity risk [99]. In case of oversupply of RECs, market price could be dived strongly and the revenue from the sales of RECs could be slumped [100]. A study shows that these uncertainties are due to the market fluctuation as well as the lack of liquidity [101]. On the supply side, the creation of futures, forwards and derivatives markets with long-term contracts are ways to limit the price volatility of RECs. On the demand side, the creation of margin, loan or banking mechanisms are some possible means to limit the price volatility that encourage proactive investors to enter the market [102]. Inconsistent REC definition and attributes, REC ownership uncertainty and lack of REC tracking system are also part of the sources of liquidity risk [103]. Enhancement of implementation is a must to reduce the uncertainties. For examples, long-term contract is a way to reduce market risk and unbundling RECs and disaggregation of attributes are ways to reduce market and liquidity risk [103], [104]. An improvement on market rules, such as ceiling prices, floor prices and ability of short selling, can also improve the situations [103], [105]. Floor price and ceiling price are set up to fix REC at an agreed range of price if supply is too substantial or shortage to reduce market risks. On the other hand, short selling is set up to allow investors to hedge their positions and improve the liquidity of REC market.

According to table 2, risks connected to capital subsidy, grant or rebates, investment tax credit and loans are mainly come from revised policies [106], [107]. The reason is because the benefits are often known to developers ahead of renewable energy developments. Therefore, new and existing developers can easily assess the risks and take necessary measures. An unanticipated increase in incentives creates comparative advantages to new entrants. To hedge the risk, existing developers can use electricity derivatives to lock in profits [108]. Electricity forwards, futures swaps are some common hedge tools. On the contrary, an unanticipated decrease in incentives will slow the growth of renewable energy developments and creates comparative advantages to existing developers. Potential entrants are advised to perform cost and benefit analysis again to make their final decisions. Undoubtedly, the market risks faced by them are increased. An additional shortcoming for loan is interest rate risk. By definition, this is the risk to the incomes from investment due to the changes in future interest rates. Firstly, if interest rates fall, existing developers will have to pay the same amount of interest which they could actually pay less. Secondly, if interest rates fall, loan rates would probably fall. As a result, new entrants enjoy comparative advantages. To hedge the interest rate risk, developers may enter an interest rate swap to pay floating rate and receive fixed rate [109]. Another major deficiency is the unsecured project performance, as the incentives are not linked with performance. To solve the problem, policymakers may impose provisions to protect themselves, such as minimal electricity generation.

Although production tax credit and energy production payment share the same objective of reducing LCOE, energy production payment provides more benefits. Firstly, production tax credit is a tax allowance and tax basis. Therefore, it is ineffective to both cash flow and debt financing [110]. Secondly, renewable energy project developers typically do not have enough taxable income to have full advantage of the production tax credits [111]. Some investors attempted to transform the tax basis benefit into cash basis. However, cost and legal status of the transaction are doubtful. Another risk caused by the policies is renewal uncertainty. Both policies provide incentives with fixed period. In general, developers usually receive incentives during the first ten years of operation. By then, the renewal is subject to the allocation and availability of funds in every subsequent fiscal year. Therefore, it often creates a boom-bust cycle of renewable energy development [12]. A study demonstrated that renewal negotiation dynamics can enlarge the influence of policy risk over corporate investment decisions [112]. Staging real option approach, scenario analysis and tax credit planning are tools to manage the risk. By diving renewable energy development in different stages, investors enlarge their flexibilities in response to the renewal uncertainty. On the other hand, investors can apply scenario analysis to minimize the renewal risk [113]. Tax credit planning can also be applied to optimize the tax credit received with consideration of different uncertainties [114].

The deficiencies caused by public competitive bidding include price dumping and shortfalls or delay in development. The problems of dumping and shortfalls or delay in development can be addressed by enacting strict development requirements [33]. Other risk concerned with the policy is the cost uncertainty. Since contract price is determined during tender procedure, cost uncertainty is usually happened [115]. Optimization based static bidding strategies have been widely adopted to reduce the market risk [116], [117]. In practical, dynamic bidding strategies, such as sequential optimization bidding strategies and game theoretic approaches, have also been used to reduce the cost uncertainty [118], [119].

4. Future Work and Conclusion

At present, a wide range of risk management instruments are offered by private and public institutions. Specific risks which are credit risk, market risk, operational risk, liquidity risk and political risk, as discussed in Section II, have created a large demand by financers and developers for risk management instruments to mitigate the risks. The main challenges for the providers of risk management tools in supporting renewable energy developments are in the following three areas:

- Further improvement of risk management instruments and innovation in their uses to make them more effective in handling the specific risks of renewable energy development
- Expansion and standardization of the use of risk management instruments, in particular to renewable energy policy risk, to promote collaboration with policymakers, financiers and developers
- Enhanced risk management assistance to financiers, developers and investors to prepare renewable energy projects and attract private and public investments

In this paper, the big picture of market risk, credit risk, liquidity risk, operational risk and political risk of renewable energy development and market has been studied. A thorough review has been provided to stakeholders in renewable energy projects, such as policymakers, financiers, developers and risk management instrument providers. Furthermore, current range of ways in which risk management instruments can diversify, hedge and transfer renewable energy policy risk has been revealed.

References

[1] A. McCrone, E. Usher, V. Sonntag-O'Brien, U. Moslener and C. Gruning, "Global Trends In Renewable Energy Investment 2013", Frankfurt School-UNEP Centre, 2013. [Online]. Available: http://fs-unep-centre.org/publications

[2] "Renewable energy coming of age", The Journal of the International Energy Agency, 2012. [Online]. Available: http://www.iea.org

[3] J. L. Sawin, "Renewables 2013: Global Status Report", REN21, 2013. [Online]. Available: http://www.ren21.net/gsr

[4] S. Carley, "State renewable energy electricity policies: An empirical evaluation of effectiveness", Energy Policy, vol. 37, pp. 3071-3081, 2009.

[5] R. C. Grace, D. A. Donovan and L. L. Melnick, "When Renewable Energy Policy Objectives Conflict: A Guide for Policymakers", National Regulatory Research Institute, 2011.

[6] G. Simpson and J. Clifton, "Picking winners and policy uncertainty: Stakeholder perceptions of Australia's Renewable Energy Target", Renewable Energy, vol. 57, pp.128-135, 2014.

[7] C. W. Lee and J. Zhong, "Top down strategy for renewable energy investment: Conceptual framework and implementation", Renewable Energy, vol. 68, pp.761-773, 2014

[8] R. Wustenhagen and E. Menichetti, "Strategic choices for renewable energy investment: Conceptual framework and opportunities for further research", Energy Policy, vol. 40, pp.1-10, 2012.

[9] Y. Glemarec, W. Rickerson and O. Waissbein, "Transforming On-Grid Renewable Energy Markets, Transforming On-Grid Renewable Energy Markets", 2011. [Online]. Available: http://www.undp.org/

[10] S. Schuman and A. Lin, "China's Renewable Energy Law and its impact on renewable power in China: Progress, challenges and recommendations for improving implementation", Energy Policy, vol. 51, pp. 89-109, 2012.

[11] A. G. Tveten, T. F. Bolkesjo, T. Martinsen and H. Hvarnes, "Solar feed-in tariffs and the merit order effect: A study of the German electricity market", Energy Policy, vol. 61, pp.761-770, 2009.

[12] T. Mezher, G. Dawelbait and Z. Abbas, "Renewable energy policy options for Abu Dhabi: Drivers and barriers", Energy Policy, vol. 42, pp.315-328, 2012.

[13] C. Klesssmann, C. Nabe and K. Burges, "Pros and cons of exposing renewables to electricity market risks – A comparison of the market integration approaches in Germany, Spain, and the UK", Energy Policy, vol. 36, pp.3646-3661, 2008.

[14] K. Sedghisigarchi, "Residential Solar Systems: Technology, Net-metering, and Financial payback", IEEE Electrical Power & Energy Conference EPEC, pp.1-6, 2009.

[15] A. J. Black, "Financial payback on California residential solar electric systems", Solar Energy, vol. 4, pp.381-388, 2004.

[16] S. Carley, "Distributed generation: an empirical analysis of primary motivators", Energy Policy, vol. 37, pp.1648-1659, 2009.

[17] Y. Yamamoto, "Pricing electricity from residential photovoltaic systems: A comparison of feed-in tariffs, net metering, and net purchase and sale", Solar Energy, vol. 86, pp.2678-2685, 2012.

[18] A. McCrone, E. Usher, V. Sonntag-O'Brien, U. Moslener, J. G. Andreas and C. Gruning, "Global Trends In Renewable Energy Investment 2011", Frankfurt School-UNEP Centre, 2011. [Online]. Available: http://fs-unep-centre.org/publications

[19] A. McCrone, E. Usher, V. Sonntag-O'Brien, U. Moslener and C. Gruning, "Global Trends In Renewable Energy Investment 2012", Frankfurt School-UNEP Centre, 2012. [Online]. Available: http://fs-unep-centre.org/publications

[20] A. Tang, N. Chiara and J. E. Taylor, "Financing renewable energy infrastructure: Formulation, pricing and impact of a carbon revenue bond", Energy Policy, vol. 45, pp. 691-703, 2012.

[21] G. Shrimali and S. Tirumalachetty, "Renewable energy certificate markets in India – A review", Renewable and Sustainable Energy Reviews, vol. 26, pp.702-716, 2013.

[22] V. Dinica and M. J. Arentsen, "Green certificate trading in the Netherlands in the prospect of the European electricity market", Energy Policy, pp.609-620, 2003.

[23] A. Ford, K. Vogstad and H. Flynn, "Simulating price patterns for tradable green certificates to promote electricity generation from wind", Energy Policy, vol. 35, pp.91-111, 2007.

[24] L. Nielsen and T. Jeppesen, "Tradable Green Certificates in selected European countries – overview and assessment", Energy Policy, vol. 31, pp.3-14, 2003.

[25] R. Wiser, S. Pickle and C. Goldman, "Renewable Energy and Restructuring: Policy Solutions for the Financing Dilemma", The Electricity Journal, vol. 10, pp.65-75, 1997.

[26] R. H. Wiser and S. J. Pickle, "Financing investments in renewable energy the impacts of policy design", Renewable and Sustainable Energy Reviews, vol. 2, pp.361-386, 1999.

[27] S. B. Darling, F. You, T. Veselka and A. Velosa, "Assumptions and the levelized cost of energy for photovoltaics", Energy & Environmental Science, vol. 4, pp.3133-3139, 2011.

[28] "Evaluating policies in support of the deployment of renewable power", International Renewable Energy Agency (IRENA), 2012. [Online]. Available: http://www.irena.org

[29] J. Badcock and M. Lenzen, "Subsidies for electricity-generating technologies: A review", Energy Policy, vol. 38, pp.5038-5047, 2010.

[30] O. Ghani, "Boosting Renewable Energy", CFA Institute Magazine, vol. 24, pp.16-17, 2013.

[31] B. G. Swezey, "The Impact of Competitive Bidding on the Market Prospects for Renewable Electric Technologies", National Renewable Energy Laboratory (NREL), 1993. [Online]. Available: http://www.nrel.gov

[32] "2013 Integrated Resource Planning Report", Hawaiian Electric Company, chapter 18, 2013. [Online]. Available: http://www.heco.com

[33] "Legal Frameworks for Renewable Energy", Deutsche Gesellschaft fur Internationale Zusammenarbeit (GIZ) GmbH, 2012. [Online]. Available: http://www.bmz.de

[34] "Principles for the Management of Credit Risk - final document", Bank for International Settlements, 2000. [Online]. Available: http://www.bis.org

[35] "Financing Solar PV at Government Sites with PPAs and Public Debt", National Renewable Energy Laboratory (NREL), 2011. [Online]. Available: http://www.nrel.gov

[36] C. Weistroffer, "Credit default swaps: Heading towards a more stable system", Deutsche Bank Research, 2010.

[37] E. S. Silva and A. A. Pereira, "Credit Risk Measures – A Case of Renewable Energy Companies", RECIPP, 2014. [Online]. Available: http://recipp.ipp.pt

[38] L. Qi, "Research on Wind Power Systems", Workshop on Power Electronics and Intelligent Transportation System, pp.408-410, 2008.

[39] S. Luthi and T. Prassler, "Analyzing policy support instruments and regulatory risk factors for wind energy deployment – a developers' perspective", Energy Policy, vol. 39,pp. 4876–4892, 2011.

[40] D. J. Swider, L. Beurskens, S. Davidson, J. Twidell, J. Pyrko, W. Pruggler, H. Auer, K. Vertin and R. Skema, "Conditions and costs for renewables electricity grid connection: Examples in Europe", Renewable Energy, vol. 33, pp. 1832-1842, 2008.

[41] M. Andor, K. Flinkerbusch, M. Janssen, B. Liebau and M. Wobben, "Rethinking Feed-in Tariffs and Priority Dispatch for Renewables", Formaet, 2010. [Online]. Available: heep://www.hks.harvard.edu

[42] "Managing the Risk in Renewable Energy", Economist Intelligence Unit, The Economist, 2012. [Online]. Available: http://media.swissre.com.

[43] L. Byrnes, C. Brown, J. Foster and L. D. Wagner, "Australian renewable energy policy: Barriers and challenges", Renewable Energy, vol. 60, pp.711-721, 2013.

[44] N. M. Bhandari, G. M. Burt, K. Dahal, S. J. Galloway and J. R. McDonald "Dispatch optimisation of renewable energy generation participating in a liberalised electricity market" International Journal of Emerging Electric Power Systems, vol. 3, pp.1-22, 2007.

[45] C. Li, Y. Sun and X. Chen, "Analysis of the blackout in Europe on November 4, 2006", Power Engineering Conference, IPEC, pp. 939-944, 2007.

[46] G. I. Maldonado, "The performance of North American nuclear power plants during the electric power blackout of August 14, 2003", Nuclear Science Symposium Conference Record, 2004 IEEE, vol. 7, pp.4603-4606, 2003.

[47] P. P. Varaiya, F. F. Wu and J. W. Bialek, "Smart Operation of Smart Grid: Risk-Limiting Dispatch", Proceedings of the IEEE, vol. 99, pp. 40-56, 2010.

[48] "Smarter Energy: optimizing and integrating renewable energy resources", IBM Sales and Distribution, Energy & Utilities, 2012. [Online]. Available: http://www-935.ibm.com

[49] C. Isakson, "Operational Risk Management During Uncertainty", Electric Light and Power (ELP), 2012. [Online]. Available: http://www.elp.com

[50] "Catastrophe Bonds — The Birth of a New Asset Class", GAM, 2012. [Online]. Available: http://www.britishchambershanghai.org

[51] J. Wiedmeyer, "Catastrophe Bonds: An Innovative Renewable Energy Risk Management Tool?", National Renewable Energy Laboratory (NREL), 2012. [Online]. Available: https://financere.nrel.gov

[52] "Wind Turbines", Local Government Association, 2012. [Online]. Available: htttp://www.local.gov.uk

[53] M. Mendelsohn, "Tapping the Capital Markets: Are REITs Another Tool in Our Toolbox?", Renewable Energy Project Finance, 2012. [Online]. Available: https://financere.nrel.gov

[54] "Risk Analysis and Recommendations on EURELECTRIC's Power Choices Study Chapter on Liquidity Risk – Definition of Liquidity Risk", Eurelectric, 2013. [Online]. Available: http://www.eurelectric.org

[55] V. Micale, G. Frisari, M. H. Mignuoui and F. Mazza, "Risk Gaps: Policy Risk Instruments", Climate Policy Initiative, 2013. [Online]. Available: http://climatepolicyinitiative.org

[56] Varadarajan, U., D. Nelson, M. Hervé-Mignucci, and B. Pierpoint, "The Impacts of Policy on the Financing of Renewable Projects: A Case Study Analysis", Climate Policy Initiative, 2011. [Online]. Available: http://climatepolicyinitiative.org/publication

[57] E. Lantz, A. Warren, J.O. Roberts, and V. Gevorgian, "Wind Power Opportunities in St. Thomas, USVI: A Site-Specific Evaluation and Analysis", National Renewable Energy Laboratory (NREL), 2012. [Online]. Available: http://www.nrel.gov

[58] A. Iimanen, "Understanding Expected Returns", CFA Institute Conference Proceedings Quarterly, pp.55-63, 2013.

[59] "International private equity and venture capital valuation guidelines", Private Equity Valuation, 2010. [Online]. Available: http://wwwprivateequityvaluation.com

[60] Z. Bodie, A. Kane and A. J. Marcus, "Investments", McGraw-Hill, chapter 18, pp.562-605, 2008.

[61] W. Pentland, "Germany's Renewable Energy Subsidies Could Threaten Economic Growth", Forbes, 2013. [Online]. Available: http://www.forbes.com

[62] "Unlocking Shareholder Value: The Keys to Success", Mergers & Acquisitions A Global Research Report, KPMG, 1999. [Online]. Available: http://www.kpmg.com

[63] A. Ness, "An Overview of the Different Types of Mergers and Acquisitions", Johnsons Corporate, 2014. [Online]. Available: http://www.johnsonscorporate.com.au

[64] C. Mikael and K. Jani, "A Procedure for the Rapid Pre-acquisition Screening of Target Companies Using the Pay-off Method for Real Option Valuation", Journal of Real Options and Strategy 4 (1): pp.117–141, 2011.

[65] S. Tongsopit and C. Greacen, "An assessment of Thailand's feed-in tariff program", Renewable Energy, vol. 60, pp.439-445, 2013.

[66] D. Jacobs, N. Marzolf, J. R. Paredes, W. Rickerson, H. Flynn, C. B. Birck and M. S. Peralta, "Analysis of renewable energy incentives in the Latin America and Caribbean region: The feed-in tariff case", Energy Policy, vol. 60, pp.601-610, 2013.

[67] "California Feed-in Tariff Design and Policy Options", KEMA, 2008. [Online]. Available: http://www.energy.ca.gov

[68] "Exploring Feed-in Tariffs for California – Feed-in Tariff Design and Implementation Issues and Options", KEMA, 2008. [Online]. Available: http://www.energy.ca.gov

[69] B. Bakhtyar, K. Sopian, A. Zaharim, E. Salleh and C. H. Lim, "Potentials and challenges in implementing feed-in tariff policy in Indonesia and the Philippines", Energy Policy, vol. 60, pp.418-423, 2013.

[70] T. Couture and K. Cory, "State Clean Energy Policies Analysis (SCEPA) Project: An Analysis of Renewable Energy Feed-in Tariffs in the United States", National Renewable Energy Laboratory (NREL), 2009. [Online]. Available: http://www.nrel.gov

[71] P. E. Morthorst, "The development of a green certificate market", Energy Policy, vol. 28, pp.1085-94, 2000.

[72] J. P. Romankiewicz, "Contrary to common knowledge, U.S. produces 64% more wind energy than China", 2012. [Online]. Available: http://sustainablejohn.com/?p=156

[73] "Impact of Spain's proposal to retroactively implement an "Industry Killing" 30% Solar Subsidy Cut on Existing Solar Plants." Green World Investor, 2010. [Online]. Available: http://www.greenworldinvestor.com

[74] F. M. Sukki, R. R. Iniguez, A. B. Munir, S. H. M. Yasin, S. H. A. Bakar, S. G. McMeekin and B. G. Steward, "Revised feed-in tariff for solar photovoltaic in the United Kingdom: A cloudy future ahead?", Energy Policy, vol. 52, pp. 832-838, 2013.

[75] S. Luthi and R. Wustenhagen, "The Price of Policy Risk – Empirical Study Looks at the Willingness of European Photovoltaic Project Developers to Invest", SolarServer, 2009. [Online]. Available: http://www.soloarsever.com

[76] C. W. Lee, S. K. K. Ng and J. Zhong, "Portfolio Optimization in Transmission Investment in Deregulated Market", IEEE Power Engineering Society General Meeting, pp. 1-8, 2007.

[77] H. Markowitz, "Portfolio selection," Journal of Finance, vol. 7, pp. 77-91, 1952.

[78] M. S. Nazir and F. Bouffard, "Risk-Sensitive Investment in Renewable Distributed Generation under Uncertainty Due to Post-Feed-In Tariff Policy", Developments in Renewable Energy Technology (ICDRET), pp. 1-5, 2012.

[79] L. Kitzing, "Risk implications of renewable support instruments: Comparative analysis of feed-in tariffs and premiums using a mean-variance approach", Energy, vol. 54, pp.495-505, 2014.

[80] X. G. Zhoa, T. T. Fend, L. Cui and X. Feng, "The barriers and institutional arrangements of the implementation of renewable portfolio standard: A perspective of China", Renewable and Sustainable Energy Reviews, vol. 30, pp.371-380, 2014.

[81] G. Wood and S. Dow, "What lessons have been learned in reforming the Renewables Obligation? An analysis of internal and external failures in UK renewable energy policy", Energy Policy, vol. 39, pp.2228-2244, 2011.

[82] R. Wiser, C. Namovicz, M. Gielecki and R. Smith, "The Experience with Renewable Portfolio Standards in the United States", The Electricity Journal, vol. 20, pp.8-20, 2007.

[83] P. Komor, "Renewable Energy Policy", Diebold Institute for Public Policy Studies, New York, 2004.

[84] J. Lipp, "Lessons for effective renewable electricity policy from Denmark, Germany and the United Kingdom", Energy Policy, vol. 35, pp. 5481-5495, 2007.

[85] A. S. Chuang, F. Wu, P. Varaiya, "A game-theoretic model for generation expansion planning: problem formulation and numerical comparisons", IEEE Transactions on Power Systems, vol. 16, pp. 885-891, 2001.

[86] J. H. Kim, J. B. Park, J. K. Park and S. K. Joo, "A market-based analysis on the generation expansion planning strategies", Intelligent Systems Application to Power Systems, pp. 458-463, 2005.

[87] X. Zhang, X. Wang, X. Wang and H. Chen, "Energy uncertainty risk management of hydropower generators", Transmission and distribution conference and exhibition: Asia and Pacific, 2005.

[88] T. E. Hoff, R. Margolis and C. Herig, "A simple method for consumers to address uncertainty when purchasing photovoltaic", Cleanpower research, 2003. [Online]. Available: http://www.cleanpower.com/research

[89] K. Dykes and R. D. Neufville, "Real options for a wind farm in Wapakoneta, Ohio: incorporating uncertainty into economic feasibility studies for community wind", World wind energy conference, 2008.

[90] E. A. M. Cesena, J. Mutale and F. R. Davalos, "Real options theory applied to electricity generation projects: A review", Renewable and Sustainable Energy Reviews, vol. 19, pp. 573-581, 2013.

[91] K. T. Yeo and F. Qiu, "The value of management flexibility – a real option approach to investment evaluation", International Journal of Project Management, vol. 21, pp. 243-250, 2003.

[92] G. E. Frances, J. M. M. Quemada and E. S. M. Gonzalez, "RES and risk: Renewable energy's contribution to energy security. A portfolio-based approach", Renewable and Sustainable Energy Reviews, vol. 25, pp.549-559, 2013.

[93] G. Hille and M. Francz, "Grid Connection of Solar PV – Technical and Economical Assessment of Net-metering in Kenya", Federal Ministry of Economics and Technology, 2011. [Online]. Available: http://kerea.org

[94] N. R. Darghouth, G. Barbose and R. Wiser, "The impact of rate design and net metering on the bill savings from distributed PV for residential customers in California", Energy Policy, vol. 39, pp.5243-5253, 2011.

[95] "Evaluation of Net Metering in Vermont Conducted Pursuant to Act 125 of 2012", Public Service Department, 2013. [Online]. Available: http://publicservice .vermont.gov

[96] "Freeing the Grid – Best and Worst Practices in State Net Metering Policies and Interconnection Standards", Network for New Energy Choices, 2008. [Online]. Available: http://www.newenergychoices.org

[97] E. Doris, S. Busche and S. Hockett, "Net metering policy development in Minnesota: Overview of trends in nationwide policy development and implications of increasing the eligible system size cap", National Renewable Energy Laboratory (NREL), 2009. [Online]. Available: http://www.nrel.gov

[98] I. Ansoff, "Strategies for Diversification", Harvard Business Review, vol. 35, pp.113-124, 1957.

[99] E. Holt, J. Sumner and L. Bird, "The role of renewable energy certificates in developing new renewable energy projects", National Renewable Energy Laboratory (NREL), 2011. [Online]. Available: http://apps3.eere.energy.gov/greenpower

[100] W. Johnston, "Australian renewable energy crisis as REC price dives", Renewable Energy Focus, 2009. [Online]. Available: http://www.renewableenergyfocus.com

[101] P. Menanteau, D. Fino and M. L. Lamy, "Prices versus quantities: choosing policies for promoting the development of renewable energy", Energy Policy, vol. 31, pp.799-812, 2003.

[102] P. Fristrup, "Forwards, Futures and Banks: Price Stability of Danish RES-E Certificates", Risø National Laboratory, 2000.

[103] E. Holt and L. Bird, "Emerging markets for renewable energy certificates: opportunities and challenges", National Renewable Energy Laboratory (NREL), 2005. [Online]. Available: http://www.nrel.gov

[104] K. Cory, J. Coughlin, T. Jenkin, J. Pater, and B. Swezey, "Innovations in Wind and Solar Financing", National Renewable Energy Laboratory (NREL), 2008. [Online]. Available: http://www.nrel.gov

[105] J. Weinstein, "Contracting for a unified RECs market", Market View, 2006.

[106] J. Cosman, "Renewable energy subsidies: lessons from Spain", Global Policy, 2012. [Online]. Available: http://www.globalpolicyjournal.com

[107] G. Black, D. Holley, D. Solan and M. Bergloff, "Fiscal and economic impacts of state incentives for wind energy development in the Western United States", Renewable and Sustainable Energy Reviews, vol. 34, pp.136-144, 2014.

[108] S. J. Deng and S. S. Oren, "Electricity derivatives and risk management", Power Systems Engineering Research Center, 2005. [Online]. Available: http://www.pserc.wisc.edu

[109] D. Skarr, "Understanding interest rate swap math and pricing", California Debt and Investment Advisory Commission, 2007.

[110] E. Kahn, "The production tax credit for wind turbine powerplants is an ineffective incentive", Energy Policy, vol. 24, pp. 427-435, 1996.

[111] M. Mendelsohn, C. Kreycik, L. Bird, P. Schwabe and K. Cory, "The impact of Financial Structure on the Cost of Solar Energy", National Renewable Energy Laboratory (NREL), 2012. [Online]. Available: http://www.nrel.gov

[112] M. I. Barradale, "Impact of public policy uncertainty on renewable energy investment: Wind power and the production tax credit", Energy Policy, vol. 38, pp.7598-7709, 2010.

[113] X. Lu, J. Tchou, M. B. McElroy and C. P. Nielsen, "The impact of Production Tax Credits on the profitable production of electricity from wind in the U. S.", Energy Policy, vol. 39, pp. 4207-4214, 2011.

[114] L. Zhang, "Optimal taxation with investment tax credit", School of Economics, Renmin University of China (RUC), 2009.

[115] C.B. Chapman, S.C. Ward and J.A. Bennell, "Incorporating uncertainty in competitive bidding", International Journal of Project Management, vol.18, pp. 337-347, 2000.

[116] M. King and A. Mercer, "The optimum markup when bidding with uncertain costs", European Journal of Operational Research, vol. 47, pp. 348-368, 1990.

[117] P. A. Naert and M. Weverbergh, "Cost Uncertainty in Competitive Bidding Models", Journal of the Operational Research Society, vol. 29, pp. 361–372, 1978.

[118] Y. Takano, N. Ishii and M. Muraki, "A sequential competitive bidding strategy considering inaccurate cost estimates", Omega, vol. 42, pp. 132-140, 2014.

[119] M. King and A. Mercer, "Problems in Determining Bidding Strategies", The Journal of the Operational Research Society, vol. 36, pp. 915-923, 1985.

7

RFIA: A Novel RF-band Interference Attenuation Method in Passive Radar

Zeinab Shamaee, Mohsen Mivehchy

Department of Engineering, University of Isfahan, Isfahan, Iran

Email address:

z.shamaee@eng.ui.ac.ir (Z. Shamaee), mivehchy@eng.ui.ac.ir (M. Mivehchy)

Abstract: Passive radars get benefit from transmitter signals in the environment for target detection. One of the most important challenges in these radars is multipath and direct path interferences that enter the target and reference antennas. This article expressed the types of weakening, and multi-step structure to mitigate the mentioned signals in passive radars. In the RF-band Interference Attenuation (RFIA) Method, at the first step the strongest interference component is mitigated by controlling the phase and amplitude. In next steps, other interference components are similarly nulled. This structure leads to separate mitigation for each interference components. Simulation results show the success of RFIA method to reduce dynamic range of analog-to-digital converter and, accordingly, the number of required bits for this converter.

Keywords: Passive Radar, Nulling, Interference Signal, Dynamic Range

1. Introduction

Passive radar has received much attention in the past decade. Conventional Radars consist of two parts including transmitter and receiver, that they are vulnerable, due to radiation [1-3]. Passive radar has one or more receivers and no dedicated transmitter and get advantage the exist signals in the environment for detecting targets. Since passive radars do not have access to modulating signal, they classified as bistatic radars. Fig. 1 shows the state of a passive bistatic radar [4].

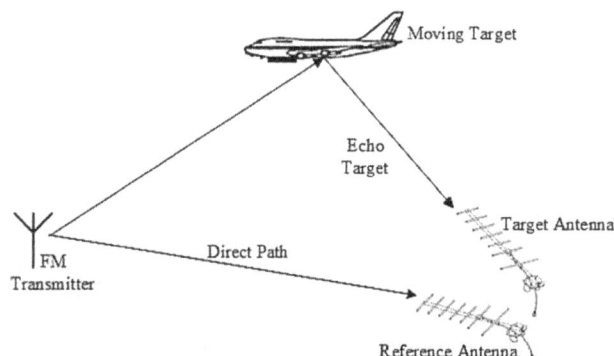

Figure 1. State of passive bistatic radar.

This radar composed of reference and target antenna. The reference antenna directly receives the signal from the intended transmitter as reference signal, and the task of target antenna is to receive the return signal from the target.

One of the most important issues in passive radar is direct path and multipath interferences (DPI & MPI) that enter mentioned antenna besides the target and the reference signals. These strong interferences cause to enlarge the required dynamic range for analog to digital converter (A/D), and for weak signal it can leads to missing of the target. Also, should be careful to saturation of amplifiers in signal paths [5].

For reduction of dynamic range of entered signals to A/D converter, this article tried to offer RFICPR method for effective mitigation of direct path and multipath interference. In the RFIA method, a structure has been used which, first, reference signal is recovered without interferences and second, the strongest component of the interference signal as DPI is canceled from the target antenna signal. Then, in the further steps, the other interference components are suppressed step by step. Interference signals step mitigation characteristics include:

- Separate mitigating for each of interference component in each step and its implication for both the target and the reference signal
- More control capabilities on the appropriate amplitude

and phase for a specific component in each step in order to further mitigation of that component.

In section 2, related literature in interference signal mitigation field are reviewed. Then, in section 3, mitigation basic concepts and the RFIA method are discussed. Besides, section 4 deals with evaluation and stimulation of the method, and section 5 presents the conclusion.

2. Literature Work

A set of available methods are trying to solve the problem in software method for reducing signal interference. In these methods, cancellation has been done in digital domain by various adaptive methods to filter out interference [6- 9]. Software methods placed after the A/D converter, and don't have effects on the required dynamic range for input of converters [10-12].

Use of physical shielding is one of the first hardware methods. Manastash Ridge Radar is the most important example for this method, in which the receiver placed on the other side of a large mountain [4].

Another hardware methods is adaptive beam forming, in which steering nulls towards the illuminator can lead to canceling the interference; although, the use of array antenna, has more nulls and finally has further canceling [13]. The main problem in these methods is increasing complexity of hardware systems which can be seen in Fig. 2.

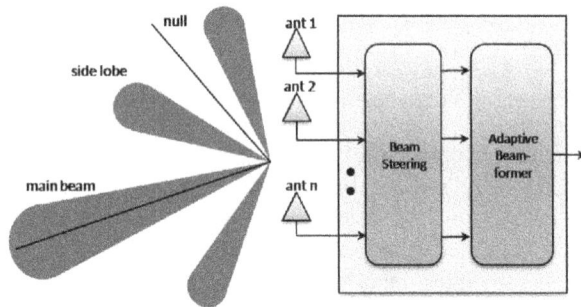

Figure 2. *Adaptive beam forming scheme [14].*

Another method is the use of high gain antenna that has a deeper null and more interference cancelling (Fig. 3), but it is impossible to build an antenna with high gain in low frequencies [15].

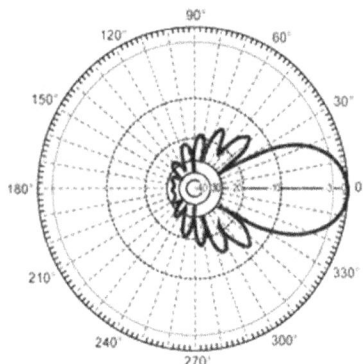

Figure 3. *A typical pattern of high gain antenna [16].*

In particular statuses, an antenna with polarization perpendicular to the polarization signal transmitter (Fig. 4) can be used to mitigate DPI for some extent. But this method is not that effective and can't have much impact on clutters [17].

Figure 4. *One of the Disk on rod antennas.*

Another hardware that can be seen in Fig. 5 is based on the difference between the reference signals from target in analog domain [18]. The reference signal and the received direct path interference are coherent, but time of arrival (TOA) and amplitude of two signals are different. Since the base of interference cancellation in this method is subtraction between the target channel and the reference channel, therefore, the reference signal and DPI in target channel must be matched in amplitude and TOA.

In [17] with using this hardware method which is placed before A/D converter, 35 dB mitigation is achieved, which is not enough. In order to improve the amount of mitigation in the proposed method, by staging cancelation of different components, totally, more mitigation can be achieved.

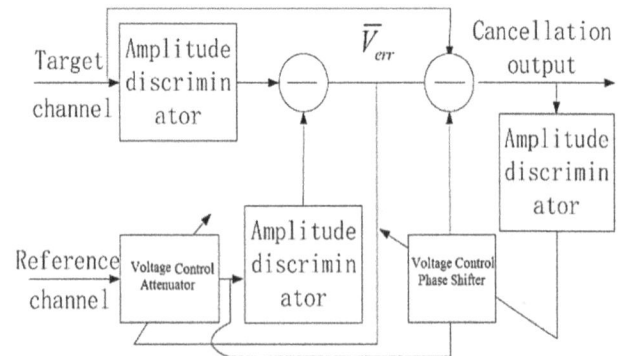

Figure 5. *DPI cancellation with conventional analog method [17].*

According to previous studies, multi-step nulling and in accordance with a specific component seems to be effective. Therefore, this paper tried to provide a method based on this structure. Details of the proposed method will be described in the next section.

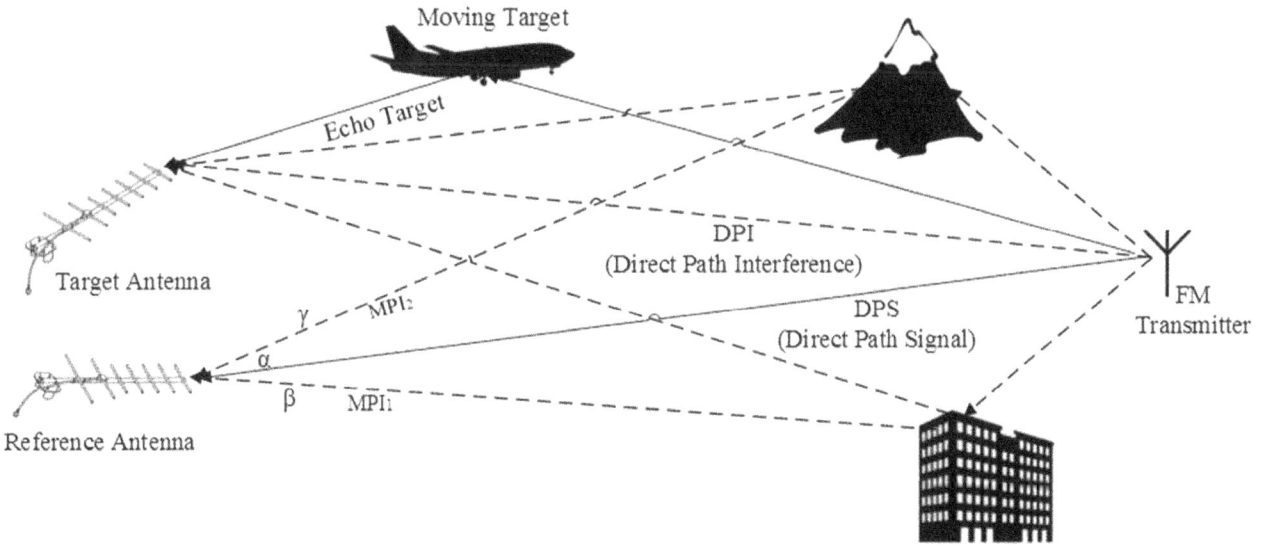

Figure 6. *Applicable signals state and interference signals in passive radar.*

3. RFIA Method

In passive radar, as seen in Fig. 6, the reference signal that carry out the process of passive radar, other interference signal unintentionally and simultaneously are received. This problem also occurs in the target antenna, where in addition to the target echo, strong interference such as DPI and returning interference are received from fixed objects placed in different distance from transmitter and receiver and have different amplitude and delays [20, 21].

All these are in situation that the radar cross section is much smaller than the DPI and multipath interference, For example, mountain ranges that make the interferences have a length of several kilometers and are distributed near the antennas. The difference between target signal power and interference signals sometimes reaches to 100 dB, have other several issues that in first step needs using the A/D converter with many bits. In a typical example to cover the dynamic range of received signals about 100 dB need A/D converter with at least 17 bits, and this estimate is in the absence of noise, because in the noise presence, the number of required bits increases [22].

In proposed method for better attenuation of a desired component in each step, it is necessary to dominate the desired component in reference antenna signal with an appropriate method. So a new reference signal is produced and gathered by received signal from the target antenna. The result of this addition is attenuation of direct path components in the received signal from the target antenna, this process is called nulling. But, if over time and under environmental conditions, the received signal changed a little, the nulling is not done properly, and even can worsen the result. But in case of proper functioning after the first step of nulling, reduction interference about 20 to 30 dB is expected. And further reduction in one step is impossible due to existing more interference components with different delay, amplitude and phase.

In fact the nulling process performs in several steps on received signals; and one of the interference components in each step of attenuation is considered. In order to have a proper process in the second step, it is needed to mitigate DPI simultaneously in the path of received signal from the reference antenna.

To simplify analysis, information signal x(t) in (1), is supposed as a sine wave with f_m frequency ($\omega_m = 2\pi f_m$). Equation (2) expresses carrier signal that frequency modulated by x(t) and f_Δ is maximum instantaneous frequency deviation around central frequency ($\omega_c = 2\pi f_c$).

$$x(t) = sin\omega_m t \tag{1}$$

$$z(t) = \sin[\omega_c + 2\pi f_\Delta co\ s(\omega_m t)]t \tag{2}$$

Signal z(t) with amplitude α and delay t_{d1} as DPS accompany with two interference signal with β and γ values as amplitude and delays t_{d2} and t_{d3}, namely R(t) in (3) where α is greater than β and γ and t_{d1} is less than t_{d2} and t_{d3}.

$$R(t) = \alpha sin[[\omega_c + 2\pi f_\Delta cos\omega_m(t - t_{d1})](t - t_{d1})]$$

$$+\beta sin[[\omega_c + 2\pi f_\Delta cos\omega_m(t - t_{d2})](t - t_{d2})]$$

$$+\gamma sin[[\omega_c + 2\pi f_\Delta cos\omega_m(t - t_{d3})](t - t_{d3})] \tag{3}$$

In (4), the signal R(t) with nr(t) noise enters to the reference antenna (Ref (t)).

$$Ref(t) = R(t) + n_r \tag{4}$$

On the other hand, transmitter signal goes to target antenna with as amplitude after hitting the moving target produces Doppler frequency fd. Finally information signal enters to target antenna with a coefficient (k) of the reference signal and ns (t) noise, as S(t), that is expressed by (5).

$$S(t) = kR(t) + T(t) + n_s(t) \tag{5}$$

a

b

c

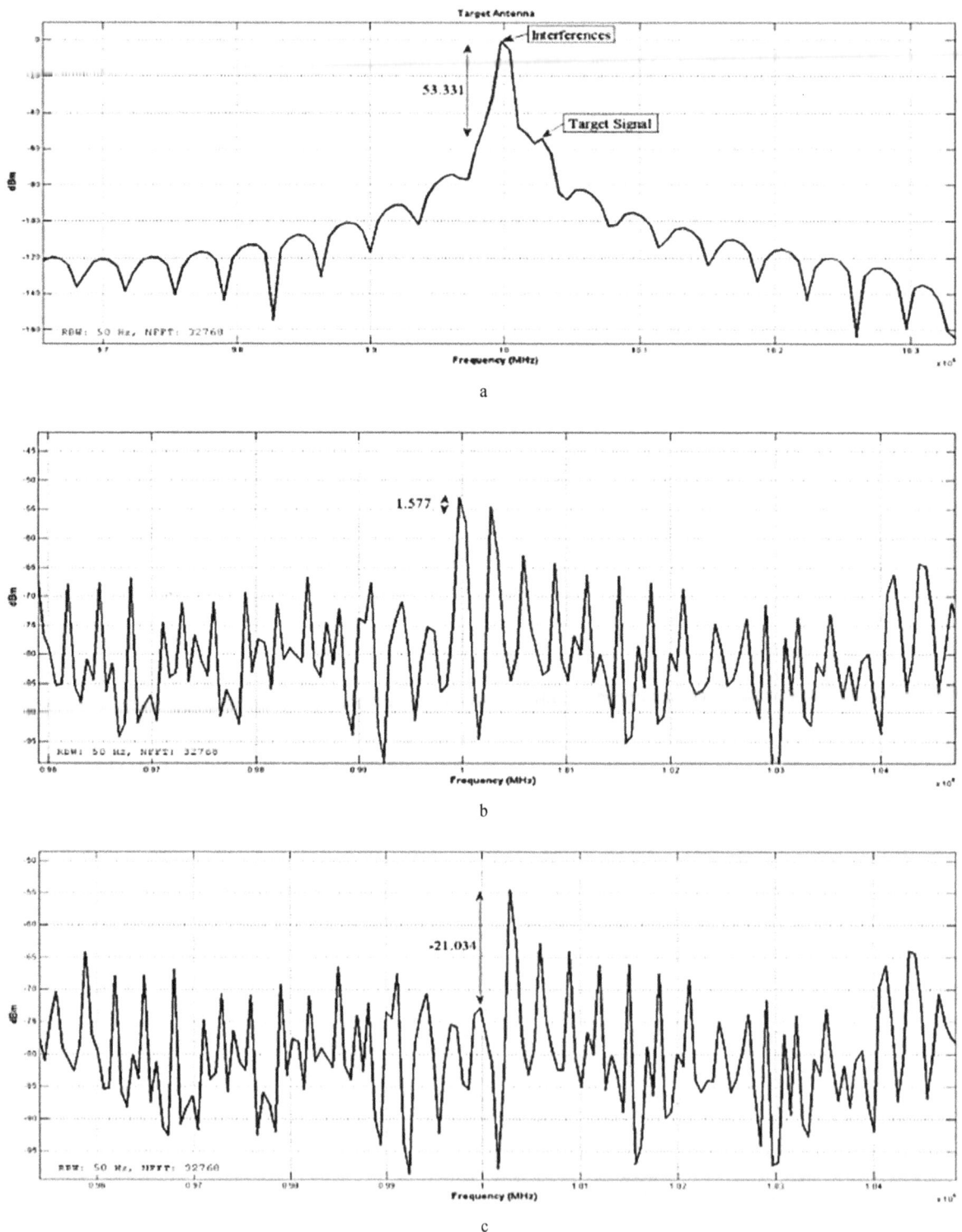

Figure 7. Frequency spectrum of target antenna signal. a. Before nulling. b. After first step attenuation. c. after second step attenuation.

4. Evaluation

Simulation of the RFIA method and also producing required signal for evaluation are done in MATLAB software. The RFIA method is implemented in two steps. In this simulation, the intended transmitter is considered from the FM type with carrier frequency of 100 MHz and the speed target is about 0.9

Mach and Doppler frequency (f_d) is about 300 Hz. Evaluation Target signal frequency spectrum with the interference components entered target antenna before nulling operations; is shown in Fig. 7.a. The target signal are placed at a frequency of 300 kHz from the carrier.

Conventional analog method that described in literature work, is based on subtracting between reference and target antenna signal with appropriate phase and amplitude. With this method, 54 dB attenuation achieved.

Spectrum of target signal after maximum attenuation by first step of RFIA method is shown in Fig. 7.b. Prior to nulling, the target component power difference with the strongest component of interference components, is about 53 dB. But after nulling this difference reached to 1.577. This amount of mitigation, is almost the highest weakening in one step. Because of presence of the next interference component that requires a separate setting of amplitude and phase shift, mitigating in this step is limited up to this rate. For further mitigations, the next steps should be passed. Frequency spectrum of target antenna signal after second step attenuation is depicted in Fig. 7.c.

The spectrum is related to the optimum gain and phase, so the amount of overall attenuation is about 73 dB. By comparing the conventional method attenuation (54 dB) and the attenuation of RFIA method (73 dB) an improvement about 20 dB is obtained as shown in Fig. 8.

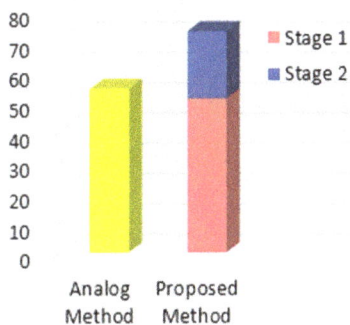

Figure 8. *Comparison between conventional and RFIA method.*

In RFIA method, About 51 dB Attenuation in one step obtained. However in the conventional analog method, the maximum attenuation was about 54 dB. The reason behind this discrepancy is that in the multi-step method, due to dominating process, in each step the strongest component itself attenuates with the highest possible amount and the power of other components does not change. But in the conventional method, with attenuating first component, the other components also slightly attenuates. But, finally, in the two step method after accomplishment of the second step, other interference component are also attenuate and generally better outcome is obtained compared to its conventional analog method.

Totally by RFIA method, the dynamic range from the target antenna dropped at least about 20 dB more than the conventional analog method, which means approximately 3 bits fewer for A/D converter input.

The distance between receiver antennas can play a major rule in the maximum limit for interference attenuation. So, the effect of delay between signals which arrives two antennas is not negligible and must be considered. The result of conventional analog method with effect of distances between receiver antennas show the maximum attenuation 39dB. With completion of the proposed method and after second step, the

amount of the overall attenuation is 50 dB which is about 11 dB higher than the conventional analog method. These results show the proposed method is successful in more attenuation of interferences.

5. Conclusion

This paper presented a multi-step structural method to weaken the direct path and multipath interference components in passive radars. In the first step of RFIA method, DPI (the strongest interference component) and in further step, the other components are attenuated. To enter the next step, mitigation first of the interference component is applied in the reference antenna in addition to the target antenna. Indeed, for each step a new reference signal generated.

In comparison with earlier works, the mitigation ability of this method is about 11 dB more; and at the same time it should be taken in to account that, this interference components mitigation is done before A/D converter. For more cancelation of the remained interference components, still software methods can be used.

The success of the RFIA method depends on this proper amplitude and phase shift proportionate for any components. Thus offering an alternative for its automatic determining for various issues is needed.

References

[1] H. Kuschel and D. O'Hagan, "Passive radar from history to future," *Radar Symp. (IRS), 2010 11th Int.*, 2010.

[2] P. Howland, "Editorial: Passive radar systems," *IEE Proc. - Radar, Sonar Navig.*, vol. 152, no. 3, p. 105, 2005.

[3] H. Kuschel and F. Fhr, "T10 — Passive bistatic radar, theory — Systems — Measurements," in *2014 IEEE Radar Conference*, 2014, pp. 37–37.

[4] H. D. G. and C. J. B. Paul E. Howland, *Bistatic Radar: Emerging Technology*. Chichester, UK: John Wiley & Sons, Ltd, 2008, pp. 247–313.

[5] J. Brown, "FM airborne passive radar-Phd Thesis," UCL (University College London), 2013.

[6] J. Palmer and S. Searle, "Evaluation of adaptive filter algorithms for clutter cancellation in passive bistatic radar," *Radar Conf. (RADAR), 2012 IEEE*, pp. 0493–0498, May 2012.

[7] F. Colone, R. Cardinali, and P. Lombardo, "Cancellation of Clutter and Multipath in Passive Radar using a Sequential Approach," in *2006 IEEE Conference on Radar*, 2006, pp. 393–399.

[8] X. Guan, D.-H. Hu, L.-H. Zhong, and C.-B. Ding, "Strong Echo Cancellation Based on Adaptive Block Notch Filter in Passive Radar," *IEEE Geosci. Remote Sens. Lett.*, vol. 12, no. 2, pp. 339–343, Feb. 2015.

[9] B. W. ; C. L. ; C. C. ; W. C. Feng, "An effective CLEAN algorithm for interference cancellation and weak target detection in passive radar," in *Synthetic Aperture Radar (APSAR), 2013 Asia-Pacific Conference on*, 2013, pp. 160–163.

[10] Z. Shamaee, M. Mivehchy, and M. F. Sabahi, "A Chi-square-based induction method for interfernce signals attenuation in passive radar," in *The Third Iranian Conference of Engineering Electromagnetics*, 2015.

[11] P. E. Howland, D. Maksimiuk, and G. Reitsma, "FM radio based bistatic radar," *IEE Proceedings - Radar, Sonar and Navigation*, vol. 152, no. 3. p. 107, 2005.

[12] B. Farhang-Boroujeny, *Adaptive Filters: Theory and Applications*, Second. Chichester, UK: John Wiley & Sons, Ltd, 2013.

[13] G. Fabrizio, F. Colone, P. Lombardo, and a. Farina, "Adaptive beamforming for high-frequency over-the-horizon passive radar," *IET Radar, Sonar Navig.*, vol. 3, no. 4, p. 384, 2009.

[14] http://www.telecomfamily.com. [Accessed: 10- May- 2016].

[15] C. Coleman, "Mitigating the Effect of Direct Signal Interference in Passive Bistatic Radar," 2009, no. red X, pp. 1–4.

[16] http://www.l-com.com. [Accessed: 10- May- 2016].

[17] R. Saini, M. Cherniakov, and V. Lenive, "Direct path interference suppression in bistatic system: DTV based radar," *2003 Proc. Int. Conf. Radar (IEEE Cat. No. 03EX695)*, 2003.

[18] C. L. Zoeller, M. C. Budge, and M. J. Moody, "Passive coherent location radar demonstration," in *Proceedings of the Thirty-Fourth Southeastern Symposium on System Theory (Cat. No. 02EX540)*, 2002, pp. 358–362.

[19] H. Wan, S. Li, and Z. Wang, "Direct Path Interference Cancellation in FM Radio-Based Passive Radar," in *2006 8th international Conference on Signal Processing*, 2006, vol. 3, pp. 4–7.

[20] H. Griffiths and C. Baker, "The Signal and Interference Environment in Passive Bistatic Radar," in *2007 Information, Decision and Control*, 2007, pp. 1–10.

[21] H. Griffiths, "Passive Bistatic Radar and Waveform Diversity," vol. 119, no. c, 2009.

[22] M. Skolnik, *Radar Handbook*, 3rd ed. McGraw-Hill Professional; 3rd edition, 2008.

8

The feature of underground channel for the wireless underground sensor networks

Farzam Saeednia[1, *], Shapour Khorshidi[2], Mohssen Masoumi[3]

[1]Department of Electrical Engineering, Kazerun Branch, Islamic Azad University, Kazerun, Iran
[2]Air-Sea Science and Technology Academic Complex, Shiraz, Iran
[3]Department of Electrical Engineering, Jahrom Branch, Islamic Azad University, Jahrom, Iran

Email address:

Farzam_2958@yahoo.com (F. Saeednia), khorshidy@yahoo.com (S. Khorshidi), maesoumi@jia.ac.ir (M. Masoumi)

Abstract: The propagation characteristics of Electro Magnetic (EM) waves in the soil and also the significant differences between the propagation in the air prevent us from obtaining one direct feature for Wireless Underground Channel. In fact, the underground environment consists of soil, rock and water instead of the air. The challenging reasons of these environments to propagate the wireless signal via the Electro Magnetic (EM) [1] waveguides are considered as: the high path loss, channel dynamic conditions and the high size of antenna. In this study, the details of Bit Error Rate (BER)[2] for 2PSK modulations, path loss and the bandwidth of the Magnetic Induction (MI) [3]Systems in the underground environment via one small induction coil were evaluated .At the end of this study, a general framework is obtained about the wireless underground communications and wireless underground sensor network. It is concluded that using the proposed framework, the transmission range in MI waves system would be raised and the path loss in that system would be declined severely.

Keywords: Channel Modulation, Electro Magnetic, Magnetic Induction

1. Introduction

The Wireless Underground Sensor Networks (WUSNs) have the wireless sensors that are buried underground. WUSNs have so many applications such as the coverage, easy to use, appropriate data, reliability and the cover density. The other applications are the control of the soil conditions, earthquake, landslides forecast, the underground substructures control, the landscapes management and security [1, 2].

It can be mentioned that the underground propagation environment consists of soil, rock and water instead of the air that confront us with three challenges if it is applied for the wireless communications via the Electro-Magnetic (EM) Waves: The high path loss, channel dynamic conditions, and the antenna size [3].

Akyilidizet.al, [2] evaluated the challenges of wireless underground sensor networks environment and Lilli et.al, [3] evaluated the features of the underground channels for the wireless underground sensor networks The magnetic induction was applied for the first time by Jack et.al, [4] in the wireless underground sensor networks [4].

The objective of the present study is to present a general framework for the wireless underground communications and WUSN that the transmission range in MI waves system would be raised and the path loss in that system would be declined severely.

The structure of this study is as follow: in the next section, the conditions of underground channel are evaluated to analyze the same channel conditions via the theory. In section three, the simulated proofs of the section two are presented and some comparisons are considered at the end. Finally, the conclusion is presented.

2. The MI Channel Properties

The propagation characteristics of Electro Magnetic (EM) waves in the underground environment (soil, water, and rock) were presented. The analyses indicate that the path loss is much higher than the ground cases and its reason is the absorption in the elements of the underground environment. The success in the communications depends on the combination of operational frequency and soil. Thus, as the operational frequency decreases, the path loss is decreased too but we need a larger antenna [3]. One suggested solution is to apply some antenna with 0.3m length to receive 300 MHz signals. Meanwhile, the signal transmission range of these antennas is about 4 meters. Recently, the Magnetic Induction method has been used as a new physical layer method for the wireless communications but it has disadvantages such as the high path loss and Low bandwidth. According to the previous studies, the MI transfer hasn't been influenced by the soil type, compositions, concentrations or the moisture rate [5].

2.1. Channel Modeling

In the MI communications, the transmitting and receiving is conducted via the induction coil (first row in fig.1). Meanwhile, a_t , a_r are the radiuses of transmitting and receiving respectively. r is also the distance between the receiver and transmitter. For example, the sinusoidal current $I = I_0 . e^{-jwt}$ is a signal that is found in a the transmitter coil whereas, $w = 2\pi f$ is the angular frequency of transmitter signal and f is the operational frequency of system. This current is able to induce another sinusoidal current in the receiver and make the connection possible. The interaction between the two paired coils is indicated by the mutual inductance. Thus, MI receiver and transmitter can be modulated as the Primary and Secondary Coil Loops of one converter (Transformer Model). As it is shown in the second row of fig.1, MI is the mutual inductance of the receiver and transmitter coil.

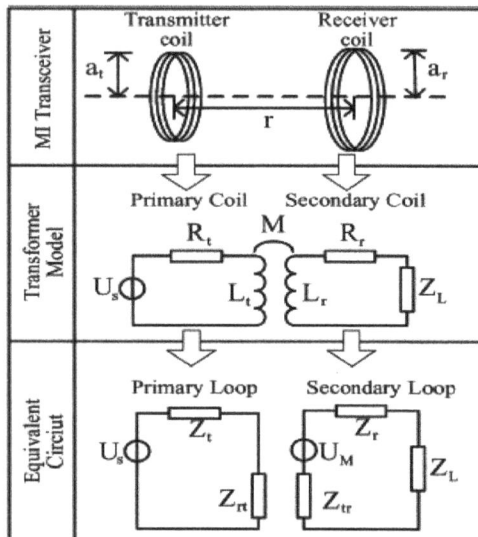

Fig. 1. *MI Communications Channel Model*

U_s is the battery voltage of the transmitter, L_T , L_r are the self-induction , R_T, R_r are resistance of coil and Z_L is the impedance of the receiver load. The third row in fig.1 indicates the analysis of the converter (transformer model) .

$$Z_t = R_t + j\omega L_t \quad * \quad Z_{rt} = \frac{\omega^2 M^2}{R_r + j\omega L_r + Z_L} \quad (1)$$

$$Z_{tr} = \frac{\omega^2 M^2}{R_t + j\omega L_t} * Z_r = R_r + j\omega L_r$$

$$U_\mu = -j\omega M \frac{U_s}{R_t + j\omega L_t}$$

Whereas, Z_t, Z_r are the impedance (impedance-based axis) of the receiver and transmitter coil .Z_{rt} is the effect of receiver on the transmitter but Z_{tr} is the effect of transmitter on the receiver .$U\mu$ is also the inductive voltage of the receiver coil loop . Both the receiver and transmitter powers are the functions with the transmission range, r.

$$P_r(r) = Re\left\{\frac{Z_L * U_M^2}{(Z_r' + Z_r + Z_L)^2}\right\} \quad (2)$$

$$P_t(r) = Re\left\{\frac{U_s^2}{Z_t + Z_t'}\right\}$$

With regard to the Transmission Line Theory, the reflection is neutralized by the same impedance. In order to increase the received power, the load impedance is equal to the conjugate of impedance in the secondary loop output obtained from the Equ.3 as follow:

$$Z_L = \overline{Z_r + Z_r'}. \quad (3)$$

According to the Equ.4, the analysis of coil resistance can be determined via the material resistance, size and the number of turns of coil.

$$R_t = N_t * 2\pi a_t * R_0 \quad (4)$$

$$R_r = N_r * 2\pi a_r * R_0$$

In the equ.4, N_t, N_r are the number of turns for the transmitter and receiver loop coils respectively.R_0 is the resistance of unit length for one loop as one magnetic dipole that is written for the various wire diameters as $3\Omega/m \cong 2 * 10^{-4}\Omega/m$ on the basis of American Standard (AWG) . The self-induction and induction can be inferred via A Magnetic Dipole Potential. The different dipole systems have been considered in [7].

$$A(r, \theta, \phi) = \frac{\mu}{4\pi r}\pi a_t^2 I_0 e^{-jwt} \sin\theta \left(\frac{1}{2} - \frac{j(2\pi)}{\lambda}\right). a_\phi \quad (5)$$

Whereas, μ is the magnetic permeability of transfer environment and λ is the signal wavelength. With regard to the Stokes Theory, the mutual inductance of two coil loops can be written as follow [7]:

$$M = (N \oint_{I_r} A . dI_r)/dI \cong \frac{\mu\pi N_t N_r(a_t^2 a_r^2)}{2r^3} \quad (6)$$

The self-inductance is calculated as :

$$L_r \cong \frac{1}{2}\mu\pi N_r^2 a_r L_t \cong \frac{1}{2}\mu\pi N_t^2 a_t \quad (7)$$

By replacing the Equations (1),(3),(4),(6) and (7) in the Equ.2 , the received and transmitted powers can be calculated. It must be mentioned that the underground transferring environment consists of various kinds of soil, water, rock, etc. Thus, the analysis of permeability differences between these materials is necessary [8]. Generally, the materials inside the underground environments can be divided in to 4 basic groups: organic materials, mineral materials, air and water. The organic materials consist of the plants and animals that live in the environment. The relative permeability in the flower, plant, air and water is nearly 1 and if the sand and clay don't have any magnetic, their relative permeability will be nearly 1. For example, the permeability of the Sedimentary rocks is about 1.009 [8]. Since most parts of the soil are without magnetic, it is supposed that the relative permeability in the underground environment is constant and it is according to the above mentioned concepts.

2.2. Path Loss

In order to have a wireless communication with the EM waveguide, the received power by the antenna will be obtained via the Frazer transport Equation [9]. The Irradiation Power is the most important consumption factor for EM waveguide transmitter. Although the transmission power $1(P_t)$ in the EM waveguide system is a constant power that isn't influenced by the receiver position, the L_{EM} path loss in the soil environment is calculated as the received power to the transmission power [3]:

$$L_{EM}(r) = -10 \log \frac{P_r(r)}{P_t} = 6.4 + 20 \log r + 20 \log \beta + 8.69 \alpha r \quad (8)$$

Attenuation constant of α waves and phase-shift constant of β waves are 1/m and Radian/m respectively. While, the transfer distance r is determined by meter, it depends on the dielectric characteristics of the soil. Also, the other air-ground interface reflection isn't considered because of high penetration [3]. If we attempt to prevent from wasting the transmission power due to radiation to the environment, we should use MI communication system because it is possible to ignore the irradiation power as the radiation resistance is negligible. It causes that the transmission power of MI system is consisted of the consumed power in MI receiver and the consumed power in the coil loop resistance .Meanwhile, since the coil loop resistance is negligible, the received power to the transmission power is close to one and the transmission power is restricted. As a result of this, the most favorable power is transferred in WUSN system but the path loss still exists. Thus, the received and transmitted power increase simultaneously as the transfer distance is raised. In order to compare the operation of EM waveguide system and MI system, the path loss in MI system with the transfer distance, r is defined as follow.

$L_{MI} = -10 \log(\frac{P_r(r)}{P_t(r_0)})$, Whereas $P_r(r)$ is the received power ratio at the receiver that is r meter away from the converter. $P_t(r_0)$ is the source transmission power when the transfer distance is very limited. $P_t(r_0) \cong \frac{U_s^2}{R_t}$ is considered when r_0 is small . When the coil resistance is low ($R_0 \ll \omega\mu$)

and the frequency is high, the path loss of MI information system is summarized as follow:

$$L_{MI}(r) = -10 \lg \frac{P_r(r)}{P_t(r_0)} \cong -\frac{10 \lg N_r a_t^3 a_r^3}{4 N_t r^6} = 6.02 + 60 \lg r + 10 \lg \frac{N_t}{N_r a_t^3 a_r^3} \quad (9)$$

In order to analyze the path loss of MI, EM waveguide systems in the underground environments, equations 8 and 9 were compared. In equation8, there are two path losses, one of them is related to the transfer environment $20 lgr$ and the other is related to the materials absorption $8.69\alpha r$ which both of them have important effects on the path loss. However, no soil dielectric materials such as α & β is found in equ.9. So, no clear effect is seen on the MI path loss. The only path loss in equ.9 is$60 lgr$ that is due to the extent of propagation environment. However, the propagation in the underground environment has no effect on the path loss. The reason is the constancy of underground materials permeability. Although the path loss due to $60 lgr$ in MI system equation is much more than the path loss due to $20 lgr$ in EM waves, it is created because of another factor in the equation of EM waves system as $8.69\alpha r$ that makes great loss from the propagation in the underground environment.

3. Numerical Analysis

3.1. Path Loss

The path loss in the MI, EM systems in Equation8 and Equation 9 were evaluated via MATLAB software. The results are shown in fig.2. The EM wave's propagation in the soil environment is influenced by the soil characteristics especially the amount of water in the soil (VWC). Thus, we indicate VWC in the soil environment as 1%, 5% and 25% [3].

In our simulation, along by VWC, the collection of soil composition is determined as follow: the percentage of sand(S) 50%, the percentage of clay(C) 15%, mass density (ρ_b)1.5 grams/cm^3, density of the solid soil particles (ρ_s) 2.66 $\frac{grams}{cm^3}$ that are considered as the normal values [3]. The permeability of the underground transmission environment is constant and applied in the air .$4\pi * 10^{-7} \frac{\mu}{m}$.

Meanwhile, other simulation parameters for EM system are also act the same. The operational frequency is set on 300MHZ. there are two reasons for selecting this frequency: 1.The path loss in this frequency is acceptable and in the frequencies lower than 300MHZ, the path loss will be raised. 2 .as the operational frequency is reduced to below 300MHZ, the size of antenna will be raised but it prevents from WUSNs implementation.

For MI system, the receiver and transmitter coil loop has a same radius (0.15) and the number of turns of coil is 5. The coil loop has been made by the cooper wire with 1.45 mm diameter. Thus, the resistance of unit length R_0 can be considered as $0.01\ \Omega$/maccording to [6] (AWG Standard). Meanwhile, the operational frequency for MI system is set on 10MHZ.This low operational frequency with the low number of turns of coils can decrease the influence of parasitic capacitance effectively [10].As it is shown in fig.2, the path loss in MI system and EM wave system has been indicated as dB against the transmission distance with the different soil, VWC. As expected, the path loss in MI system is not influenced by the environment because the permeability, μ remains the same in the whole propagation environment. On the other hand, as VWC increases, the path loss in EM waves system is also raised severely. When the soil is very dry (good insulation), VWC=1%, the path loss of EM waves lesser than MI system. When the soil is very wet (good conductor), VWC=25%, the path loss of EM waves is much more than MI system [12]. When the moisture of soil is as VWC=5%, the path losses are same for two systems. The main reason of path loss in EM wave system is the absorption of EM transmission waves inside the underground materials. When the amount of water in the soil is VWC=5%, the EM wave system has the lowest path loss in the range of 0.5m to 3m. In relatively remote areas, r>3m, MI system has the lesser path loss than EM system .Even in MI system; it has the lesser path loss than EM wave system in remote distances.

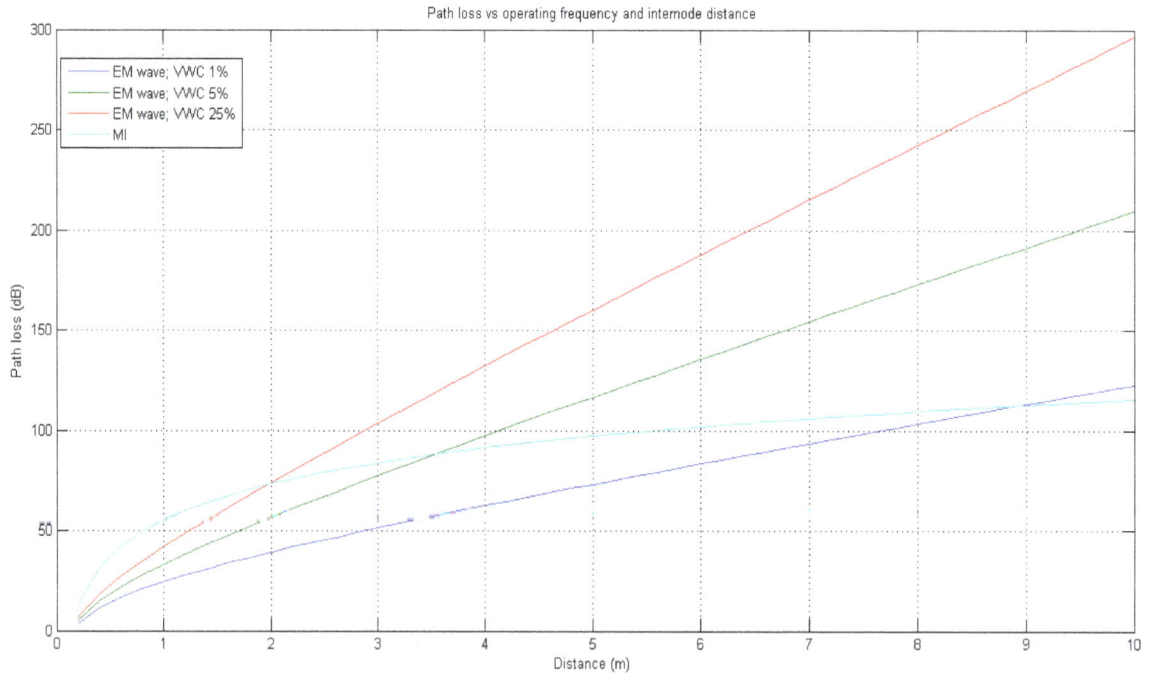

Fig. 2. The path loss of EM waveguide system and MI system is different from the soil water content.

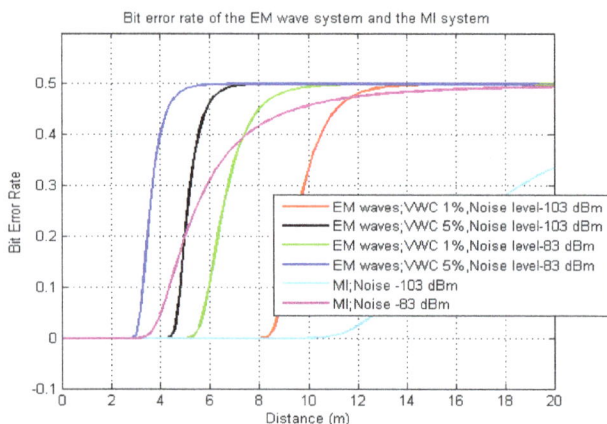

Fig. 3. Bit Error Rate for EM wave system with different soil water content and MI system with noise.

3.2. Bit Error Rate

As it was evaluated, the characteristics of BER(fig.3) depends on three factors: 1. Path Loss 2. Parasite Level 3.Modulation Method used by system. The path loss in MI system and EM wave system has been considered in Equation 8 and Equation 9. The parasite power in the soil was measured by wireless spectrum analyzer (BVS yellow jacket) [3] and [11]. The mean of noise level P_n was considered as about -103d B m and -83d Bm. The signal to noise (SNR) [1] can be calculated by SNR=P_t-L-P_n whereas P_t id the transmission power and L is the path loss (Equations 8 and 9) . In the simulation, P_t=10 dBm was considered. 2PSK modulation with the simple design was used while it is applied extensively. BER was obtained on the basis of a SNR function: BER = 0.5er fc($\sqrt{SNR}$ (in this function, erfc($\sqrt{SNR}$) is the error function [24]. In fig.3, BERs in MI and EM systems were shown with the different VWC soils as a function of transmission distance, r. In the low noise scenario, the transmission range in MI system is more than EM wave system (regardless of VWC). This scenario is explained via the following reasons:

A. In the low noise, the path loss below 100d B has no effect on the operation of BER.

B. MI system has more path loss than EM system in the sections that are below 100d B. But the path loss in this system is lower in the sections that are more than 100d B.

C. In the high noise scenario, since the transmission range of MI system is between the range of EM wave system in the dry and wet soil conditions, the path loss more than 80d B can influence on the operation of BER.

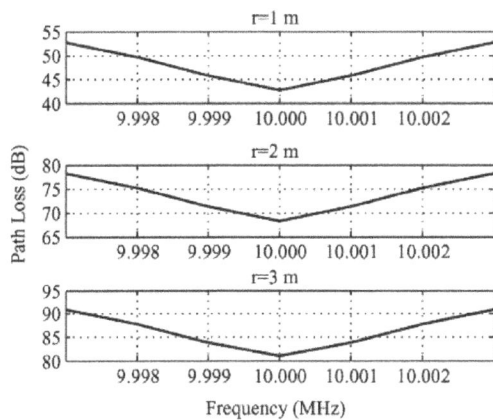

Fig. 4. The Frequency Response of MI Waveguide System with Different Wire and Remote Relay Resistances.

3.3. Bandwidth

It must be mentioned that the path loss in MI system is created because of the following factors: only one central frequency can make the load impedance matching possible but the output impedance is originated from resistance and reactance. The load impedance should be equal with the impedance conjugation of output in the secondary loop to prevent from any deviation in the central frequency that causes the reflection of transmission power and the increase of path loss. According to the analyses, the bandwidth ratio in MI system with the different transmission distances (fig.4) is 3-dB that is about 2KHZ. When the operational frequency is 10MHZ, the bandwidth isn't influenced by the transmission distance. Thus, MI system provides the wider range (10m) than EM waveguide system (about 4m). The advantage of MI system is that its operation is not influenced by the environment characteristics of soil especially the water content. But the transmission range of both systems is still too short to use for the scientific applications in the underground environment [2].

4. Discussion and Conclusion

In the underground wireless communications via EM waves, there are three major challenges: the high path loss because of the material absorption, the channel dynamic conditions because of the various properties of the soil and the very high size of the antenna.

MI is an alternative method with the same channel conditions and it can accomplish the communications with the small loops but recently, no detailed analysis performed to evaluate the path loss and the bandwidth of MI system in the underground soil environment. At the present study, one analytic model was shown that indicated the

communicational underground channel characteristics of MI. According to the channel analysis, we presented one MI wave method that increased the transmission range. Our analysis indicated that MI method had the constant conditions of channel because the path loss only depends on the permeability of propagation environment but the materials absorption is one of the most important parts in the path loss of EM system that may change in the various soil conditions. In the underground environments, the path loss of MI system is slightly lower than EM wave system in the natural and wet soils.

References

[1] I. F. Akyildiz, W. Su, Y. Sankarasubramaniam, and E. Cayirci, "Wireless sensor networks: A survey," *Comput. Netw. J.*, vol. 38, no. 4, pp. 393–422, March 2002.

[2] I. F. Akyildiz and E. P. Stuntebeck, "Wireless underground sensor networks: Research challenges," Ad Hoc Networks (Elsevier), vol. 4, pp. 669–686, Jul. 2006.

[3] L. Li, M. C. Vuran, and I. F. Akyildiz, "Characteristics of underground channel for wireless underground sensor networks," presented at the Med-Hoc-Net'07, Corfu, Greece, Jun. 2007.

[4] N. Jack and K. Shenai, "Magnetic induction IC for wireless communication in RF-impenetrable media," presented at the IEEE Workshop on Microelectronics and Electron Devices (WMED 2007), Apr. 2007.

[5] A. R. Silva and M. C. Vuran, "Development of a testbed for wireless underground sensor networks," EURASIP J. Wireless Commun.Netw. (JWCN) [Online]. Available: http://cse.unl.edu/~mcvuran/ugTestbed.pdf

[6] Standard Specification for Standard Nominal Diameters and Cross- Sectional Areas of AWG Sizes of Solid Round Wires Used as Electrical Conductors, ASTM Standard B 258-02, ASTM International, 2002.

[7] D. R. Frankl, Electromagnetic Theory. Englewood Cliffs, NJ: Prentice- Hall, 1986.

[8] W. M. Telford, L. P. Geldart, and R. E. Sheriff, Applied Geophysics, 2nd ed. New York: Cambridge Univ. Press, 1990.

[9] J. D. Kraus and D. A. Fleisch, Electromagnetics, 5th ed. New York: McGraw-Hill, 1999.

[10] L. A. Charles and W. A. Kenneth, Electronic Engineering, 3rd ed. New York: Wiley, 1973.

[11] "YellowJacket Wireless Spectrum Analyzer," Berkeley Varionics Systems, Inc. [Online]. Available: www.bvsystems.com

[12] Zhi Sun, Student Member, IEEE, and Ian F. Akyildiz, Fellow." Magnetic Induction Communications for Wireless Underground Sensor Networks," IEEE TRANSACTIONS ON ANTENNAS AND PROPAGATION, VOL. 58, NO. 7, JULY 2010

9

Evaluation criteria for reliability in computer systems

Esmail Kaffashi[1], Soheyla Amirian[1], Sahar Hematian Bojnourdy[2], Soheila Zahraee[2]

[1]Information Technology and Computer Engineering Department, Amirkabir University of Technology (Tehran Polytechnic),Tehran, Iran
[2]Industrial department, University of Applied Science, SID Bojnourd Center, Bojnourd, Iran

Email address:
Esmail2001201@Gmail.com (E. Kaffashi), Soh.amirian@Gmail.com (S. Amirian), Saharhematian2012@Gmail.com (S. H. Bojnourdy),
Zahraees@Gmail.com (S. Zahraee)

Abstract: Numerical methods, particularly finite element methods, are widely used in solving different problems. Since these methods are approximate, having a real understanding of the distribution of errors is extremely important. With the increasing number of users, the number of cause dfailuresin creases by their fault. In this article we will discuss per formance evaluation system, the performance evaluation of computer and communication systems for quality research that is finding its profit goals for the number of ways to predict the behavior of the system. One of the main parameters in determining the performance evaluation is reliability and because some complex systems cannot be easily modeled by hybrid methods (RBD), we use Markov method.

Keywords: Performance Evaluation, Reliability, Reliability Block Diagram, Markov Method

1. Introduction

The size and complexity of computer systems have increased with faster than our ability to design, test, implement and maintain it in the last decade. Computer systems are increasingly used in various applications. This will certainly continue in the future. By using computer in all devices, services, and activities of daily life, its importance has increased. With the increasing number of users, the number of failures increases due to their fault, so we need to evaluate the system performance. In this paper we discuss and analyze the reliability and Markov method. Assessing system performance targets in a number of ways isto predict the system behavior. When new systems are built or existing systems will be added or configured, the performance evaluation can be used to predict architectures collision or comprehensive changes in the system efficiency. An important aspect of evaluation and performance is performance measurement and monitoring. By monitoring, operating system informs more about real and important events of the system. Note that the requirement for performance measurement is system availability that can be observed and measured. Therefore, we can better understand performance measurements on systems where it may arise gradually improved with newer systems. Another measure of the most important aspects of the system that is measured to check for changes and to implement special codes need to be time stamps are the event logs writing. Obviously, these changes affect system performance. Monitoring models (software, hardware, and hybrid) is one of the basic models for performance evaluation, but up slightly from replicating or expressing. In the performance evaluation, an abstract described model, which is based on (Mathematics) the concept of a system that is clearly a part of math and interactions, along with theimpact that well-expressed. Most of the time this part of model called the system model screw. Stresses specified in the performance evaluation based model, which means it is in fact the best model for computer communications systems, the task is challenging. In fact, the performance models require a lot of engineering skills that still is not enough [1,2].

2. Reliability Concepts

Definition of system reliability must be defined on the basis of precise concepts. Since the set of same systems that operate under identical conditions, may fail at any point of time, the failure phenomenon can be explained in terms of the follow in gpotential methods[7].

2.1. Check and Calculate the Reliability of a Simple System

Reliability is the probability that a system is intended to mandate ata specific time and specific performance under particular condition to success. System reliability can be used as a measure of success in carrying out their duties properly. For more explanation and understanding of reliability, we need to know, how to calculate the reliability of a simple system, and will continue to obtain a general formula.

System reliability R (T) is the probability that a system meets its functional specifications in the interval [0, t] failsafe function. There is a simple example for calculating the reliability. In this example, we want to estimate the average number of failures in a device or system in the interval [0, t] that is the failure rate. If at time zero, the same number N of normal components (of a sort) have been starting to work. After time t, some components are corrupted. And damaged areas until t (in the interval [0, t]) called F (t) and the normal component of the residual at time t as S (t). The reliability of these components will be equal to [5,4]:

$$R(t) = \frac{S(t)}{N} = \frac{S(t)}{S(t) + F(t)} \quad (1)$$

And so the unreliability of these components will be equal to:

$$Q(t) = \frac{F(t)}{N} \quad (2)$$

Therefore, at any time t, we would have:

$$R(t) + Q(t) = 1 \Rightarrow R(t) = 1 - Q(t) \Rightarrow R(t) = 1 - \frac{F(t)}{N} \quad (3)$$

To calculate the failure rate, the above equation would be derived with respect to time.

$$\frac{dR(t)}{dt} = -\frac{1}{N} \frac{sF(t)}{dt} \Rightarrow -N \frac{dR(t)}{dt} = \frac{dF(t)}{dt} \quad (1)$$

$\frac{dF(t)}{dt}$ is an instantaneous failure rate, means the rate attime t components fail. At time t, $S(t)$ the number of components remains intact. In the follow in gequation:

$$Z(t) = \frac{\frac{dF(t)}{dt}}{S(t)} \quad (5)$$

The Z (t) is called the failure rate or hazard represent the instantaneous failure rate at time t relative to the number of components in the areas of time t.

$$Z(t) = \frac{1}{S(t)}[-N \frac{dR(t)}{dt}] = \frac{-1}{R(t)} \frac{dR(t)}{dt} \Rightarrow$$
$$Z(t) = \frac{1}{1 - Q(t)} \frac{dQ(t)}{dt} \quad (6)$$

$$\Rightarrow Z(t) = -\frac{\frac{dR(t)}{dt}}{R(t)} = \frac{\frac{dQ(t)}{dt}}{1 - Q(t)} \quad (7)$$

Z (t) is time dependent. But experience has shown that for many systems, Z (t) at some time intervals is relatively healthy. Usually hardware component failure rates follow an experimental plot of the bathtub curve. As seen in the chart below, the system life span is divided to three infancy, young, and oldcategories. Intervals failure rate is very high in infancy and old age, but youth has almost constant rate.

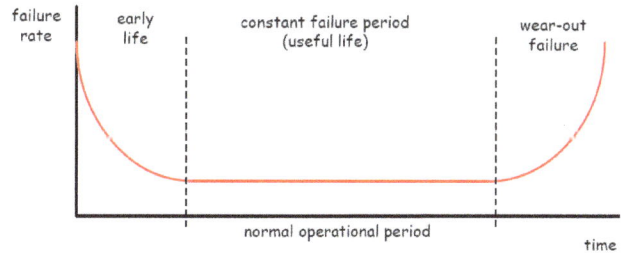

Figure 1. Bathtub diagram

Infancy typically in heat and stress tests is done at the factory pass. One can usually assume that the systems are in use in the period of youth. The following equation:

$$Z(t) = \lambda = \frac{-1}{R(t)} \frac{dR(t)}{dt} \Rightarrow \lambda dt = -\frac{dR(t)}{R(t)} \quad (8)$$

Integrating both sides of equation we have:

$$\lambda \int_0^t dt = -\int_1^{R(t)} \frac{dR(t)}{R(t)} \Rightarrow \lambda t \mid_0^t = LnR(t) \mid_1^{R(t)} \Rightarrow$$
$$\lambda t = LnR(t) \Rightarrow R(t) = e^{-\lambda t} \quad (9)$$

This relationship is known as the Exponential failure law. It means with a constant failure rate, reliability varies as a power function of time [5,4].

2.2. Reliability Modeling

Reliability, which can be calculated using various methods, is divided into two main groups: model-based and measurement based. The first method is widely used to evaluate the reliability of complex software / hardware systems is used, which is based on creating a model suitable for systems that briefly, with a sufficient level of detail to provide the aspect of interest to evaluate. Analysis models are classified into two types: Hybrid and state-based. The Hybrid model provides system structure in a logical connecting operation of elements in order to determine the success or failure of the system. The state model offer the behavior of system based on states tracks and available states. Models are used extensively to assess the characteristics of trust, especially reliability analysis. However, with some possible input value, the actual behavior of the system provider may not be accurate enough. Measurement-based approach may provide better results because they are based on actual operating data and application of statistical techniques. However, because the actual data may not be available using this method is not always possible [7].

3. Evaluation of Other Criteria

Reliability is not the only efficient criteria for evaluating the system. The following criterions also are used for this purpose:

3.1. Mean Time to Failure (MTTF)

The mean time to failure is the mean time that the system works before you experience a failure. Consider N systems work identical. If we show the deterioration of system i with ti, we have:

$$MTTF = \frac{1}{N}\sum_{i=1}^{N} t_i \qquad (10)$$

We know $\frac{dQ(t)}{dt}$ is the failure density function and the MTTF is in fact expected time of failure. Following:

$$MTTF = \int_0^\infty t\frac{dQ(t)}{dt}dt = -\int_0^\infty t\frac{dR(t)}{dt}dt \Rightarrow MTTF = \int_0^\infty R(t)dt \quad (11)$$

This equation shows that the area under the curve R (t) is compared to the MTTF. However, if the system obeys the exponential failure law, we have:

$$MTTF = \int_0^\infty e^{-\lambda t} dt = \frac{-1}{\lambda}[e^{-\lambda t}]_0^\infty = \frac{1}{\lambda} \qquad (12)$$

Means that, MTTF is inversely related to the rate of failure [3].

3.2. Mean Time to Repair (MTTR)

Mean time between failures is mean time between successive failures of the system. The mean time to repair (MTTR) cannot be easily estimated, and based on many parameters such as the type of defect, skills and experience to repair work, repair tools, system testing capabilities, and so on. Usually this time is estimated with the injection error to system and investigatation required time to fix the error. In mostcases, in order to simplify the evaluation, we show μ for the repair rate, fixed and equal to reciprocal of MTTR.

$$\mu = \frac{1}{MTTR} \qquad (13)$$

Now, it can be stated maintainability of the system. Maintainability of the system is probability of a system could be damaged in the interval [0, t] to be repaired and re-used. With this definition we have:

$$M(t) = 1 - e^{-\mu t} \qquad (14)$$

3.3. Mean Time between Failures (MTBF)

Mean time between failures, is the distance between two successive failures of a system. In other words, after a system crash, the system is repaired and back in operation until the next failure occurs. Based on this definition, the MTBF can be obtained as follows [5,7].

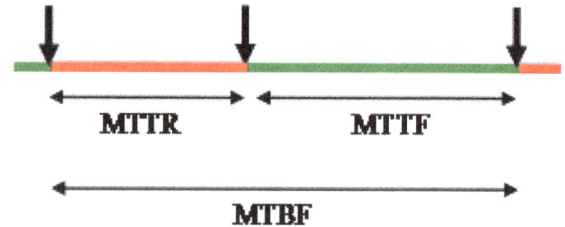

Figure 2. Mean time between failure

$$MTBF = MTTR + MTTF \qquad (15)$$

Now based on these criteria, availability can be expressed. Availability is the probability that the system is functioning correctly at a specified time t. As you can see the availability is defined at the moment, but reliability is defined at the time interval.

The system can be accessed but fails repeatedly, and on the other hand have low reliability. If N is the number of failures occurred during the execution time, Steady-state availability is calculated as follows:

$$A_{Steady-State} = \frac{N.MTTF}{N.MTTF + N.MTTR} = \frac{MTTF}{MTTF + MTTR} = \frac{MTTF}{MTBF} \qquad (16)$$

If the system abide failure rate and exponential repair, we have:

$$A_{Steady-State} = \frac{\frac{1}{\lambda}}{\frac{1}{\lambda}+\frac{1}{\mu}} = \frac{1}{1+\frac{\lambda}{\mu}} \qquad (17)$$

4. Evaluation of the System Using a Markov Model

Some complex systems cannot be easily modeled by hybrid methods like Reliability Block Diagram (RBD), such as feature detection and fault coverage, to repair and replace a module, the dynamic changes in the configuration of thesystem. The methods used for modeling should alsoconsider the state of the system. RBD shows how system components work properly and is effective in the correct working of the entire system. In this model, each piece of the system is a block that contains the input and output terminals. Block operation is the same as a switch.If the device is in operation mode switch is closed, but if the device is disabled, the switch will open. Given that the failure of any component of the system, some of the blocks are closed and some are open, if there is a path between input and output, the entire system is working well, but if there is no path then the system is disabled. RBD reliability of complex systems is divided in series, parallel, and M out of N. If the above not be solved, we solve by Markov method [5].

4.1. TheMarkov Model

For the Markov process, we divide the time into three periods: past, present, and future. The future of the process does not depend on the direction of the past but only present. For example, the Poisson process is a Markov process because the number of events that occur after a certain time is regardless of the events that happened before that. Markov chains is a special case of Markov process where the parameter T and the system selects only discrete values. Accordingly, a series of random variables X1, X2, X3, ...,Xn is called a Markov chain. Markov model is used for the analysis of stochastic systems based on the notions of states and transitions between states of the system, the system state is a combination of distinct modules, andthe system shows the healthy and faulty. If the system has n modules and each module has a health condition or damaged, then 2n can be considered for the position. State transitions model changes between states of the system that this changes can be result of a malfunction, or the repair of a faulty module, such as a TMR, when the system works properly that the 2 or 3 module works correctly [5,6].

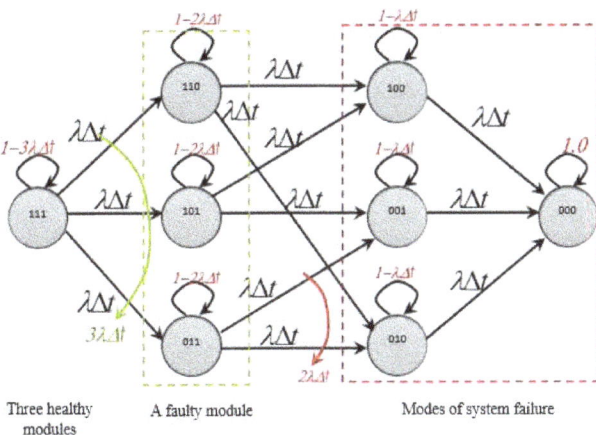

Figure 3. *Markov model for a TMR system*

In general, the probability of the system at time t + 1 in a given case depends on the state of the system at time t, t-1, t-2 and so on. The first order Markov model, the state of the system at time t + 1 depends the state of the system at time t not on the state of the system to chain all transitions between state before time t, in this model because there is only a failure between t and to t + 1 to describe many systems, is sufficient with memory length 1. If the system state at time t + 1 is independent of the previous state of the system, the Markov model is zero-order, such as polynomial is a probability

distribution, and this model does not have a memory. In m-order Markov model, the order of a Markov model is equal to the amount of memory that it is possible to calculate the next state. For example, the next state in a second-order Markov model depends on two previous states.

The possibility that one module at a time breaks at $t + \Delta t$ time, provided that the module worked properly at t time (healthy) is equal to:

$$P = 1 - \frac{R(t + \Delta t)}{R(t)} \quad (18)$$

If the failure of modules subordinate Exponential failure law, we have:

$$R(t) = e^{\lambda t} \Rightarrow P1 - \frac{e^{-\lambda(t+\Delta t)}}{e^{\lambda t}} \quad (19)$$

$$e^{-\lambda\Delta t} = 1 + (-\lambda\Delta t) + \frac{(-\lambda\Delta t)^2}{2!} + ... \Rightarrow P = 1 - e^{-\lambda\Delta t}$$
$$= 1 - P[1 + (-\lambda\Delta t) + \frac{(-\lambda\Delta t)^2}{2!} + ...] \Rightarrow P = (-\lambda\Delta t) - \frac{(-\lambda\Delta t)^2}{2!} - ... \quad (20)$$

In Markov model, the possibility of a module failure rate λ in the considered time step Δt is considered $\lambda\Delta t$.

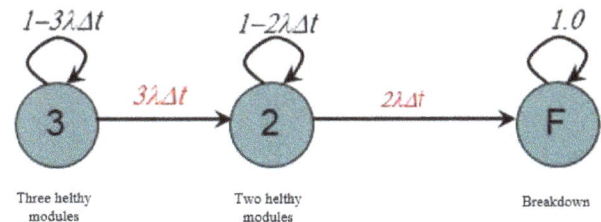

Figure 4. *Simplified model of a TMR system*

Since the first order Markov model is used, the system state at time $t + \Delta t$ depends only on the system state at t time. Thus we have:

$$P(t + \Delta t) = A.P(t) \quad (21)$$

The probability of being in different states of the system can be showed in matrix:

$$\mathbf{P}(t + \Delta t) = \mathbf{A}.\mathbf{P}(t) \quad (22)$$

$$\begin{bmatrix} P_3(t + \Delta t) \\ P_2(t + \Delta t) \\ P_F(t + \Delta t) \end{bmatrix} = \begin{bmatrix} 1 - 3\lambda\Delta t & 0 & 0 \\ 3\lambda\Delta t & 1 - 2\lambda\Delta t & 0 \\ 0 & 2\lambda\Delta t & 1 \end{bmatrix} . \begin{bmatrix} P_3(t) \\ P_2(t) \\ P_F(t) \end{bmatrix}$$

To calculate the probability of each state of the system at time t, we open up above matrix equation and sides of each share would be divided by Δt to a differential equation then solving the probability of being in each state of the system at time t is calculated. TMR, system reliability is equal to:

$$P_{TMR}(t) = P_3(t) + P_2(t) = 1 - P_F(t) \Rightarrow 3e^{-2\lambda t} - 2e^{-3\lambda t} \quad (23)$$

The result is identical with the result obtained from the TMR model helping the block diagram reliability [1].

Various types of Markov can be noted discrete Markov and Hidden Markov processes. If we have a system at any moment in a distinct states S1, ..., SN, at discrete times or regular intervals, the system state changes according to a set of possibilities. Appropriate to describe the current system needs to know the current status along with all the previous cases is that for the special case of a first order Markov chain, describing the probability is determined only by the current state and the previous state [2].

$$P(q_t = S_j \mid q_{t-1} = S_i, q_{t-2} = S_k, ...) = P(q_t = S_j \mid q_{t-1} = S_i) \quad (24)$$

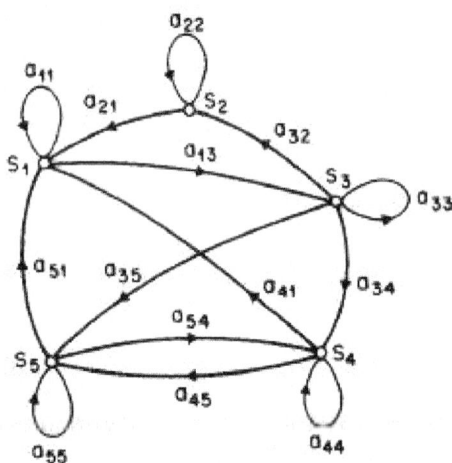

Figure 5. *A Markov chain with 5 states*

We consider only processes in which the right side of the above equation is independent of time and that's why we have a set of transition probabilities between states.

$$a_{i,j} = P(q_t = S_j \mid q_{t-1} = S_i) \ 1 \le i, j \le N \quad (25)$$

The above random processis calleda Markov model, because the output is a set of states that their exposure is associated with a view. We can produce the sequence of observations required to calculate the probability that the Markov chain. In this model, each state corresponds to an observable event, but the hidden model of the observation is probabilistic functions of states. The resulting model is a stochastic model is underlying (hidden) and only by a set of random processes that produce the sequence of observations is observed [2].

5. Discussion and Conclusion

This paper presented the system performance goals and evaluation criteria including reliability. The system performance evaluation purposes is the number of ways for predicting the behavior of the system, monitoring performance measurement, andpredicting changes in the system concept in new architectures. In this paper, we examined the reliability criterionand because complex systems cannot be easily combined with hybrid techniques such as RBD model, we usedMarkov methods. Markov models are used for theanalysis of stochastic systems based on the concepts of states of transition between states of the system, wherethe unknown probabilities can be obtained by iteration.

References

[1] BOUDEWIJN R.HAVERKORT, PERFORMANCE OF COMPUTER COMMUNICATIN SYSTEM: A Model – Based Approach. John Wiley & Sons, Inc. New York, NY, USA ©1998, ISBN:0471972282.

[2] Piet van Mieghem, Performance Analysis Of communications Networks And Systems, 2006. Book DOI:http://dx.doi.org/10.1017/CBO9780511616488

[3] M. Danielle Beaudry, performance-related reliability measures for computing system, IEEE TRANSACTIONS ON COMPUTERS, VOL. c-27, NO. 6, JUNE 1978

[4] Dhirajk.Pradhan".Fault-Tolerant Computer System design." by Prentice Hall PTR, A Simon &Sxhuster Company Upper Saddle River, New Jersey 07458,1996

[5] Zarandi, HR. Markov systems and evaluate the system performance ", Amirkabir University (Tehran Polytechnic), second edition. September 2011.

[6] Moatar, Mohammad Hussain. "Hidden Markov models and training algorithms", Amirkabir University of Technology (Tehran Polytechnic), 2006.

[7] Abbas Abadi, S. Reliable, Humayun. "Component-based software reliability prediction using stochastic Petri nets and Markov chains". International Symposium on Modeling and Optimization, Tehran, 2012.

10

Performance of Superconducting Synchronous Generator for Three-Phase Short-Circuit at Its Terminals

H. Lale Zeynelgil

Department of Electrcal Eng., Electric & Electronics Faculty, Istanbul Technical University, Maslak Istanbul, Turkey

Email address:

zeynelgil@itu.edu.tr

Abstract: In this paper, three-phase short-circuit at the terminals of a superconducting synchronous generator is simulated by using MATLAB. In this simulation, a single screened superconducting generator directly connected to power system is considered and two operating cases are investigated: the first case is that the fault is permanent and the second case is that the fault is cleaned by a protection system. To show the effects of this short-circuit and the performance of the superconducting synchronous generator for both cases, the deviations of armature momentary currents, angular speed and field current are plotted. In order to compare the obtained results, the present simulations for the both cases mentioned above are repeated for a conventional synchronous generator having the same rating as considered superconducting counterpart, which is directly connected to power system, and the deviations of the same variables as considered for the superconducting synchronous generator are also given in the paper.

Keywords: Superconducting Synchronous Generator, Screen, Conventional Synchronous Generator, Three-Phase Short-Circuit

1. Introduction

Electrical energy has very important role for human life and it is essential to supply qualified electrical energy to the consumers. The transient performances of the synchronous generators in the power system are very important for this aspect. Since the requirement of electrical energy increases more and more, the ratings of generating units must be increased and the distribution networks must become relatively complex. However, the ratings of generators are limited due to technological constraints such as electrical, mechanical and transportation problems. Superconducting synchronous generators have various advantages over conventional ones to overcome the maximum rating problem beside improving the transient performance [1-3].

One of the most familiar methods to investigate the transient performance of synchronous machines is the three-phase short-circuit test since the transient current mostly occurs for this fault. When the three-phase short-circuit occurs in a power network, the fault is cleaned by protection system during the generator is on. Following this, very high currents pass through the armature windings. Especially, when there is a three-phase sudden short-circuit at the terminals of a synchronous generator the transient current is much higher than its nominal current [2, 4, 5].

A superconducting synchronous machine consists of a rotating cryogenic superconducting field winding and stationary armature winding at ambient temperature. Superconducting materials are not preferred for armature windings in the stator because it is difficult to use these as alternative current windings [6]. Also, an exceed flux is produce at superconducting field winding, so the iron core in rotor reach to the saturation point and overheat and noisy operation are occurred. Therefore, iron is used only to stator yoke as shield [7].

There are one or two screens between rotor and stator in superconducting synchronous machines. The screens have two fundamental tasks. One of them is to protect the superconducting field winding form magnetic field deviations and the other is to damp rotor angle oscillations especially for short circuits. Therefore construction and material selection are restraint factors for the screens [8].

In this paper, three-phase short-circuit at the terminals of a superconducting synchronous generator directly connected to power system is simulated by using MATLAB. Both the cases of the fault is permanent and cleaned are investigated. To show the behaviour of superconducting synchronous generator when there is a three-phase short circuit at its terminals the deviations of armature momentary current, rotor angle, angular speed and field current are plotted for the cases which are considered. Finally, this investigation is repeated for a conventional synchronous generator, which is directly connected to power system in order to compare the behaviours of the superconducting and the conventional synchronous generators for the considered fault.

2. Mathematical Model Used in the Simulation

2.1. Modelling of Single Screened Superconducting Generator

A superconducting synchronous generator having superconducting field windings and single rotor screen is considered, and a three-phase short-circuit at the terminals of the generator is investigated in the simulation. It is assumed that the generator is directly connected to power system. To obtain the mathematical model of the generator, armature windings are assumed to be static and they are transformed to the rotating and orthogonal windings with the same speed as the rotor windings by using Park Transformation [3, 9, 10]. The transformation matrix equation is given as,

$$\begin{bmatrix} u_d \\ u_q \\ u_o \end{bmatrix} = u_{dq} = T u_{ph} = T \begin{bmatrix} u_a \\ u_b \\ u_c \end{bmatrix} \tag{1}$$

where

$$T = \frac{2}{3} \begin{bmatrix} \cos\theta & \cos\left(\theta - \frac{2\pi}{3}\right) & \cos\left(\theta + \frac{2\pi}{3}\right) \\ -\sin\theta & -\sin\left(\theta - \frac{2\pi}{3}\right) & -\sin\left(\theta + \frac{2\pi}{3}\right) \\ \frac{1}{2} & \frac{1}{2} & \frac{1}{2} \end{bmatrix} \tag{2}$$

From above expression, the mutual inductances between armature and rotor windings are constant, and thus the mathematical model having constant coefficients is obtained for real superconducting generator, which is a system varying in time.

When the Park transformation mentioned above is applied to flux-current relations for armature windings, the following equations for armature and rotor windings of the machine are given as [9, 10],

$$\begin{bmatrix} \lambda_d \\ \lambda_{kd} \\ \lambda_f \end{bmatrix} = \begin{bmatrix} L_d & L_{akd} & M \\ \frac{3}{2}L_{akd} & L_{kd} & L_{fkd} \\ \frac{3}{2}M & L_{fkd} & L_f \end{bmatrix} \begin{bmatrix} -i_d \\ i_{kd} \\ i_f \end{bmatrix} \tag{4}$$

$$\begin{bmatrix} \lambda_q \\ \lambda_{kq} \end{bmatrix} = \begin{bmatrix} L_q & L_{akq} \\ \frac{3}{2}L_{akq} & L_{kq} \end{bmatrix} \begin{bmatrix} -i_q \\ i_{kq} \end{bmatrix} \tag{5}$$

$$\lambda_0 = L_0 . i_o \tag{6}$$

The relations for transformed stator voltages in d and q axes are obtained as,

$$V_d = -r_a i_d + \frac{d\lambda_q}{dt} - \omega\lambda_q \tag{7}$$

$$V_q = -r_a i_q + \frac{d\lambda_d}{dt} + \omega\lambda_d \tag{8}$$

where $\omega = \frac{d\theta}{dt}$.

and transformed relation of the power is obtained as,

$$p(t) = \frac{3}{2}v_d . i_d + \frac{3}{2}v_q . i_q + \frac{3}{2}v_0 . i_0 \tag{9}$$

By substituting the relations for stator voltages in d and q axes in Eq. (9), it is obtained instantaneous power as,

$$p(t) = \omega . \frac{3}{2}\left(\lambda_d i_q - \lambda_q i_d\right) + \frac{3}{2}\left(\frac{d\lambda_d}{dt}i_d + \frac{d\lambda_q}{dt}i_q\right) + 3\frac{d\lambda_0}{dt}i_0 \tag{10}$$

According to above given equation, the instantaneous power consists of shaft power and the variation in stored energy. Thus, the shaft torque is defined as,

$$T_e = \frac{3}{2}\left(\lambda_d i_q - \lambda_q i_d\right) \tag{11}$$

2.2. The Model in Per-unit Values

The per-unit values are used to investigate the performance of power systems. A base value X_B is determined to represent any variable X in per-unit values and per-unit value for this variable is defined as,

$$x \triangleq \frac{X}{X_B} \tag{12}$$

Generally, nominal values of the considered system are used as base quantities. In this respect, the base values for power, impedance, flux and torque are determined, respectively, as

$$P_B = \frac{3}{2}V_B I_B \tag{13}$$

$$Z_B = \frac{V_B}{I_B} \tag{14}$$

$$\lambda_B = \frac{V_B}{\omega_0} \tag{15}$$

$$T_B = \frac{P}{\omega_0}P_B \tag{16}$$

Since the frequency is different for any winding, the base quantities for voltage and current can have different values Therefore, the flux-current relations for d axis in per-unit values are obtained as,

$$\begin{bmatrix} \psi_d \\ \psi_{kd} \\ \psi_f \end{bmatrix} = \begin{bmatrix} x_d & x_{akd} & x_{ad} \\ x_{akd} & x_{kd} & x_{fkd} \\ x_{ad} & x_{fkd} & x_f \end{bmatrix} \begin{bmatrix} -i_d \\ i_{kd} \\ i_f \end{bmatrix} \tag{17}$$

where

$$x_d = \omega_0 \frac{I_{dB}}{V_{db}} L_d \tag{18}$$

$$x_{kd} = \omega_0 \frac{I_{kB}}{V_{kb}} L_{kd} \tag{19}$$

$$x_f = \omega_0 \frac{I_{fB}}{V_{fb}} L_f \tag{20}$$

$$x_{akd} = \omega_0 \frac{I_{kB}}{V_{db}} L_{akd} = \frac{3}{2}\omega_0 \frac{I_{dB}}{V_{kb}} L_{akd} \tag{21}$$

$$x_{ad} = \omega_0 \frac{I_{fB}}{V_{db}} M = \frac{3}{2}\omega_0 \frac{I_{dB}}{V_{fb}} M \tag{22}$$

$$x_{fkd} = \omega_0 \frac{I_{kB}}{V_{fb}} L_{fkd} = \frac{\omega_0 I_{fB}}{V_{kb}} L_{fkd} \tag{23}$$

and

$$\frac{3}{2} V_{dB} I_{dB} = V_{fB} I_{fB} \tag{24}$$

$$\frac{3}{2} V_{dB} I_{dB} = V_{kB} I_{kB} \tag{25}$$

$$V_{fB} I_{fB} = V_{kB} I_{kB} \tag{26}$$

Flux-current relations for q axis are written as similarly to those obtained for d axis as follows:

$$\begin{bmatrix} \psi_q \\ \psi_{kq} \end{bmatrix} = \begin{bmatrix} x_q & x_{akq} \\ x_{akq} & x_{kq} \end{bmatrix} \begin{bmatrix} -i_q \\ i_{kq} \end{bmatrix} \tag{27}$$

The same base power value is assumed for all of windings since the base values for voltages and currents of armature windings are taken as effective values while those of rotor windings are taken as DC values in the particular relations. If it is assumed that the base values for voltages and currents of all windings are same the mutual inductances between armature windings and, field windings and screen in each axis are equal as,

$$x_{akd} = x_{ad} \tag{28}$$

$$x_{akq} = x_{aq}$$

Also, the fact that the rotor screen is a thin, continuous shell leads to the shielding constraint:

$$x_{fkd} = x_{kd} \tag{29}$$

Thus, flux-current relations given by Eqs. (17) and (27) are obtained as [2, 9, 10],

$$\begin{bmatrix} \psi_d \\ \psi_{kd} \\ \psi_f \end{bmatrix} = \begin{bmatrix} x_d & x_{ad} & x_{ad} \\ x_{ad} & x_{kd} & x_{kd} \\ x_{ad} & x_{kd} & x_f \end{bmatrix} \begin{bmatrix} -i_d \\ i_{kd} \\ i_f \end{bmatrix} \tag{30}$$

$$\begin{bmatrix} \psi_q \\ \psi_{kq} \end{bmatrix} = \begin{bmatrix} x_q & x_{aq} \\ x_{aq} & x_{kq} \end{bmatrix} \begin{bmatrix} -i_q \\ i_{kq} \end{bmatrix} \tag{31}$$

The relations for stator voltages in per-unit values are given as,

$$v_d = \frac{1}{\omega_0} \frac{d\psi_d}{dt} - \frac{\omega}{\omega_0} \psi_q - r_a i_d \tag{32}$$

$$v_q = \frac{1}{\omega_0} \frac{d\psi_q}{dt} + \frac{\omega}{\omega_0} \psi_d - r_a i_q \tag{33}$$

where r_a is armature windings resistance in per-unit values. The relations for rotor windings in per-unit values are also given as,

$$v_f = \frac{1}{\omega_0} \frac{d\psi_f}{dt} + r_f i_f \tag{34}$$

$$v_{kd} = \frac{1}{\omega_0} \frac{d\psi_{kd}}{dt} + r_{kd} i_{kd} \tag{35}$$

$$v_{kq} = \frac{1}{\omega_0} \frac{d\psi_{kq}}{dt} + r_{kq} i_{kq} \tag{36}$$

Finally, the relations for rotor dynamics in per-unit values are given as,

$$\frac{2H}{\omega_0} \frac{d\omega}{dt} = T_m - T_e \tag{37}$$

$$\frac{d\delta}{dt} = \omega - \omega_0 \tag{38}$$

where

$$H = \frac{\frac{1}{2} J \left(\frac{\omega_0}{p} \right)^2}{P_B} \tag{39}$$

2.3. Simulation Model

In order to investigate the dynamic performance of a superconducting synchronous generator in the case of a three-phase short-circuit at the terminals of the generator, a model containing the differential equations for voltages behind sub-transient reactance are used in this paper [9, 10].

The mathematical model used for the simulation model consists of number of ten first order differential equations given as:

$$\frac{de_d''}{dt} = -\frac{x_q - x_q''}{x_q''} \frac{\psi_q}{T_{q0}''} - \frac{x_q}{x_q''} \frac{e_d''}{T_{q0}''} \tag{40}$$

$$\frac{de_q''}{dt} = \frac{x_d - x_d''}{x_d''} \frac{\psi_d}{T_{d0}''} - \frac{x_d}{x_d''} \frac{e_q''}{T_{d0}''} + \frac{e_q'}{T_{d0}''} \tag{41}$$

$$\frac{de_q'}{dt} = (\alpha - 1) \frac{e_q''}{T_{d0}'} - \alpha \frac{e_q'}{T_{d0}'} + \frac{e_f}{T_{d0}'} \tag{42}$$

$$\frac{d\psi_d}{dt} = -\frac{\psi_d}{T_a} + \omega \psi_q + \frac{e_q''}{T_a} + v_d \tag{43}$$

$$\frac{d\psi_q}{dt} = -\frac{\psi_q}{T_a} - \omega \psi_d + \frac{e_d''}{T_a} + v_q \tag{44}$$

$$\frac{d\psi_f}{dt} = -\omega_0 r_f i_f + \omega_0 v_f \tag{45}$$

$$\frac{d\psi_{kd}}{dt} = -\omega_0 r_{kd} i_{kd} \tag{46}$$

$$\frac{d\psi_{kq}}{dt} = -\omega_0 r_{kq} i_{kq} \tag{47}$$

$$\frac{d\omega}{dt} = \frac{\omega_0}{2H} \left[T_m + \frac{\psi_d e_d''}{x_q''} + \frac{\psi_q e_q''}{x_d''} + \psi_d \psi_q \left(\frac{1}{x_q''} - \frac{1}{x_d''} \right) \right] \tag{48}$$

$$\frac{d\delta}{dt} = \omega - \omega_0 \tag{49}$$

where

$$\alpha = \frac{x_f}{N_f \cdot N_{kd}} = \frac{x_d - x_d''}{N_d' \cdot N_d''} \qquad (50)$$

3. Simulation Studies

Three-phase short-circuit is assumed as one of the biggest fault that a generator must withstand. Especially, when short-circuit occurs at the terminals of the generator very high currents pass through armature windings of the generator.

In this paper, three-phase short-circuit at the terminals of a superconducting synchronous generator directly connected to power system is simulated. The parameters for the synchronous generator are given in Table A1 at Appendix [3]. It is assumed that the considered generator has superconducting field windings on rotor and single rotor screen. Then, this investigation is repeated for a conventional synchronous generator directly connected to power system for comparison to behaviours of these generators. The parameters for the conventional synchronous generator are also given in Table A2 at Appendix [3]. For the simulation, a programme is written by using MATLAB and the mathematical models reflecting real behaviour of the system are used to represent both of the generators.

In these simulations, firstly, it is assumed that each machine considered operates in steady state conditions and then a three-phase short-circuit occurs at the terminals of the considered machine at arbitrary time $t = t_0$. For comparison, steady state operation conditions for the superconducting and the conventional synchronous generators are given in Table 1 and 2, respectively. The strategy of simulation study is same for each case, except the differences in mathematical models of the generators resulting from physical differences between these generators.

Table 1. *Steady state operation conditions for superconducting synchronous generator.*

Terminal Voltage of the machine: V_t (p.u.)	1
Terminal Voltage of the machine in d axis: V_d (p.u.)	0.2122
Terminal Voltage of the machine in q axis: V_q (p.u.)	0.9772
Stator current: I_t (p.u.)	1
Stator current in d axis: I_d (p.u.)	0.6951
Stator current in d axis: I_q (p.u.)	0.7189
Field winding current: I_f (p.u.)	6.0155
Voltage induced in the machine: E_f (p.u.)	1.1851
Voltage behind transient reactance: e_{q0}' (p.u.)	1.1239
Voltage behind sub-transient reactance in d axis: e_{d0}'' (p.u.)	0.1301
Voltage behind sub-transient reactance in q axis: e_{q0}'' (p.u.)	1.060
Rotor angle: δ	12°.25
Armature winding flux in d axis: ψ_d (p.u.)	0.9786
Armature winding flux in q axis: ψ_q (p.u.)	−0.2135
Field winding flux: ψ_f (p.u.)	2.5159
Rotor screen flux in d axis: ψ_{kd} (p.u.)	1.1612
Rotor screen flux in q axis: ψ_{kq} (p.u.)	−0.1416

Two operating cases for both superconducting and conventional synchronous machines are investigated in this study; the considered short-circuit is permanent and cleaned by protection system at 0.2 seconds. For both cases, to show

behaviour of superconducting synchronous generator when a three-phase short circuit occurs at its terminals, the deviations of armature momentary current, angular speed and field current are plotted in the simulation. The graphics for the first case are given in Figs. 1 to 3 and for the second case are given in Figs. 4 to 6.

Table 2. *Steady state operation conditions for conventional synchronous generator.*

Terminal Voltage of the machine: V_t (p.u.)	1
Terminal Voltage of the machine in d axis: V_d (p.u.)	0.6447
Terminal Voltage of the machine in q axis: V_q (p.u.)	0.7645
Stator current: I_t (p.u.)	1
Stator current in d axis: I_d (p.u.)	0.9507
Stator current in d axis: I_q (p.u.)	0.3102
Field winding current: I_f (p.u.)	1.3567
Voltage induced in the machine: E_f (p.u.)	2.8761
Voltage behind transient reactance: e_{q0}' (p.u.)	0.9890
Voltage behind sub-transient reactance in d axis: e_{d0}'' (p.u.)	0.5940
Voltage behind sub-transient reactance in q axis: e_{q0}'' (p.u.)	0.7973
Rotor angle: δ	40°.14
Armature winding flux in d axis: ψ_d (p.u.)	0.7656
Armature winding flux in q axis: ψ_q (p.u.)	−0.6483
Field winding flux: ψ_f (p.u.)	1.0563
Amortisseur winding flux in d axis: ψ_{kd} (p.u.)	0.8607
Amortisseur winding flux in q axis: ψ_{kq} (p.u.)	−0.6173

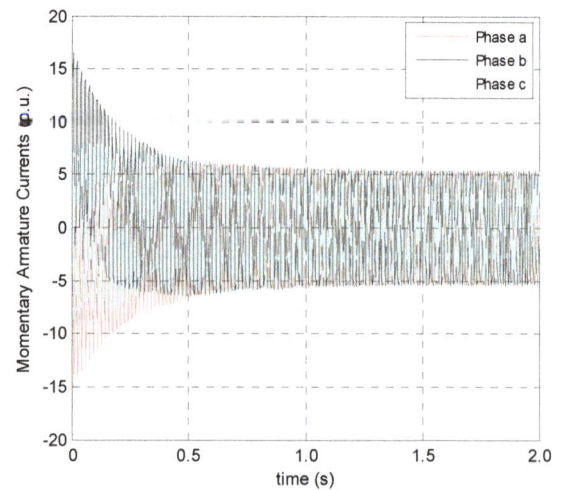

Figure 1. *Deviations of momentary armature currents for superconducting synchronous generator for the case of permanent fault.*

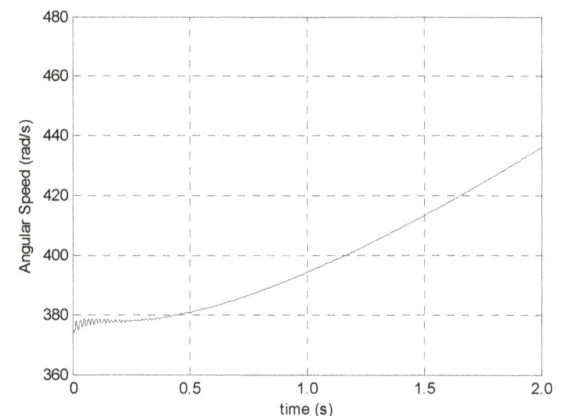

Figure 2. *Deviation of angular speed for superconducting synchronous generator for the case of permanent fault.*

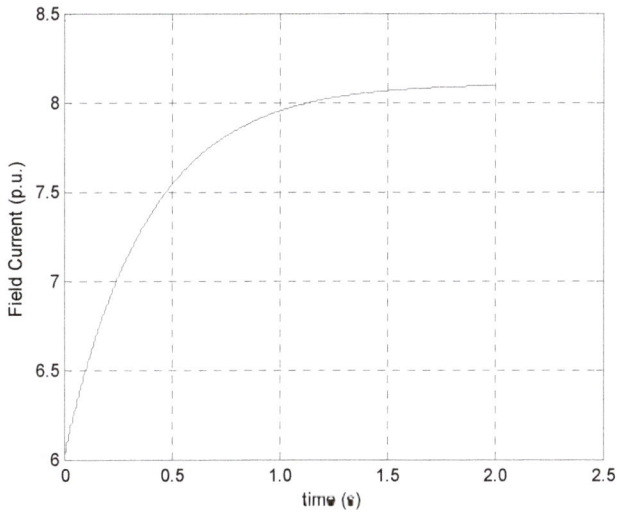

Figure 3. *Deviation of filed current for superconducting synchronous generator for the case of permanent fault.*

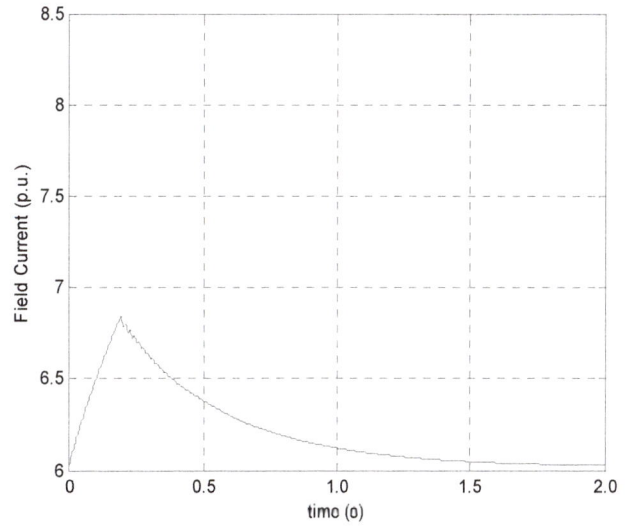

Figure 4. *Deviations of momentary armature currents for superconducting synchronous generator for the case of removing the fault.*

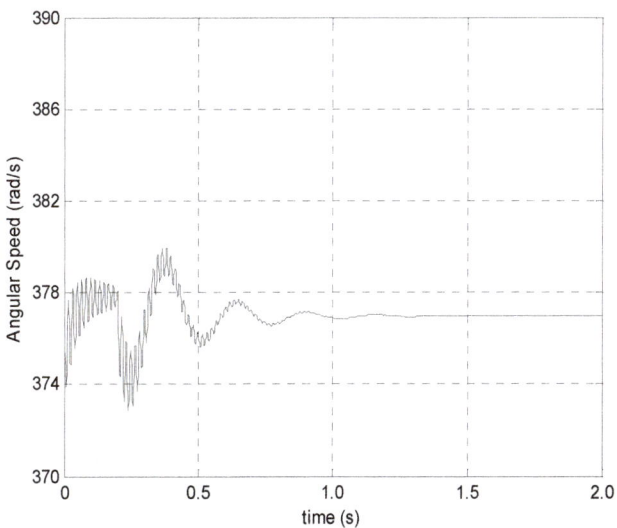

Figure 5. *Deviation of angular speed for superconducting synchronous generator for the case of removing the fault.*

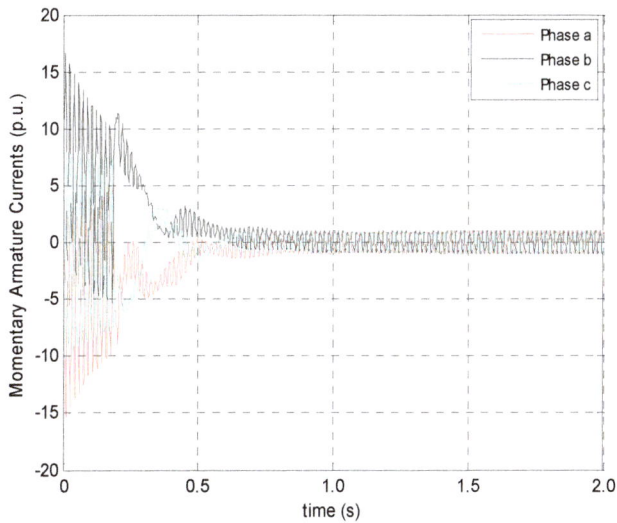

Figure 6. *Deviations of momentary armature currents for superconducting synchronous generator for the case of removing the fault.*

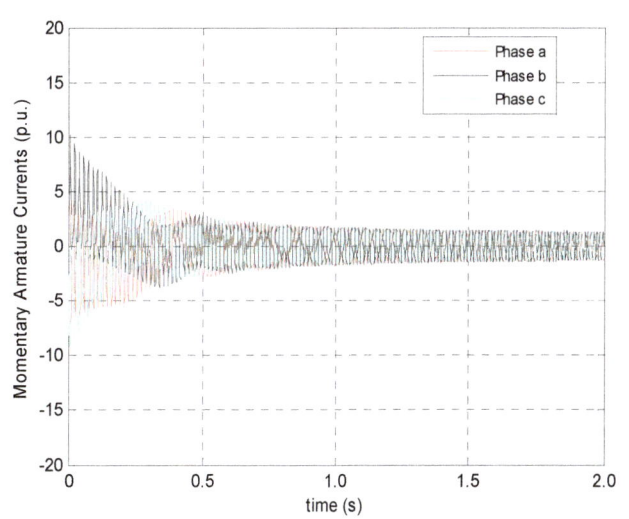

Figure 7. *Deviations of momentary armature currents for conventional synchronous generator for the case of permanent fault*

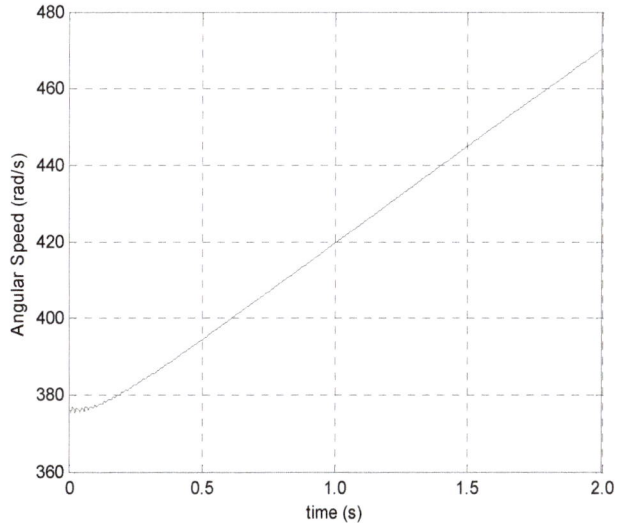

Figure 8. *Deviation of angular speed for conventional synchronous generator for the case of permanent fault.*

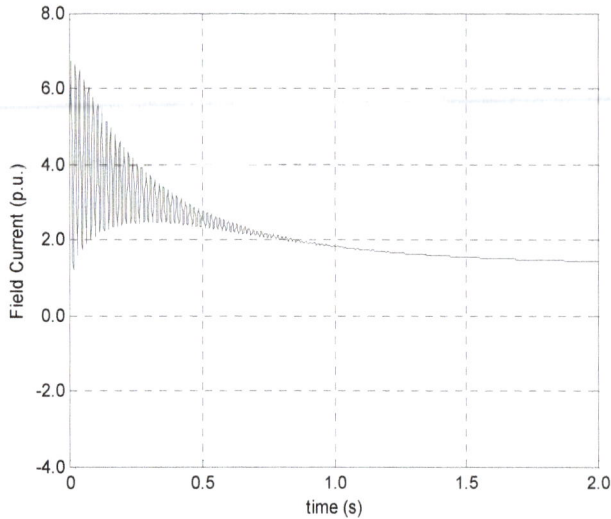

Figure 9. *Deviation of filed current for conventional synchronous generator for the case of permanent fault.*

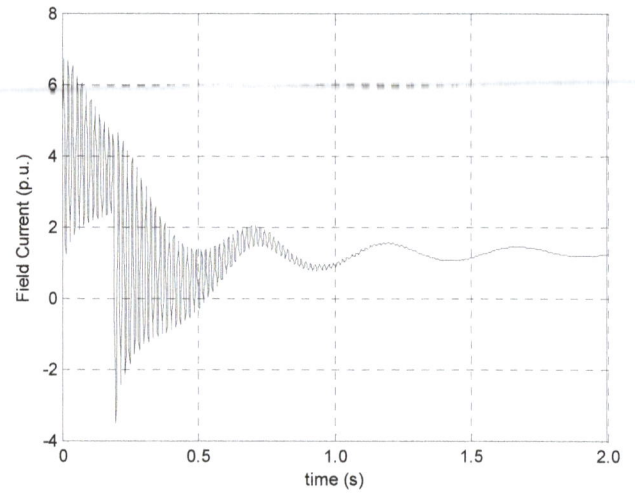

Figure 10. *Deviations of momentary armature currents for conventional synchronous generator for the case of removing the fault.*

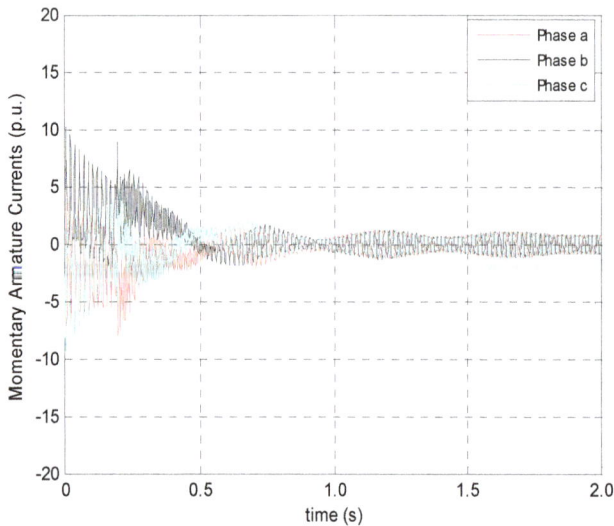

Figure 11. *Deviation of angular speed for conventional synchronous generator for the case of removing the fault.*

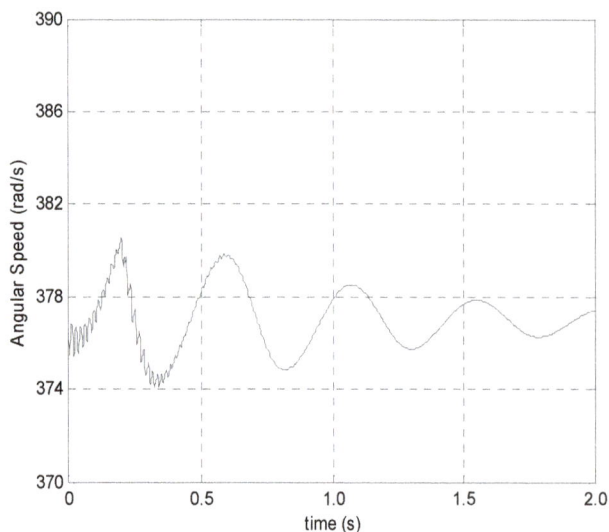

Figure 12. *Deviation of filed current for conventional synchronous generator for the case of removing the fault.*

4. Conclusions

One of the most familiar methods to investigate the transient performance of synchronous generators is three-phase short-circuit test because the most transient current occurs for this fault. In this paper, three-phase short-circuit at the terminals of a superconducting synchronous generator is simulated by means of a simulation programme using MATLAB. In this simulation, a single screened superconducting generator directly connected to power system is considered. To show the effects of the short-circuit the deviations of armature momentary current, angular speed and field current are also plotted in the simulation. Finally, this simulation is repeated for a conventional synchronous generator directly connected to power system for comparison.

Form the results, for the permanent fault case it seen that settling time of the momentary armature currents for superconducting synchronous machine is shorter than those of conventional synchronous machine but steady state currents for superconducting synchronous machine is bigger than those of conventional synchronous machine because synchronous reactance of superconducting synchronous machine is smaller than that of conventional one. On the other hand, momentary armature current rises to 17 p.u. and then oscillates at the interval of (-5;5) p.u. for superconducting synchronous machine. For conventional synchronous machine, these values are obtained as 10 p.u. and (-1.5;1.5) p.u., respectively. Angular speed arises continuously for both machines because the considered fault doesn't removed. Deviations of field currents for both machines are as expected for the permanent fault case.

For the case that the fault is removed it is seen that the momentary armature currents for conventional synchronous machine cannot settled to steady state values in the simulation period while for superconducting synchronous machine are settled at about 1.0 seconds. For deviations of angular speeds and field currents for both machines are obtained similar results to deviations of momentary armature currents. So, angular speed for conventional synchronous machine cannot settled to steady state values in the simulation period while for superconducting synchronous machine are settled at about 1.35 seconds, and deviations of field currents for conventional synchronous machine cannot settled to steady state values in the simulation period while for superconducting synchronous machine are settled at the end of the simulation period.

Finally, from the simulation results it is seen that the superconducting synchronous machine has more stability than the conventional synchronous machine for a three-phase short circuit at its terminal at both cases of the fault is permanent and cleaned by protection system. Especially, it is clearn that the superconducting synchronous generator is superior to the conventional one in the case of removing the fault.

On this subject, it can be investigated a stability analysis can be made by using equal area criterion. Also, a similar investigation can be carried out for a double-screened superconducting synchronous generator.

Appendix

Table A1. Parameters for superconducting synchronous generator used in the simulation.

Rating of the machine: S (MVA)	907
Terminal Voltage of the machine: V_t (kV)	26
Armature resistance: r_a (p.u.)	0.0019
Field winding resistance: r_f (p.u.)	$7.867 \cdot 10^{-7}$
Rotor screen resistance in d axis: r_{kd} (p.u.)	$4.2465 \cdot 10^{-4}$
Rotor screen resistance in q axis: r_{kq} (p.u.)	0.0037
Armature reactance in d axis: x_d (p.u.)	0.297
Armature reactance in q axis: x_q (p.u.)	0.297
Transient armature reactance in d axis: x_d' (p.u.)	0.209
Transient armature reactance in q axis: x_q' (p.u.)	0.209
Sub-transient armature reactance in d axis: x_d'' (p.u.)	0.116
Sub-transient armature reactance in q axis: x_q'' (p.u.)	0.116
Field winding reactance: x_f (p.u.)	0.441
Rotor screen reactance in d axis: x_{kd} (p.u.)	0.2158
Rotor screen reactance in q axis: x_{kq} (p.u.)	0.2144
Mutual reactance in d axis: x_{ad} (p.u.)	0.197
Mutual reactance in q axis: x_{aq} (p.u.)	0.197
Energy Time Constant: H ( )	2.456
Armature winding time constant: T_a (s)	0.1619
Field winding time constant: T_{d0}' (s)	1487
Sub-transient time constant for d axis: T_{d0}'' (s)	0.798
Sub-transient time constant for q axis: T_{q0}'' (s)	1.557

Table A2. Parameters for conventional synchronous generator used in the simulation.

Rating of the machine: S (MVA)	907
Terminal Voltage of the machine: V_t (kV)	26
Armature resistance: r_a (p.u.)	0.0038
Field winding resistance: r_f (p.u.)	0.0012
Amortisseur winding resistance in d axis: r_{kd} (p.u.)	0.0269
Amortisseur winding resistance in q axis: r_{kq} (p.u.)	0.3918

Armature reactance in d axis: x_d (p.u.)	2.22
Armature reactance in q axis: x_q (p.u.)	2.09
Transient armature reactance in d axis: x_d' (p.u.)	0.235
Transient armature reactance in q axis: x_q' (p.u.)	0.235
Sub-transient armature reactance in d axis: x_d'' (p.u.)	0.175
Sub-transient armature reactance in q axis: x_q'' (p.u.)	0.175
Field winding reactance: x_f (p.u.)	2.2642
Amortisseur winding reactance in d axis: x_{kd} (p.u.)	2.2887
Amortisseur winding reactance in q axis: x_{kq} (p.u.)	2.0679
Mutual reactance in d axis: x_{ad} (p.u.)	2.12
Mutual reactance in q axis: x_{aq} (p.u.)	1.99
Energy Time Constant: H ( )	3.134
Armature winding time constant: T_a (s)	0.122
Field winding time constant: T_{d0}' (s)	5.2
Sub-transient time constant for d axis: T_{d0}'' (s)	0.03
Sub-transient time constant for q axis: T_{q0}'' (s)	0.014

References

[1] Grant PM (1997) Superconductivity and Electric Power: Promises, Promises ... Past, Present and Future, IEEE Trans. on Applied Superconductivity, 7(2): 112-113

[2] Umans SD (2009) Transient performance of a High-Temperature-Superconducting Generator, IEEE Electric Machines and Drives Conference, 451-457

[3] Bozdag HO (2011) Simulation of Three-phase Short-circuit at the Terminals of Superconducting Synchronous Generator, M.Sc Thesis, Istanbul Technical University, Institute of Science and Technology

[4] Liu X, Liu D, Huang Y (2008) Simulation of Three-phase Short-circuit at the Terminals of Synchronous Machine, ICEM, 115-117

[5] Shenkman, Arieh L. (2009) Transient analysis of Electric Power Circuits Handbook", Springer

[6] Goddard, K. F., Kukasik, B., Sykulski, J. K. (2009) Alternative designs of High-Temperature Superconducting Synchronous Generator, IEEE Applied Superconductivity, .19(6): 3805-3811

[7] Kalsi, Swarn S. (2002) Development status of Superconducting rotating machines" IEEE Pes Meeting

[8] Furuyama, M., Kirtley, J. L., 1975, "Transient Stability of Superconducting Alternators", IEEE PAS-94, 320-328

[9] Kirtley JL (1993) Large System Interaction of Superconducting Generators, IEEE Proc. on Power Systems, 81(3): 449-461

[10] Kirtley JL (1981) Armature Motion Damping of High Temperature Superconducting Generators, IEEE Tras.. on PAS, 100(6): 2870-2879

11

A new attack on link-state database in open shortest path first routing protocol

Esmail Kaffashi[1], Ahmad Madadi Mousavi[1], Hamid Rezaei Rahvard[2], Sahar Hemmatian Bojnordi[3], Forough Khademsadegh[1], Soheila Amirian[1]

[1]Information Technology and Computer Engineering Department, Amirkabir University of Technology (Tehran Polytechnic), Tehran, Iran
[2]Pardazeshgaran Gam Aval, Mashhad, Iran
[3]Industrial department, University of Applied Science, SID Bojnourd Center, Bojnourd, Iran

Email address:

Esmail2001201@Gmail.com (E. Kaffashi), A.madadi@Gmail.com (A. M. Mousavi), HamiD.Rezaei04@Gmail.com (H. R. Rahvard), Saharhematian2012@Gmail.com (S. H. Bojnordi), Forooghkhadem@Gmail.com (F. Khademsadegh), Soh.amirian@Gmail.com (S. Amirian)

Abstract: Open Shortest Path First (OSPF) protocol known as interior gateway of routing protocol is a major competitor for Cisco's EIGRP of a special routing protocol. Most attacks on this protocol are based on LSA fake router which the attacker has control over it. These attacks can affect the part of the routing domain or cause severe damage based on the strategic location of the router in the AS to bring domain routing. Attacks that cause much damage to a network security mechanism and enables fight-back will not have effect on routing domain. In this paper we will describe an attack that can arbitrarily change the routing domain routing table with harmfully threats without fight back mechanism enabled.

Keywords: Security, Routing Protocols, Link-State, OSPF, New Attack

1. Introduction

Open Shortest Path First (OSPF) is a complexity routing protocol of link-state [8, 10]. Link-State routing protocol is much faster and more convenient way to reach your destination and obtains in comparison to Distance Vector protocol Msyrbaby [1]. Routers using Link-State algorithm creates a map of a network that allow them to choose the best path accurately. OSPF uses of Link-State algorithm to calculate the shortest path to all destinations and select the best-known [1]. In this process, we analyzed the algorithm.

When the situation of a link was changed and the device detected a link change began to publish an LSA message about links. After the publication of the LSA message, router send it to all neighboring devices via a special multi-cast message [4]. As depicted in Figure 1, the schema of LSA message sending in a hypothetical network is presented.

Each router makes up to date own by using the received LSA or Link-State Database (LSDB), furthermore, the router sends LSA to its other neighbors. When the database of each router is completed, the shortest path to the destination is measured in a tree that Dijkstra's algorithm uses it to calculate the shortest path tree destinations, costs and next hope to reach its destination in routing table form [3, 13].

If no changes happen in the OSPF net such as change in link cost or adding and omitting, an OSPF net continues smoothly. Any changes which are announced through link-state packets, Dijkstra algorithm finds the shortest route to be calculated again.

An area is defined as a group of contiguous networks and host. All routers in the same area share a common area ID. All routers within the same area have the same topology table as well.

As mentioned previously, OSPF LSA uses a Link-State updates for exchange uses. Any change in routing information is sent to all network routers, area concept was introduced in order to limit the part of the Link-State updates explosion occur. Dijkstra's algorithm to compute a function of a router is limited to changes within an area [3]. All routers within an area of Link-State databases are complete and accurate.

This protocol is designed to be used within a single Autonomous System or AS and can be divided into different groups of the network called area. Each area has its own database; the topological database for each area will be hidden from other areas, which will reduce traffic on the network. All areas must be connected to an area called Backbone.

Routers belonging to a single area are called Internal Router. Routers which belong to more than one area are called Area Border Routers (ABRs). The router will start to exchange routing information with an external AS Autonomous System Boundary Router (ASBR) is called. Figure 2 shows the layout of the routers in the area [6].

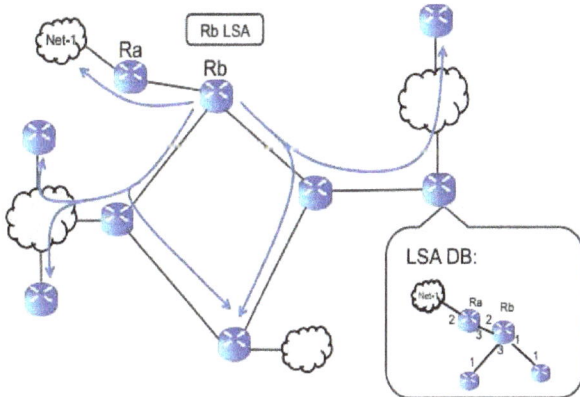

Figure 1. An example of LAS sending.

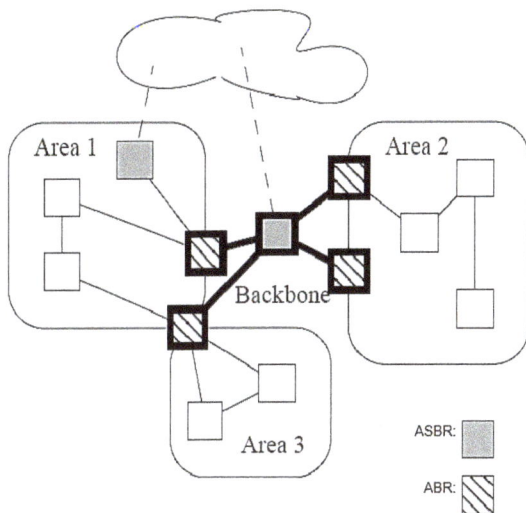

Figure 2. The role of routers in area [6]

This project will include the following milestones:
1 Careful reading of RFC 2328 (OSPF v2).
2 Research on known vulnerabilities of attacks on traffic of this protocol.
3 A new method of attack on the protocol.
4 Conclusions from the findings.

2. Vulnerability and Risk

Protect infrastructure and network infrastructure is very important. The mean of the Internal / Insider routers are the trusted network and has been accepted as a router in the network and the company is in the process of information

exchange and the face of the routers External / Outsider reverse the Internal / Insider will [13].

2.1. Basic OSPF Vulnerability

OSPF risks arise mainly from the following vulnerabilities:
• Distance attack: even if the OSPF routing protocol to be used as the range, but in many situations and conditions routers OSPF unicast packets that will be sent to the address it receives and accepts (in paragraph 8.1 of RFC2328) [8].
• The fight-back mechanisms: in OSPF, this mechanism can be explained as follows: when a router LSA has the right to see their fake just to send LSA and to inform the public that the LSA has been true and the other false. But this mechanism is not very effective because in RFC2328 OSPF v2 is related to the mechanism, there are no words to and there is no official word fight. Procedures for the inactivation mechanism during attacks there [9, 10].
• Abusing the fight-back mechanisms: Can help to implement a DOS attack. Update an LSA storm may make the mistake of trying to make the LSA wrong. Even though this is an acceptable and effective response.
• External routes: Routes that are received from external sources such as other area or AS other, we can not validate them in areas that are not defined as stub is reproduced [13].

2.2. Vulnerabilities in Protocols

This protocol has more security mechanisms for authentication, such as checksam received packets over the communication link. OSPF is generally 5 types of messages:
1 Hello
2 Database Description
3 Link-State Request
4 Link-State Update
5 Link-State Acknowledgement [9]

Message type attacks can be performed on each of the 5 and the fields that are defined for these messages can use to your advantage to get a more sophisticated ground attack [8]. The JiNao, 4 presented the attacks that can be said is essentially a denial of service attack. Below the name of four attacks are mentioned [7]:
1 Max Age
2 Sequence++
3 Max Sequence#
4 Bogus LSA

Age of the field in the packet header of the attack 1 LSA message is abuse or episodes 2 and 3 of the Sequence Number field of the message header of the previous abuse [1, 3].

2.3. Sources of Vulnerability

These attacks are the main ways of taking system resources such as CPU, RAM or disk space, or network connections by which they aim to avoid a system or network is usually associated with a particular sector or other networks. Because sometimes hackers use to destroy your network Prdarnd resources such attacks are also known to attack non-symmetric or asymmetric attack [13].

3. Understanding the Attack

Suppose that, a remote attacker could leverage the insider to attack the router, Some vulnerabilities have been reported as CVE-2010-0581, CVE-2010-0580, CVE-2009-2865 CVE-2010-0581 that an attacker can execute code with a SIP packet to and routers with Broadcom chipsets or just a new bug has been discovered by a team DefenseCode which allows an attacker without authentication to run their code on the root surface [2, 13].

To attack the protocols that are stable, we can use other poor protocols. For example, we can use the RIP protocol, which is completely unsafe and send the OSPF or BGP message in wrong routes, and circumvent their natural protection. Another critical point in OSPF networks is DR routers that can be attacked and disrupt the network [10].

Two new attacks on this protocol are designed to follow the rules of the protocol is as follows:

1 Wrong adjacent from the remote: it allows attackers to trick a remote router and by fake LSAs define a new link to the router that this link does not exist in real. This attack focuses on the routers adjacency process. This attack can create a Block-Hole for a particular subnet.

2 Fake LSA messages: this attack is of those high damage attacks on the network and will also enable the fight-back security mechanism. But what is more important is that the mechanism itself is used in the attack, in attack implementation focus point is on the creation and transfer of the fight-back messages [9].

In the following description, we will describe attack which is optimized version of the fake LSA.

3.1. New Attack on Routing Tables

The header of a LSA instance is shown in Figure 3:

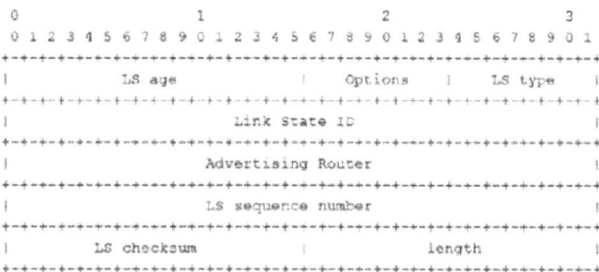

Figure 3. *LSA Header Format [7]*

Structure descriptions are:

- LS Age: elapsed time of the creation of the LSA indicates that on seconds.
- Options: used for optional feature.
- LS Type: specifies the type of LSA (e.g. Router, Network, Summary, etc.) will be discussed further on Router LSA.
- Link State ID: AS topology is part of the LSA to describe it will specify section.

- Advertising Router: Router ID of the router that created the LSA.
- LS Sequence Number: number of LSA that each LSA specifies its own unique number.
- LS Checksum: Checksum refers to the entire contents of the LSA.
- Length: shows LSA content length to bits [7].

In attack we do need to focus on two fields, here are two fields when creating Router LSA:

- Link State ID: each router creates the LSA, should be assigned LS ID the same as the router ID of the router.
- Advertising ID: specifies the router LSA's original creation (i.e. the origin of creation LSA).

Based on OSPF regulation each router creates own LSA and is not expected to cause the router to other routers to generate router LSA, as described above, then the two fields must have the same value [3]. In OSPF to check the same for these two fields, the specific operation is not performed and this will send the LSA so that the two fields have different values [5]. Based on section 13.4, a router can activate fight-back when you receive a fake LSA:

"The Advertising Router is equal to the router's own Router ID"

This means that the Advertising Router from fake LSA equal to victim router's Router ID, fight-back victim will not be activated by the victim router even if the LS ID is equal to the Router ID of the victim router. Explanation may be that we assume the attacker will send Router LSA from some of Rv victims:

1 LS ID: is equal to the ID of Rv Router.
2 Advertising Router: every value except the ID of Rv Router.

According to the rules of OSPF it can be assured that fight-back will not be activated even in other routers in the AS with Rv Router and they put a fake LSA in their LSDB, but we have a problem. In section 12.1 of RFC is determined based on the following three fields identifies uniquely the LSA or not:

1 LS Type
2 Advertising Router
3 LS ID

Therefore fake LSA is not valid in the LSDB because they are different (Advertising Router is not equal) and cannot trust until fake LSA are not deleted from LS DB [2, 3]. OSPF as discussed in RFC where there is doubt that it can be used and have a successful attack [8]. Section 16.1 is said route calculation on LSDB is based on the Vertex ID:

"This is a lookup ... based on the Vertex ID".

The description of OSPF Link State ID field is the same as Vertex ID. When routers create routing table they act on the basis of this field. This will be an ambiguity in the description of the protocol and on the other side of the LSA are identified and described three fields before the other calculations will be based on the Link State ID field. This ambiguity raises the question: When the LSA fetch from LSDB for the fetch calculation then which LSA is fetched, authentic and original LSA or fake LSA? Remember that both the fake and the original LSA all are in the LSDB as there. Both LSA and Link

State ID field has the same value. OSPF is not able to answer the question, so the answer depends on the implementation [3, 8].

Most networks that have implemented OSPF, is based on Cisco IOS. According to infonetrics research, almost 75 percent enterprise networks use Cisco's in world [11]. To implement attack, we use GNS3 and SCAPY with the latest stable version of IOS on the C7200 router and we test it with M1-150 version provided by Cisco.

We send fake LSA with higher sequence number than the original LSA. Fake LSA is not only in the LSDB but also in the entire LSDB will be replaced within the AS. All routers have the victim router. In Figure 4, the result of running attack can be seen and the whole process is shown in Figure 5-10.

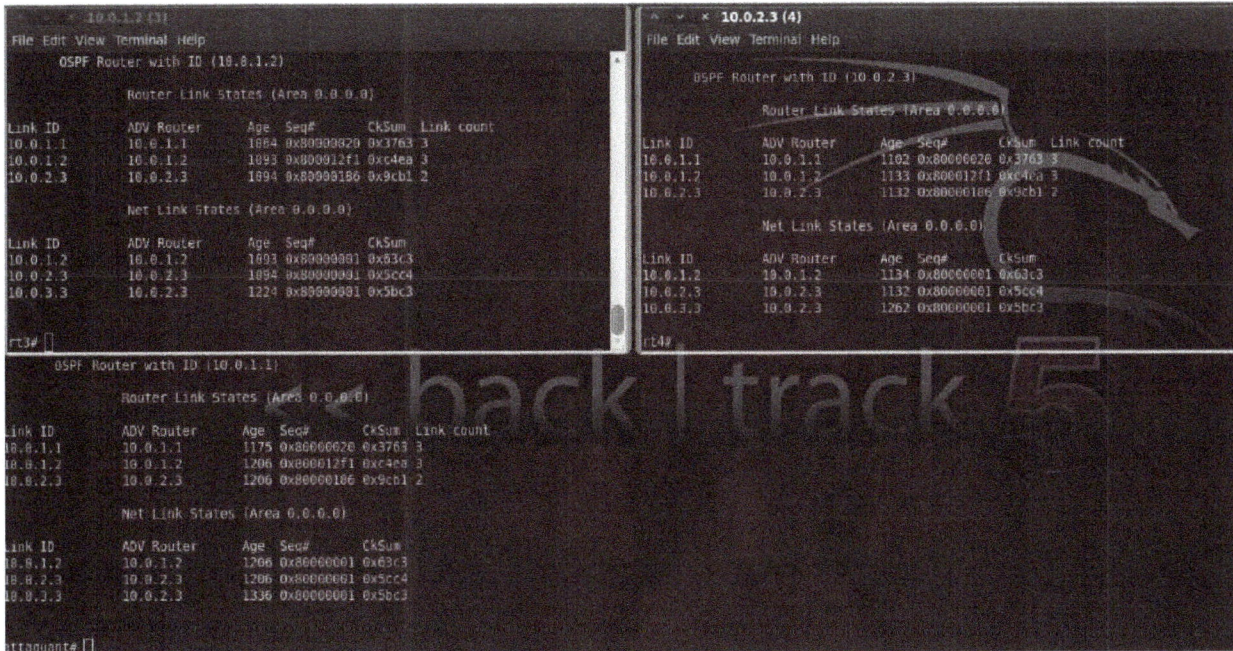

Figure 4. *OSPF database before the attack*

Using Wireshark, we follow OSPF: first appears the tigger packet use to force a response the rt4 router (id = 10.0.2.3). Sequence is set to: 80005200 and the metric (30) is false, it should provoke a fight-back from rt4.

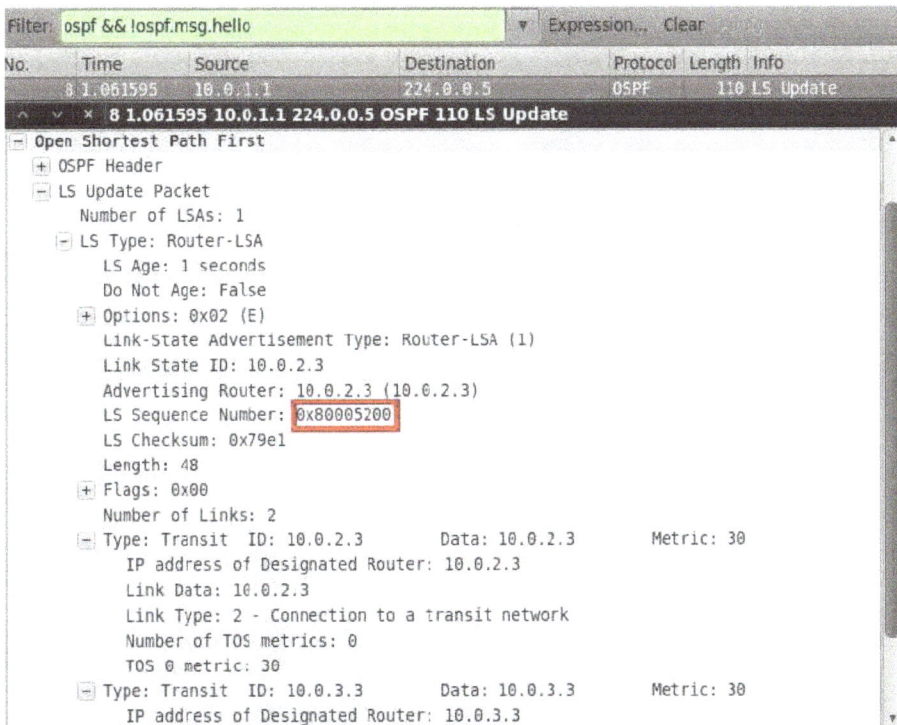

Figure5. *Sending a packet specially crafted to spoof an R1 LSA.*

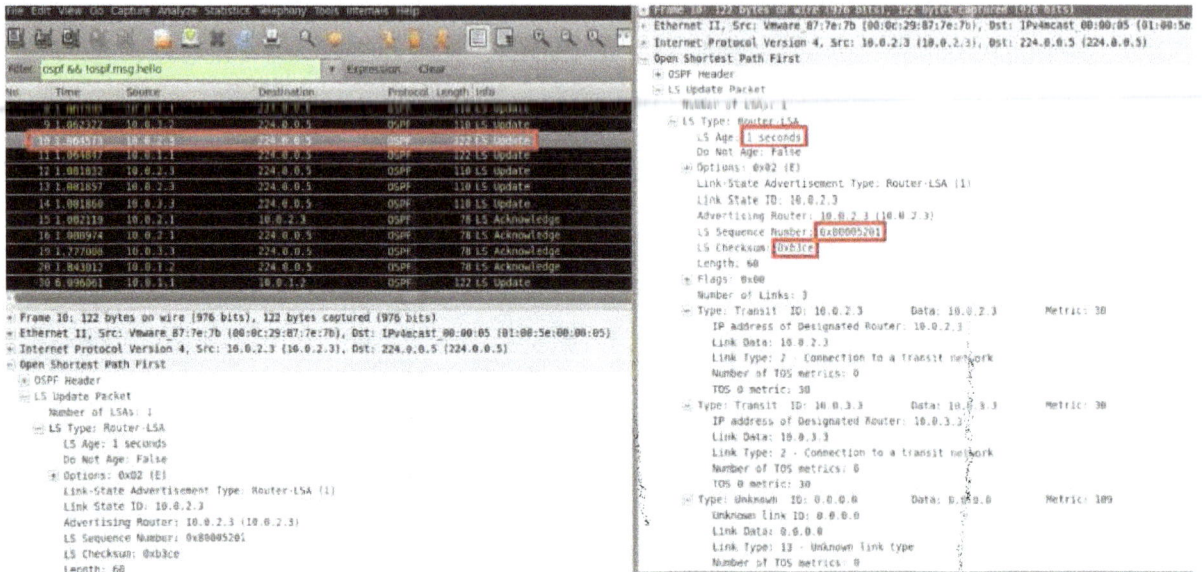

Figure 6. *Sending the disguised LSA craft to match the LSA fight-back from rt4*

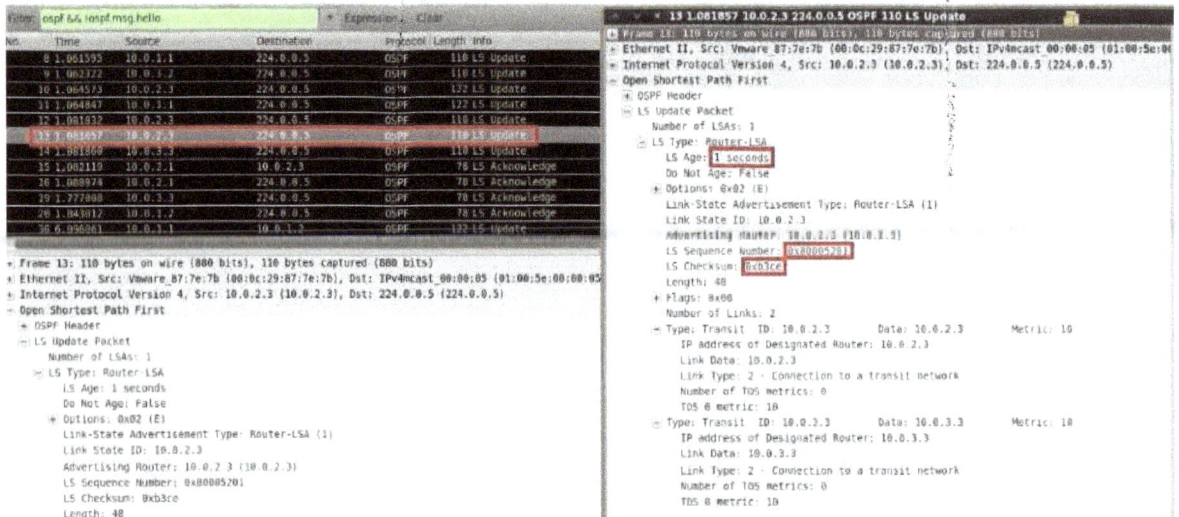

Figure 7. *R1 sends the fight-back that will be rejected due to the previous packet craft at step 2.*

Comparison of two packets:

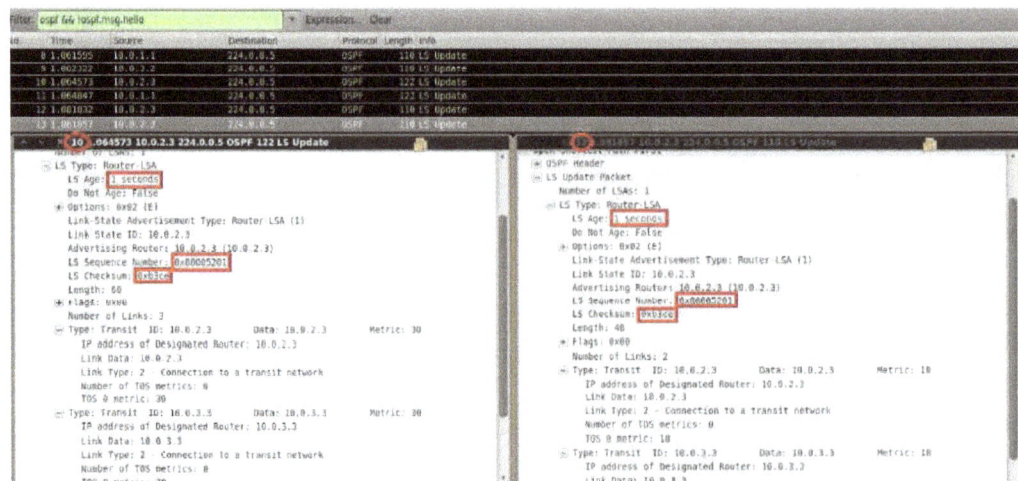

Figure 8. *both packets contains the same sequence number, checksum and age (+/- 15 min)*

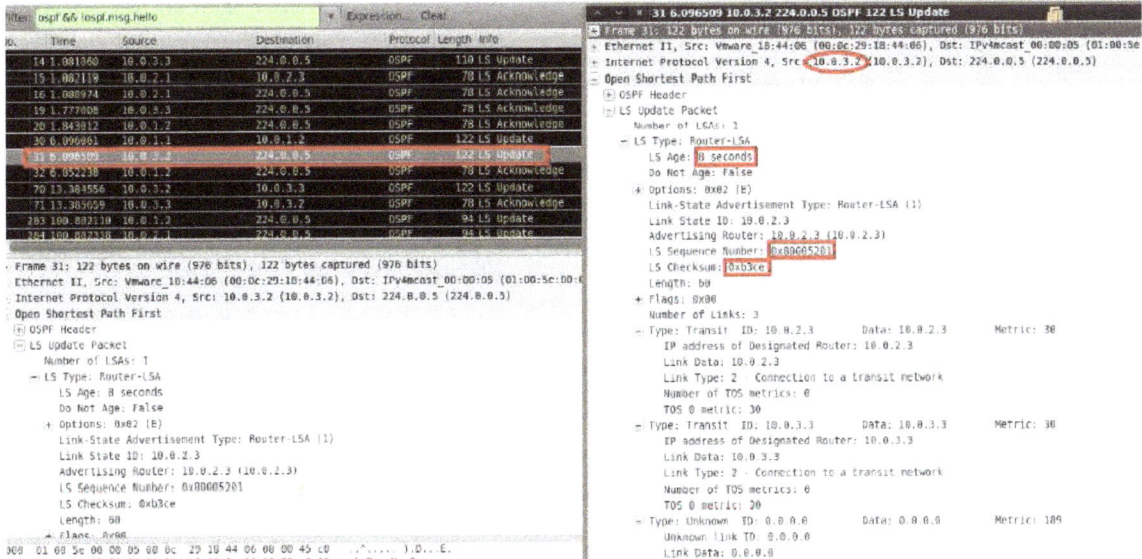

Figure 9. *R2 flood the disguised LSA, R1 receives it and drops the packet, seeing it as the one it has sent as step 3.*

We then check the rt3 router' database that should be corrupted:

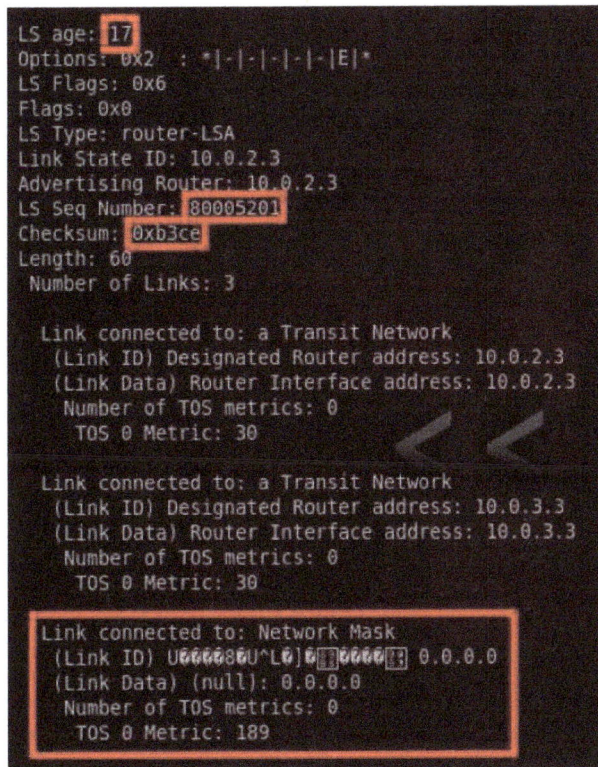

Figure 10. *OSPF database after the attack*

4. Discussion and Conclusion

Attacks were according to RFC2328 and new attack due to ambiguity in the definition of the protocol is achieved. These attacks may be the basis for the creation of more efficient attacks and destructive affects, however Cisco devices are vulnerable to the attack. All beliefs are broken about the attacks to this protocol and by the presented attacks a router can easily control all routing domains without enabling fight-back security mechanism.

References

[1] Russell Chris, Security of IP Routing Protocols, SANS Institute, Global Information Assurance Certification Paper, October 7, 2001.

[2] Andrew A. Vladimirov, Konstantin V. Gavrilenko, Hacking Exposed Cisco Networks, McGraw-Hill Companies, 2006.

[3] Michael Sudkovitch, David I. Roitman, OSPF Security Project – Technion Institute of Technology, 2010.

[4] Vanessa Antoine, Raymond Bongiorni et al, Router Security Configuration Guide, National Security Agency [C4-040R-02], 2005.

[5] Faraz Shamim, Zaheer Aziz, Troubleshooting IP Routing Protocols (CCIE® Professional Development), Cisco Press, 2002.

[6] Brian Vetter, Feiyi Wang, S. Felix Wu, an Experimental Study of Insider Attacks for the OSPF Routing Protocol, In 5th IEEE International Conference on Network Protocols, 1997.

[7] S. F. Wu, H.C. Chang, F. Jou, F. Wang, F. Gong, C. Sargor, JiNao: Design and Implementation of Scalable Intrusion Detection System for the OSPF Routing Protocol, DARPA Information Survivability Conference and Exposition. DISCEX'00, 1999 , Pages 69-83, IEEE Article, 1999.

[8] John Moy, OSPF Version 2, IETF RFC 2328, April 1998. https://www.ietf.org/rfc/rfc2328.txt

[9] Emanuele Jones, Olivier Le Moigne, OSPF Security Vulnerabilities Analysis, Internet Draft: draft-jones-ospf-vuln-01.txt, IETF 58 –RPSEC Working Group, November 2003.

[10] Daniel Mende, Rene Graf, Enno Rey, Christopher Werny, Burning Asgard, an Introduction to the Tool Loki, Black Hat Digital Self Defense Conference USA, 2010 Jul 05.

[11] Wendell Odom, CCNP ROUTE 642-902, Cisco Press, 2010.

[12] Infonetics Research, "Enterprise Routers Quarterly Market Share, Size, and Forecasts", May 2012

[13] Esmail Kaffashi, Hamid Rezaei rahvard, "Discovered a new security hole in OSPF routing protocol", 16th conference of National Association of Electrical Engineering, Iran, Kazeroon, August 2013.

Design of a PID feed-forward controller for controlling Output fluid temperature in shell and tube hear exchanger

Navid Khalili Dizaji[1, *], Aidin Sakhvati[2], Seyed Hossein Hosseini[3]

[1]Department of Mechatronics Engineering, Tabriz Branch, Islamic Azad University, Tabriz, Iran
[2]Department of Electrical Engineering, Tabriz Branch, Islamic Azad University, Tabriz, Iran
[3]Department of Electrical & Computer Engineering, University of Tabriz, Tabriz, Iran

Email address:
navidkhalili@yahoo.com (Navid K. D.), aidin_sakhavati@iaut.ac.ir (Aidin S.), hosseini@tabrizu.ac.ir (Seyed H. H.)

Abstract: The most important part in chemical processes, which is directly related to energy consumption, is heat exchanger. Main purpose of heat exchanger is transferring heat from hot fluid to cold fluid. There are different heat exchangers in industry, which their common types are shell and tube heat exchangers. In these heat exchangers, one fluid flows in tubes and the other in shell around tubes. In heat exchangers, one of the important issues is reaching output fluid temperature to given temperature in the least possible time. In this paper, PID controller along with forward-feeding controller was designed to control output fluid temperature of a shell and tube heat exchanger. First, process mathematical modeling is done using experimental data, and then the controller is designed. Designed controller regulates output temperature of heating fluid to a desired point in the least possible time without considering a non-linear process. Then, controller performance is evaluated by unit step response analysis and performance indicators related to the control system. Modeling all processes and designing controllers are done in MATLAB software Simulink.

Keywords: PID Controller, Feed-Forward Controller, Shell and Tube Heat Exchangers

1. Introduction

In a chemical process, heat exchangers are used in order to transfer heat from one hot fluid through the solid wall to a cold fluid. There are various types of heat exchangers in industry. Certain type of these heat exchangers, which are known as shell and tube heat exchangers have been produced in different configurations as well as various sizes and can operate in high pressure. These heat exchangers apply in fields such as condensation, electricity production, chemical processes, medicine, etc. one shell and tube heat exchanger is a set of parallel tubes, which are enclosed in a cylindrical cover. In heat exchangers, one fluid flows in tubes and other fluid around tubes [1].

Researches about controlling heat exchangers have dealt with other heat exchangers like plate exchangers and control plans like neural network, and fuzzy network in order to control output fluid temperature of these heat exchangers. Therefore, there is less attention to controlling these heat exchangers' temperature. It is necessary to mention that regarding domestic research, there is no valid plan in order to control temperature of these heat exchangers and this work in unique in this regard.

This paper deals with a shell and tube heat exchanger and one-input-output system using experimental data. Output fluid temperature of heat exchanger should be kept in an optimized point regarding process need. Then a PID controller runs with feed-forward controller in order to reach control purpose. Also analyzing designed controller performance is done in the time domain. In order to evaluate designed controller performance, unit step response analysis and performance indicators were used. Processes modeling and controllers' design indicators steps have implemented in MATLAB software Simulink.

2. Mathematical Modeling

A common chemical process for heating includes chemical reactors and heat exchangers system. Cold fluid heat exchanger is prepared by super hot vapor 180^0C from boiler. Reservoir provides cold fluid in 32^0c using one pump and one-way valve. Super hot vapor comes out from boiler and flows through tubes in heat exchanger; while process fluid flows through inter-shell and tubes. After heating fluid was done by super hot vapor, dense vapor exits system in 93°c. Heat exchanger warms fluid to 52°c in fact, the purpose is controlling output fluid temperature that enters heat exchanger.

Figure (1) shows schematic of one shell and tube heat exchanger.

There are different assumptions for shell and tube heat exchanger. First assumption is that rate of input and output fluid stream is identical. Second assumption is that heat storage capacity of walls is trivial. Shell walls are fully insulated.

Figure (2) shows PID control system with feedback controller for shell and tube heat exchanger.

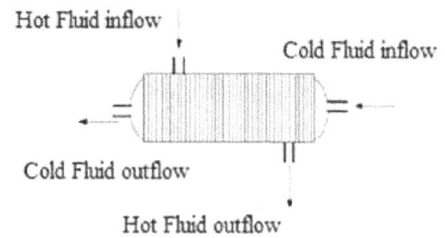

Fig (1). shell and tube heat exchanger

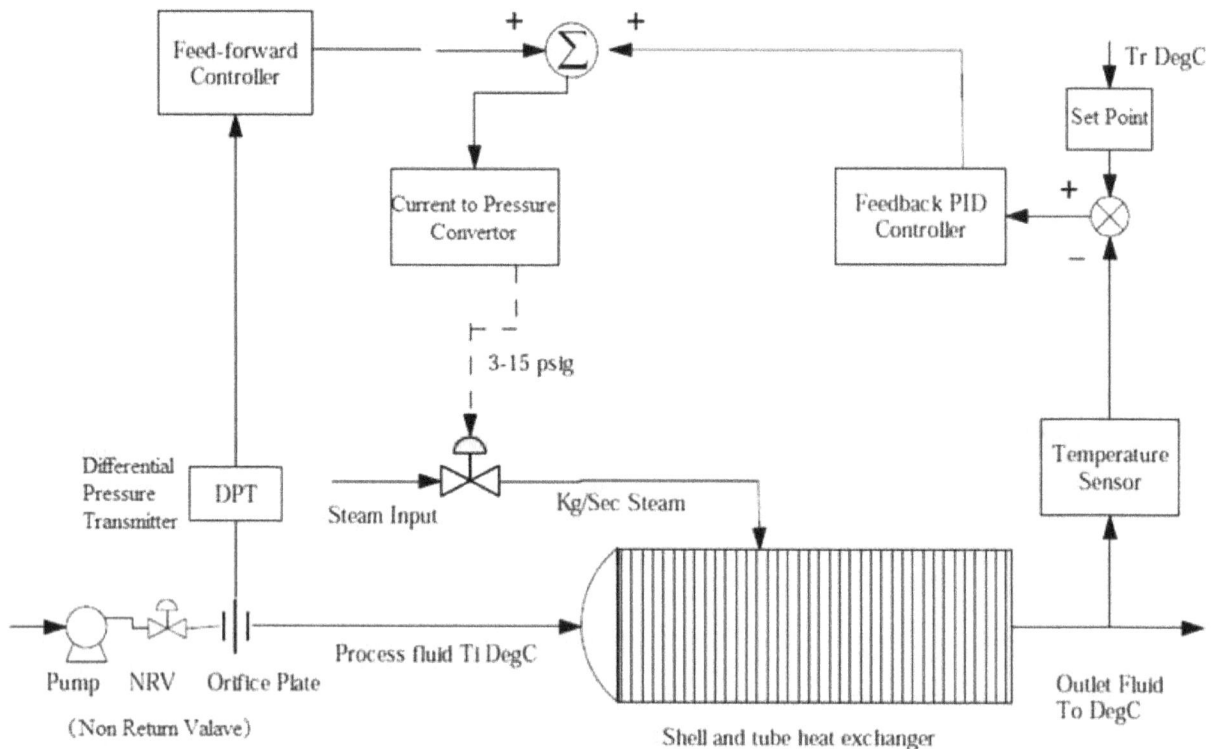

Fig (2). Shell and tube heat exchanger control system

In figure (2) used, control valve is opening with air. Used sensor in this design is a thermocouple with a transmitter. Thermocouple is used to measure temperature in control architecture feedback path. Output fluid temperature is measured by the thermocouple, and its output is sent to the transmitter. Finally, temperature output is converted to a standard signal in 14-20 mA range. This transmitter unit output is given to the PID controller unit. There is an orifice in this sheet, which creates pressure difference. In fact, it reduces pressure and this pressure difference can be converted to standard signal 4-20 mA using a transmitter. In this control system, controller estimates error feed-forward and changes the control variable before distortion can influence output.

Controllers run control algorithm and send a combined signal in order to control output temperature of shell and tube heat exchanger to drive. Then command for final control is given from drive. Drive unit is current to pressure pneumatic heat exchanger, and it is the ultimate control unit for pneumatic valve. Drive converts controllers combined output signal to one standard pressure unit in the 3-15psig range. Drive decides to open and close valve considering sent signals. As a result, vapor stream enters heat exchanger and develops final output temperature using other inputs. Generally, there are two distortions in this process, one is changes in the input fluid stream, and the other is changes in input fluid temperature. However, in practice, changes in input fluid stream create significant distortion than temperature changes.

Furthermore, we show PID control along with feed-forward controller in figure (2) as a diagram block. Figure (3) shows closed loop diagram, including process, controllers, drive, control valve, sensor, optimized point, and distortion.

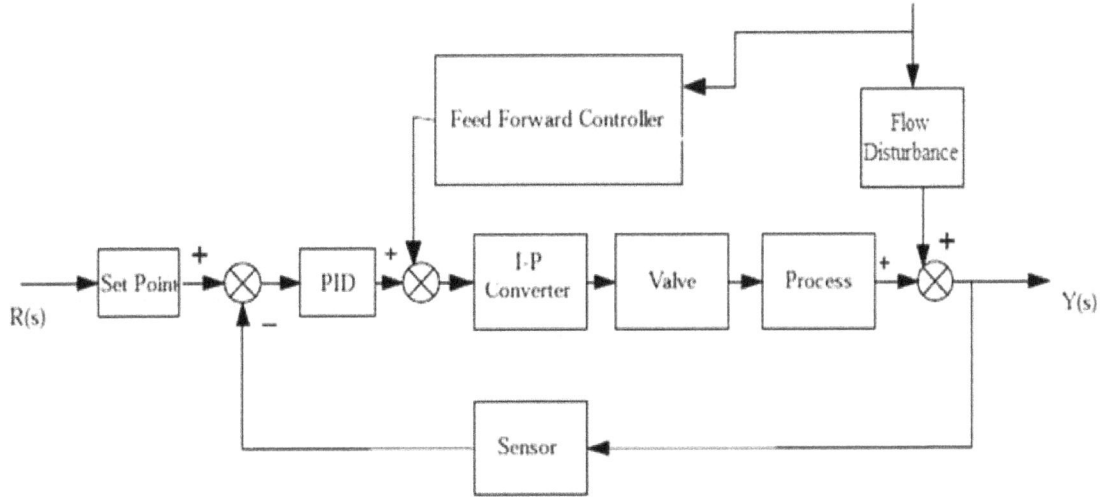

Fig (3). diagram block PID controller with feed-forward controller for shell-tube heat exchanger

Fig (3): diagram block PID controller with feed-forward controller for shell-tube heat exchanger.

Controller imposes control using the final control elements through the drive. Sensor receives output and feedbacks it to the controller. In fact, temperature is measured by sensor and output signal is sent to the controller, where it is compared with the optimal point. Output signal of the controller enters the flow to pressure pneumatic exchanger that creates a pneumatic signal to regulate control valve and vapor stream enters. This stream enters heat exchanger and creates final output temperature with other inputs.

Experimental data for process are presented in table (1) [3]:

Table (1). experimental data for heat exchanger control process

Control valve capacity	1.6kg/sec	Exchanger response to vapor	50^0C/(kg/sec)
Control valve time constant	3sec	Process time constant	30 sec
Sensor temperature	$50\text{-}150^0C$	Drive current range	4-0 mA
Sensor time constant	10 sec	Drive pressure range	3-15psig

Transformation functions related to different system elements and process, and their rates are gathered in table (2).

Table (2). transformation function and rates of different system components

Process transformation function	30s+1
Valve function	3s+1
Drive rate	0.75
Distortion function	30s+1
Sensor function	10s+1

In this control plan, feed-forward controller regulates distortion by the input fluid stream. Feed-forward transformation function is shown by $G_{cf}(s)$, which is expressed as equation (1) [4]:

$$G_{cf}(s) = -\frac{G_d(s)}{G_p(s)} \qquad (1)$$

In equation (1), Gp(S) is process transformation function and $G_d(s)$ is distortion transformation function for input stream, which are shown by equations (2) and (3), respectively.

$$G_p(s) = \frac{5e^{-s}}{90s^2 + 33s + 1} \qquad (2)$$

$$G_d(s) = \frac{1}{30s + 1} \qquad (3)$$

By substituting current distortion transformation function and process transformation function in equation (1), feed-forward controlling function is expressed as equation (4):

$$G_{cf}(s) = \frac{-18s^2 - 6.6s - 0.2}{27s^2 + 30.9s + 1} \qquad (4)$$

Using experimental data, transformation functions and rates in shell and tube heat exchanger control system are shown in figure (4):

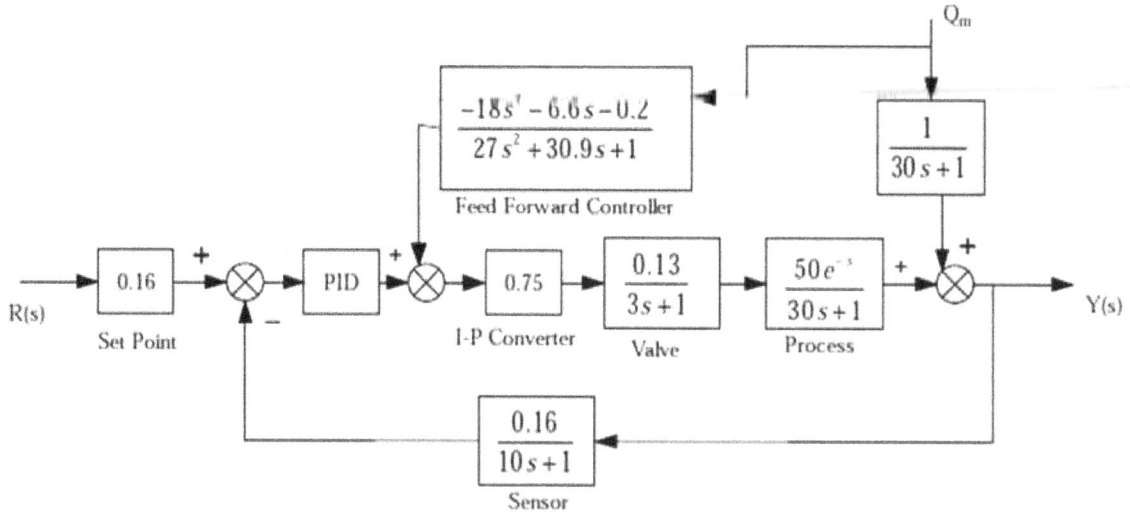

Fig (4). *transformation functions and rates in sell and tube heat exchanger control system*

In order to regulate PID controlling parameters Ziegler–Nichols closed loop regulation method was used, which is an automatic regulation method. In this method, K_u is the final rate. High increase in this rate pushes the system to instability. In this method, system fluctuates and frequency (w) and vibration period (P_u) are calculated from fluctuations. Table (3) shows regulation laws based on Ziegler–Nichols closed loop fluctuation [5].

Table (3). *PID parameters' regulation law*

K_c	τ_i	τ_d	method
K_u 0/6	P_u 0/5	P_u 0/125	Ziegler–Nichols

3. Results of Computer Simulation

Using experimental data, PID controller Simulink with feed-forward controller for shell und tube heat exchanger is shown in figure (5):

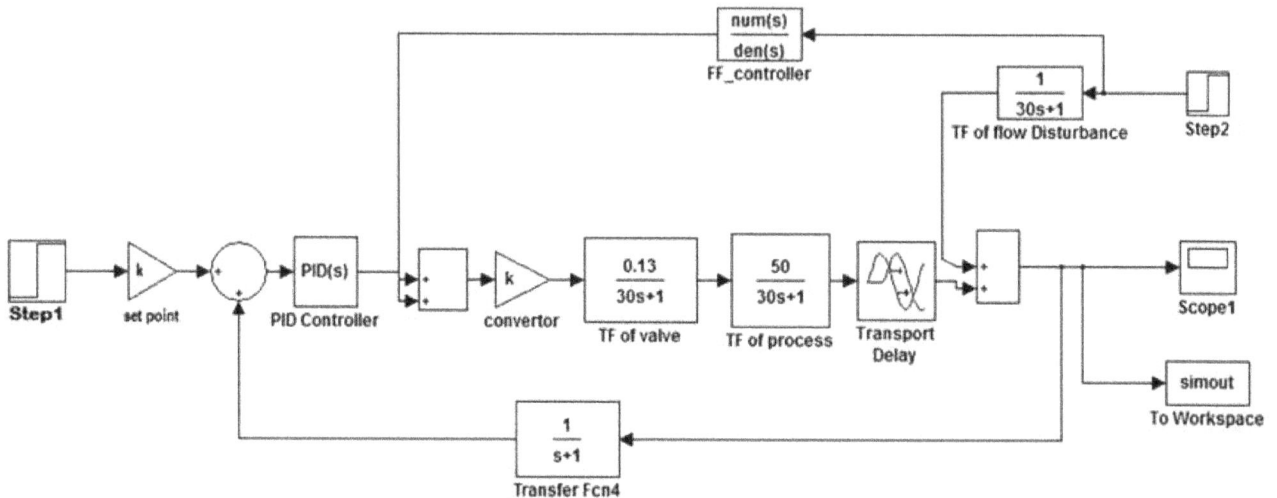

Fig (5). *PID controlling Simulink with feed-forward controller for shell and tube heat exchanger*

For PID controller, the amount of K_c , τ_d τ_i parameters are obtained by Ziegler–Nichols closed loop regulation method, which is shown in table (4).

Table (4). *PID parameters regulation*

K_c	τ_i	τ_d	method
14/28	14/395	3/59	Ziegler–Nichols

Purpose of designing control systems in time range is using

features and characteristics of system time domain. These characteristics are often expressed as standard values in ascending time, settlement time, maximum jump, peak time, and stable mood error by unit step response. In this designed controller, settlement time and maximum mutation were evaluated using unit step response analysis.

Unit step response PID controller with feed-forward controller for shell and tube heat exchanger system is shown in figure (6):

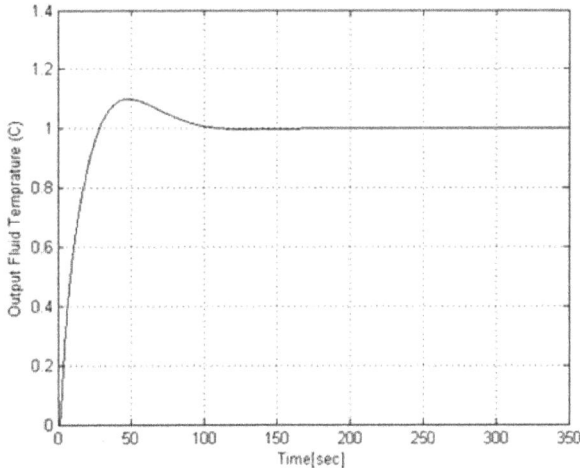

Fig (6). Unit step response PIN controller with feed-forward controller in shell and tube heat exchanger

Maximum mutation is the between time response's Peal and sustainable output. This relation can be defined as equation (5) [6]:

$$\%M_P = \frac{C(t_p) - C(\infty)}{C(\infty)} * 100\% \qquad (5)$$

Settlement time (t_s) is required time for system to reach and stays final value range (about 2%-5% of final value) [6]. In this article, it is considered as 5% of final value.

By analyzing a unit step response of this control system settlement time is 90. 23 seconds, and maximum mutations are 9.1.

Performance indicators were used in order to evaluate controller performance. Performance indicators for evaluating this control system are integral absolute error (IAE), integral square error (ISE), integral time absolute error (ITAE) and integral time square error (ITSE). Performance indicator is a quantitative amount of system performance. A system is considered optimal control system when its performance indicator reaches to the minimum. Performance indicators can be obtained by following relations [6]:

$$IAE = \int_0^\infty |e(t)| dt \qquad (6)$$

$$ISE = \int_0^\infty e^2(t) dt \qquad (7)$$

$$ITAE = \int_0^\infty t|e(t)| dt \qquad (8)$$

$$ITSE = \int_0^\infty te^2(t) dt \qquad (9)$$

Performance indicators for PID controller with feed-forward controller for shell and tube heat exchanger are shown in table (5):

Table (5). PID controller performance indicators with feed-forward controller

controller	IAE	ISE	ITAE	ITSE
PID controller with feed-forward	3/528	0/331	80/66	3/474

4. Discussion and Conclusion

In this paper, we designed a PID feed-forward controller for controlling output fluid temperature in shell and tube heat exchanger. Maximum mutation in this control system was 9.1% and settlement time is 90. 23 seconds, which in chemical process's settlement time about few minutes are ideal. Performance indicator's values for this system are in a satisfactory range. Regarding obtained results, we can control output fluid temperature in shell and tube heat exchanger in the least possible time without considering non-linearity of the process. By controlling output fluid temperature in shell and tube heat exchanger; we can have significant savings in energy consumption. For future research, other control systems like fuzzy controller, internal model and neural networks can be used.

References

[1] Ramesh, K., Dusan, R., *Fundamentals of heat exchanger design*, John Willey and Sons, New York, 2003.

[2] Kaddour, N., *Process modeling, and control in chemical engineering*, Marcel Dekker Inc, New York, 1989.

[3] Gopal, M., *Control systems principles and design*, Tata McGraw-Hill Education, New York, 2008.

[4] Shue L., Feng, L., *"Feed forward Compensation Based the Study of PID Controller"*, Advances in Intelligent and Soft Computing, Vol. 149, pp. 59-64, 2012.

[5] Kiam, A., Gregory C., Yun, L., *"PID control system analysis, design and technology"*, Control System Technology, Vol. 13, pp. 559-576, 2005.

[6] Kenneth M., Wim, V., *Management Control Systems: Performance Measurement, Evaluation and Incentives*, Prentice Hall, New York, 2007.

13

Evaluation of uncertainties on the performance of distance relay

Seyyd Mahmood Mosavi[1, 2]

[1]Khorasan Regional Electric Company, Mashhad, Iran
[2]Department of Electrical Engineering, College of Ebne Yamin, Sabzevar, Iran

Email address:
mahmod.mosavi487@yahoo.com

Abstract: Global demand growth for electric energy has speed up development in designing power systems in order to meet users' needs for providing reliable, cheap and high quality electric energy. In power networks the first zone of distance relay is considered as main protection for transmission lines. Zones 2 and 3 of this relay are considered as supporting protection for next lines. Distance relay can be affected by many factors which are not clear and definite. These factors which are considered as uncertainties are fault resistance, fault in measurement transformers, ground resistance, and change in structure and exploitation conditions. In this article we prove their considering uncertainties approaches zone 1 distance relay extent to 100 percent of protected line provides more reliable protection for transmission lines.

Keywords: Distance Relay, Uncertainty, Zone, Protection

1. Introduction

Global demand growth for electric energy has speed up development in designing power systems in order to meet users' needs for providing reliable, cheap and high quality electric energy. From the beginning of this industry the problem of electric energy production, transmission and distribution was always accompanied with likely faults and reliability problem [3]. Task of protection system is fast and timely detection of unusual conditions in power system and running revisions in order to restore power system to usual conditions [4]. One of the important elements of power system is transmission line which transfers energy from producer to consumer. Protection of transmission line is done by distance relay and high current [3]. These relays not only act as main protection but also they are used as supporting protection [1]. Distance relay detects fault location through measuring path impedance between relay location and short circuit. However there are factors which have undesirable effect on distance relay and disrupt their performance; Therefore, investigating them and presenting suitable solution has special importance in preventing these factors [3].

2. Effective Uncertainties on Distance Relay Performance

2.1. Fault Resistance

Fault resistance has undesirable effect on distance relay performance as an unknown phenomenon and usually is considered as an important uncertainty. If fault resistance was zero, measured impedance in relay point is only dependent on the part of tool which is located between relay and line but if resistance was non-zero, measured impedance is not equal with part of line which is located between relay and fault point. Therefore, fault tolerance results in uncertainty in distance relay [2].

2.2. Fault in Measurement Transformers (PT, CT)

Regarding that distance relay is an impedance element, it needs sampling current and voltage to calculate impedance using these values. This current and voltage sampling was done via CT (current transformer) and PT (voltage transformer). These equipment's have certain accuracy class and in the fault moment this sampling may be disturbed by problems CT saturation or CVT transient fault (capacitor voltage transformer) and it will influence distance relay performance.

2.3. Ground Resistance

Ground resistance is another uncertainty which is considered for single phase to ground faults. In this case, measured impedance by relay will be equal with sum, line impedance between relay and fault location and fault resistance and ground resistance. As ground resistance in each zone is different from other zones because of ground especial resistance, soil type and humidity, it will influence correct relay performance [3].

2.4. Change in Structure and Exploitation Conditions

Since distance relay settings are those settings which are stable and impose offline for worst case fault in system and relay, therefore, by changing structure and exploitation of network distance relay performance will suffer. In fact when network structure changes, network impedance and short circuit in network will change as well. Change in exploitation condition creates this problem. By increasing or decreasing load or production of plants network voltage will change [3].

3. Related Works

In a research study, Sidha al,[9] presented a new approach for calculating zone 2 distance relay settings and its application in comparative protection. They presented two comparative and non-comparative algorithms for calculating zone 2 settings in response to change in exploitation condition. In non-comparative algorithm, impedance is calculated for maximum and minimum and production and zone 2 as worst case but in comparative method online awareness from exploitation is presented and new settings impose in relay per each change. This method provides better support and coverage for zone 2 [9].

In another study, Jamali and Shateri [4] calculated the maximum fault resistance regarding measured impedance in relay point for which distance relay with tetrahedron characteristics has desired performance.

Furthermore Benteghi et. al, [11] presented comparative distance relay based on cvt transient fault estimation. They presented the following issues:

Measurement impedance is obtained by dividing ct and cvt voltage and current. Therefore, it influences cvt transient voltage fault influences this impedance and leads it to saturation point. Measured impedance includes ohm and self-parts. Based on distance relay parameters, reaching saturation point was caused by reactance fault. Therefore, reactance problem is solved with combining old distance relay with comparative reactance relay [11]. In 2008, Joy Mooni presented an article entitled "distance relay under ct saturation condition". He studied waveform of ct output current in secondary and showed that how saturated ct caused sub-saturation point and up-saturation point and studied effect of ct saturation on zones and direction and distance

relay time [12].

In addition Taqizade et. al, [13], studied the effect of single phase to ground fault with arc-resistance on distance relay performance. They used THD factor (harmonic disturbances) to detect faults along with arc and a comparative distance relay was presented by this factor.

4. Methodology

First we are seeking to determine effect of each uncertainty factor on distance relay performance. For this, we design a simple 8 buses network according to IEEE standard as depicted in Figure 1. Then we regulate distance relays of this network based on status quo (without uncertainty), single phase and two phase and three phase faults were simulated in different locations (different zones) and measured impedance by distance relay for each condition was recorded. In the next step, we simulate uncertainties which were used separately for each fault. This time measured impedance by relay was recorded considering uncertainties. It is necessary to mention that some uncertainties are related to each other and influence each other. For example, exploitation conditions influences fault resistance and we should consider it in modeling and model both uncertainties synchronously or some uncertainties have more influence on certain zones. We should consider them to determine their effects. Therefore, effect of each on uncertainties on distance relay performance were determined and prioritized according to importance of these uncertainties.

4.1. Design of a Standard 8-Buses Network

The Mont-Carlo method was used regarding type and nature of problem, purposes and questions. In simulations instead of creating formula for solving problem we try to analyze and test a model for several times in likely conditions and results were recorded each time. Then by analyzing obtained statistics we gain reliable results for real performance of the system.

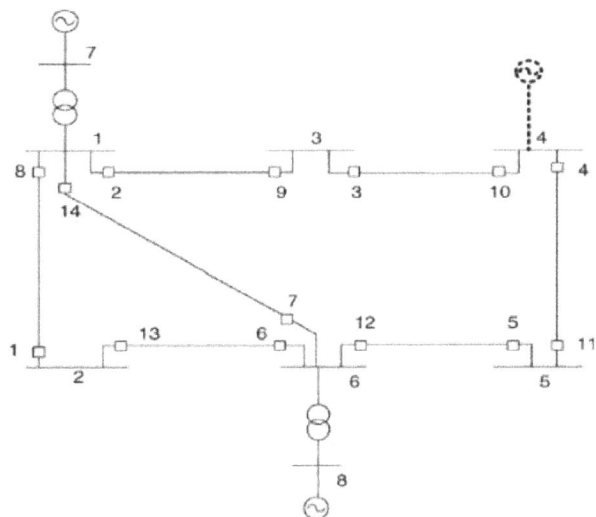

Figure 1. 8-buses standard network

4.2. Research Methodology

4.2.1. Setting of First Zone

It is expected that relay R in ideal state covers ijth line by its protection i.e. for each fault between buses i and j, relay should be able to detect fault and cut ij circuit. For this purpose, at first glance it seems that if relay R set by ij line impedance this line will be under protection but uncertainties influence measured impedance by relay R and may increase or decrease this impedance.

Assume that there is a fault near bus j (k_1). In this case, current will increase and voltage reduces. If impedance measured by relay increases because of uncertainties, in this case relay R detects this impedance in zone 2 (k_2).

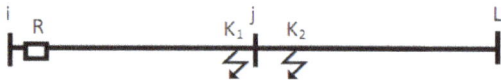

Figure 2. *Simple three buses network with distance relay for introducing setting zone 1 and 2*

However if fault occurs in next line i.e. jL line near bus j and impedance measured by relay decreases because of some uncertainties, relay R detects wrongly fault in zone 2 as zone 1 and cut ij circuit. Therefore, relay R cannot be set by ij line impedance but it is clear that impedance measured by relay R is in following interval because of uncertainties.

$$Z_{min} < Z_{meas} < Z_{max} \qquad (1)$$

$$Z_{meas} \in [Z_{min}, Z_{max}] \qquad (2)$$

In which Z_{min} is minimum impedance Z_{means} is impedance measured by transformers and Z_{max} is maximum impedance. Since zone of one Relay R per faults in next line should not have performance, impedance measured by this relay should be higher than relay set impedance i.e.

$$Z_{min} > Z_{1set} \leftarrow [Z_{min}, Z_{max}] > Z_{1set} \leftarrow Z_{meas} > Z_{1set} \qquad (3)$$

In which z_{1set} is set impedance for distance relay.

Therefore, minimum impedance measured by relay R considering uncertainties should be higher than set impedance; Therefore, if three phase fault simulated in end point of line ij and on bus j and repeat it thousand times, minimum impedance measured by relay R in these repetitions can be considered as zone 1 setting in relay R.

$$Z_{min} = Z_{1set} \qquad (4)$$

Now we should see that how much percent of line ij is covered by setting relay R without uncertainties i.e. measured impedance by relay in this condition should be lower that set amount to place in relay performance scope. Therefore, we should displace fault on ij line to farthest point for which fault be always $Z_{means} < Z_{1set}$. Therefore, we have:

$$Z_{max} < Z_{1set} \leftarrow [Z_{min}, Z_{max}] < Z_{1set} \leftarrow Z_{meas} < Z_{1set} \qquad (5)$$

Therefore, we change fault point on line ij and it is

simulated thousand times in different locations to obtain maximum impedance in each location considering uncertainties. This will be repeated until maximum impedance measured in one place should be equal with zone 1 set by relay which shows that relay covers it as a zone considering uncertainties.

4.2.2. Setting of Zone 2

In order to set zone 2 relay R, we consider following network:

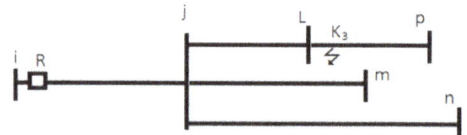

Figure 3. *Simple three buses network with distance relay*

In ideal condition, relay R should cover all lines connected to bus j as zone 2. Perhaps we consider that of zone 2 was set by jn line impedance we can achieve this but if a short circuit happens on Lp relay wrongly detects it as zone 2 while fault has occurred in zone 3. Therefore, in order to set zone 2 relay R we should consider shortest line connected to bus j and as we discussed about zone 1 relay we should simulate a three phase fault at the end of shortest line with uncertainties. Impedance measured by relay is in $z_{meas} \in [z_{min}, z_{max}]$ interval. Therefore, the following equation should be hold in order that relay has not performance. Therefore, we have:

$$Z_{min} > Z_{2set} \leftarrow [Z_{min}, Z_{max}] > Z_{2set} \leftarrow Z_{meas} > Z_{2set} \qquad (6)$$

We repeat this fault several thousand times. Minimum impedance measured by relay R was considered as zone 2 setting. In following we consider that by setting zone 2 this relay covers some percents of jL line completely considering uncertainties. For this purpose, we should displace fault on line iL to farthest point in which $z_{meas} < z_{2set}$. Therefore, we have:

$$Z_{max} < Z_{2set} \leftarrow [Z_{min}, Z_{max}] < Z_{2set} \leftarrow Z_{meas} < Z_{2set} \qquad (7)$$

Therefore, we change fault point on line jL and it is simulated thousand times in several locations to obtain maximum impedance in each location considering uncertainties. His will be repeated until maximum measured impedance become equal with zone 2 setting. In this case, this location shows percent of jL line which relay covers it completely as zone 2 considering uncertainties

4.2.3. Setting of Zone 3

In order to obtain zone 3 relay we consider following network:

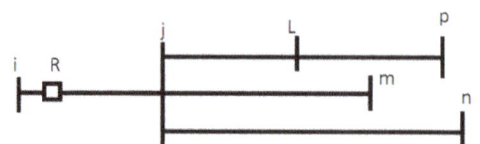

Figure 4. *Three buses simple network with distance relay*

At the end of longest line connected to bus j (on bus n) we simulate three phase fault. We assume that in this fault there is uncertainty in measurement device, fault in calculating line impedance and lines' resistance. We repeat this simulation several thousand times. Maximum measured impedance in this repetition is considered as zone 3 setting.

Of course, we should consider that uncertainties mentioned in simulations occur in short circuit or fault not others or it may decrease or increase impedance measured by relay; therefore, in all simulations we consider uncertainties randomly and its value by normal distribution.

In simulations we considered uncertainties of CT and PT faults and faults in calculating line impedance and fault resistance. We considered CT and PT fault as 2% and fault in lines impedance as 1 percent and fault resistance two times of line resistance.

5. Setting Algorithms

Regarding the setting algorithms used for calculating zone setting and obtaining maximum covered area the following details are provided:

5.1. Zone 1 Setting Algorithm

1. Entering network information including introducing buses, line impedance, generators production power, load consumption power.

2. Entering relay number in which zone 1 is placed and line which this relay protects it as zone 1.

3. Imposing one percent fault in lines impedance matrix using normal distribution.

4. Forming network impedance and admittance

5. Raphson-Newton load distribution (for calculating buses voltage before short circuit).

6. Simulation three phase fault on bus j (worst case scenario)

7. Imposing fault resistance randomly in zero to double amount of ij line resistance using normal distribution.

8. Calculating short circuit and obtaining relay voltage and current after short circuit.

9. Imposing 2% fault in relay voltage and current using normal distribution.

10. Calculating seen impedance by relay using following formula and storing in Z_{sss}

For Impedance Relay: $Z_R=V_R/I_R$ and for Admittance Relay: $Z_r=(V_R/I_R)/(\cos(\varphi_r-\varphi_x)$

11. Step 9 repeats thousand times.

12. Calculating minimum Z_{sss} and storing in Z_{tttt}

13. Step 7 repeats thousand times.

14. Step 3 repeats thousand times

15. Calculating minimum Z_{tttt} as zone 1 setting

16. end

If R is the impedance relay : $Z_R=V_R/I_R$

if the relay R is moho: $Z_r=(V_R/I_R)/(\cos(\varphi_r-\varphi_x))$

5.2. Zone 2 and Zone 3 Setting Algorithms

Zone 2 and zone 3 setting algorithms is similar to zone 1 setting algorithm with this difference that in step 2 for zone 2 setting enter 2 and line relay is on it (j,i) and shortest next line (L) and zone 3 setting as -2 and longest next line.

5.3. Algorithm for Calculating Maximum Covered Area by Zone 2 Setting

1 Entering network information including buses, lines impedance, power generated by generators, loads consumption power..

2 Entering relay number and line in which relay locates and shortest next line and zone 2 setting.

3 Place a as 1.

4 Change bus 1 location

5 Impose one percent error in impedance matrix using normal distribution.

6 Forming network impedance and admittance

7 Newton load distribution (for calculating buses voltage before short circuit)

8 Simulation three phase fault on bus j (worst case scenario)

9 Imposing fault resistance randomly in zero to double amount of ij line resistance using normal distribution.

10 Calculating short circuit and obtaining relay voltage and current after short circuit.

11 Imposing 2% fault in relay voltage and current using normal distribution.

12 Step 9 repeats thousand times.

13 Calculating maximum Z_{sss} and storing in Z_{tttt}

14 Step 7 repeats thousand times.

15 Step 3 repeats thousand times

16 Calculating maximum Z_{tttt} as zone 1 setting

17 If $Z_{max}=Z2_{SET}$ go to step 20

18 Place a=a-(0-01) and go step 4

19 a is percent of jl line which relay R covers it/

20 end

6. Discussion & Conclusion

Settings imposeding as distance relay are stable settings which are considered by dominant condition of systems. In these settings effects of factors like fault resistance, measurement devices fault, change in structure and exploitation conditions are not considered which are introduced as uncertainty. In this study we obtained settings for distance relay which cover uncertainties and increase relay accuracy. By considering security marging several setting groups are considered for different conditions of network and by changing condition certain setting groups impose on relay.

For this purpose, we used the Mont-Carlo method with MATLAB software to analyze and simulate faults without interference of uncertainties and obtain accurate and reliable setting for distance relay. With these settings zone 1 distance relay approaches 100% protection and increases reliability and stability of transmission lines and power network.

References

[1]　Shateri, H., & Jamali, S. ideal characteristic for distance relay. International power conference, Tehran, Iran, 1990.

[2]　Shateri, H., Jamali, S. fault border resistance for setting distance relay with tetrahedron chaaracteristics. 19th international electricity conference, Tavanir company Tehran. Iran, 2004.

[3]　Catalogues, instructions for power networds, Deputy exploit, Khorasan Regional Electricity Company, 2012.

[4]　Hashemi, S.M., Tarafdarhaqh, M., & Seyyedi, H. protecting transmission lines equipped with IPFC. Iran. Tabriz University. 6th power systems conference, , 2012.

[5]　Tarlochan S.Sidhu , David Sebastian Baltazar , Ricardo Mota Pahomino , and Mohindar S.Sachdev, A New Approach for Calculating Zone-2 Setting of Distance Relays and Its Use in an Adaptive Protection System , IEEE Transactions on Power Delivery, January ,v01.19 , No.1 , 70, 2004.

[6]　H.Shateri , S. Jamali . Measured Impedance by Distance Relay Elements in a Single Phase to Ground Fault , IEEE , 5th International Conference on Electrical and Computer Engineering, 20 _ 22 December , 978_1_4244 , 2008.

[7]　Benteg He , Yiquan Li , Zhiqian Q.Bo ,An Adaptive Distance Relay Based on Transient Error Estimation of CVT , IEEE TRRANSACTIONS ON POWER DELIVERY, OCTOBER , VOL.21 , NO.4 , 2006.

[8]　Joe Mooney, P.E., Distance Element Performance Underconditions of CT Saturation, IEEE, Schweitzer Engineering Laboratories, Inc., 2008.

[9]　Majid. Taghizadeh , Javad.Sadeh, Mohsen.Bashir , Ebadollah. Kamyab , Effect ofSingle Phase to Ground Fault With ArcResistance on the Performance of Distance Relay , IEEE, 1_4244_8782 , 97, 2011.

14

Production of iridium metal thin films for application as electrodes in DRAMs and FRAMs

Sakine Shirvaliloo[1, 3, *]**, Hale Kangarloo**[2]

[1]Department of Medical Physics, Iran University of Medical Sciences, Tehran, Iran
[2]Faculty of Science, IAV, Branch Urmia, Urmia, Iran
[3]Young Researchers Club, Mahabad, Iran

Email address:
Sakine.shirvaliloo@gmail.com (S. Shirvaliloo), h.kangarloo@yahoo.com (H. Kangarloo)

Abstract: Thin films of Noble metals such as Iridium have several potential applications in ICs. Electronic devices are fabricated on the integrated circuits that include transistor, capacitor and resistance. They can be used as electrodes in DRAMs and FRAMs, and as gate electrodes in MOSFETs. Noble metals are excellent metals for electrode fabrication because chemical stability, highly electrical resistance, highly work function and many of them can withstand highly oxidizing conditions. In this paper, emphasis is on reaction mechanisms and limits, leakage currents, electrodes and electrode interfaces and deposition techniques.

Keywords: Atomic Layer Deposition, Precursor, Gate Electrode, Dynamic and Ferroelectric Memories, Capacitors

1. Introduction

Atomic layer deposition (ALD), also called atomic layer epitaxy (ALE), is a thin film deposition technique developed in Finland in the mid-1970s. Initially, ALD was developed for deposition of materials used in electroluminescent displays but during the years, the selection of materials and their potential applications has expanded [1]. Iridium is chemically very stable, and can withstand highly oxidizing conditions [2]. Therefore, this metal is applicable to ICs as electrodes in dynamic random access memories (DRAMs) and ferroelectric random access memories (FRAMs). Other applications in ICs include gate electrodes in MOSFETs, and seed and barrier layers for copper in interconnect metallization. In addition to ICs, noble metals have applications for example in catalysis, sensors, and magnetic data storage [3-5].

2. Data and Material

In this work, we used Iridium thin films. Iridium is the chemical element with atomic number 77, and is represented by the symbol Ir. Iridium is a very hard, brittle, silvery-white transition metal of the platinum family. Iridium is noble metal similar to silver and gold, and the Platinum group metals ruthenium, rhodium, palladium, osmium, and platinum [2]. Noble metals are called noble because of their chemical stability; they have Positive standard reduction potentials, and many of them are highly resistant to oxidation, even at elevated temperatures [3].

3. Research Methodology

In our research study, we have used atomic layer deposition. ALD is a gas phase thin film deposition method which can be regarded as a special modification of the more widely used chemical vapor deposition (CVD) technique [1]. In ALD, The precursors are led into the reaction in alternate pulses. While in CVD, precursors are led into the reaction chamber simultaneously ALD processes for iridium have been reported so far only for Ir(acac)$_3$ and oxygen, and For Ir(acac)$_3$ and hydrogen. Table1 presents a summary of these iridium ALD processes [5].

Ir(acac)$_3$ is solid at room temperature and has a relatively low vapor pressure,195 which, however, is not a problem in ALD. In the oxygen-based process, Ir(acac)$_3$ was sublimated at 150 °C. Air with a flow rate of 5–40 sccm or pure oxygen with a flow rate of 5–20 sccm was used as the oxygen source [2]. The film growth took place on Al$_2$O$_3$ surface at temperatures of 225°C and higher. At 400°C, signs of thermal self-decomposition of Ir(acac)₃ were observed, and therefore, 375 °C was considered the upper temperature limit for the process. The temperature range for the film growth is in good accordance with the reported thermal stability of Ir(acac)₃ in vacuum and in oxygen atmosphere.

Table 1. Summary of the reported iridium ALD processes.

Metal precursor	T vaporization (°C)	Reaction gas	Flow rate (sccm)	T growth (°C)
Ir(acac)₃	150	Air	5-40	225–400
	150	O₂	5-20	225–300
	N/A	H₂	separate reduction step	200–250 / 275–450

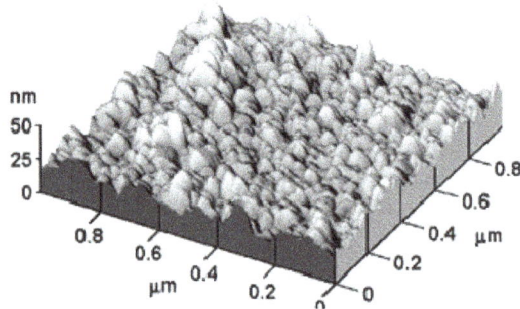

Figure 1. AFM image of a 85-nm iridium film grown at 350°C from Ir(acac)₃ and air.

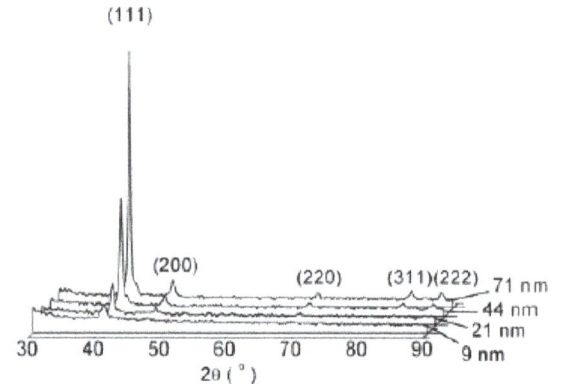

Figure 2. XRD patterns measured from iridium films of different thicknesses. The films were grown at 300 °C.

Resistivities of about 70-nm thick iridium films grown from Ir(acac)₃ and air at temperatures between 225 and 400 °C were lower than 12 μΩ·cm. The film resistivity decreased with increasing growth temperature which was most probably due to increased crystallinity of the films. At 400 °C, there was a small increase in the resistivity, possibly due to the increased impurity contents of the film. The increase in the resistivity with decreasing film thickness is related to an increased effect of grain boundary and surface scattering [2].

The iridium films grown from Ir(acac)₃ and oxygen at temperatures of 225–400 °C were metallic iridium with a preferred (111) crystal orientation as studied by XRD. The films showed a strong tendency to orient the (111) crystal plane, which has the lowest surface energy, parallel to the surface. Similar tendency was also observed with platinum, which is a fcc metal, too. The iridium films had a preferred (111) crystal orientation even at the lowest growth temperatures, implying high surface mobility of the iridium atoms. Figure 14 shows that the (111) crystal orientation is favorable already at the beginning of the growth of an iridium film at 300 °C; in the XRD pattern measured from a 9-nm film, the intensity ratio of the iridium (111) and (200) reflections is 4.6 : 1 while in a randomly oriented powder, the intensity ratio would be 2 : 1.21.

Iridium has been deposited by ALD from Ir(acac)₃ and hydrogen on large surface area Al₂O₃ and SiO₂-Al₂O₃ supports for catalytic applications. The reaction temperature during Ir(acac)₃ exposure was 200 °C [2]. The results imply that when Ir(acac)₃ adsorbs on the support surface, one of the three acac ligands is released in reaction with the surface hydroxyl groups. The remaining ligands were removed by a separate reduction step at temperature of 275–450 °C.

4. Results and Analysis

4.1. Dynamic Random Access Memory Electrodes

The data storage in DRAMs is based on introducing a charge into the memory capacitor. The charge must be repeatedly refreshed (thereof the term 'dynamic'), the time interval between the refreshments depending on the leakage current through the capacitor. The charge storage capacity of the capacitor is determined by its capacitance, which can be increased by decreasing the thickness of the dielectric layer, by using a dielectric material with a higher dielectric constant, or by increasing the active capacitor area [6]. Three-dimensional structures provide larger capacitor areas even if the lateral dimensions of the capacitors decrease, and therefore, three-dimensional structures are needed in order to further increase the number of capacitors per unit area, and thus, the storage density of the DRAMs. In stacked DRAM memory cells, the capacitors are fabricated above the silicon substrate. Examples of stacked capacitor structures are shown in Figure [5].

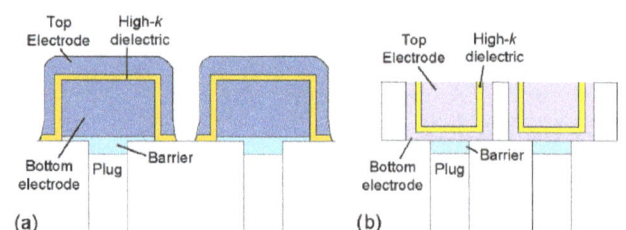

Figure 3. (a) An example of a simple stacked DRAM capacitor structure, and (b) an example of a concave stacked DRAM capacitor structure.

Noble metals are suitable electrode materials because they remain conductive in highly oxidizing conditions; platinum is highly resistant to oxidation, whereas iridium and ruthenium form conductive oxides (IrO_2 and RuO_2) when they become oxidized. Furthermore, noble metals have high work functions, and thus high barrier for leakage by Schottky emission, that is, thermionic emission in the presence of an applied electric field. The bottom electrode affects the dielectric and electrical properties of the high-k oxide, and plays therefore a significant role in the performance of the memory capacitor. The following sections summarize the noble metal electrode structures studied for DRAM capacitors with Ta_2O_5 and BST as the dielectric materials [6].

4.1.1. Electrodes for (Ba, Sr)TiO₃

Iridium is a promising electrode material for BST because good dielectric properties have been reported for BST films grown on iridium electrodes. BST films having a high dielectric constant have been obtained on iridium electrodes. The top electrode material affects the dielectric constant of the BST film as well; higher dielectric constants have been obtained when iridium is used as the top electrode instead of platinum. This difference was explained by a large compressive stress in the BST film induced by the iridium top electrode. The leakage currents reported for BST films grown on iridium are, however, higher than for BST films grown on platinum. High-temperature post-annealing in O_2 atmosphere has been reported to lower the leakage currents of BST films grown on iridium electrodes [5].

An iridium oxide layer forms on the surface of the iridium electrode during BST processing at temperatures of 600 °C or higher. This iridium oxide layer can act as an oxygen diffusion barrier. Iridium films annealed at 700 °C in N_2 atmosphere have been reported to prevent oxygen diffusion better than non-annealed films.85 this is because the number of grain boundaries reduces during the high temperature annealing. Thus, there are fewer pathways for oxygen diffusion. The stability of iridium in contact with silicon is moderate; iridium reacts with silicon when annealed in O_2 atmosphere at 650 °C but is not completely consumed in the reaction with silicon [7].

4.1.2. Electrodes for Ta₂O₅

Iridium is a potential bottom electrode material for capacitors with Ta_2O_5 as the dielectric because high dielectric constants and low leakage currents have been reported for Ta_2O_5 grown on ruthenium electrodes. For crystalline Ta_2O_5 films, dielectric constants in the range of 45–110 have been reported whereas for amorphous films, the dielectric constant is about 25–32. Thus, in order to obtain higher dielectric constants, the Ta_2O_5 films need to be crystallized. The crystallization is usually performed by post-annealing at a temperature of 700 °C or higher. If the high-temperature annealing is performed in an O_2 atmosphere, the ruthenium bottom electrode is likely to become oxidized to ruthenium oxide [2]. The RuO_2 formation causes roughening of the Ta_2O_5 –electrode interface which leads to higher leakage currents. Therefore, in order to prevent ruthenium oxidation,

the high-temperature annealing of Ta_2O_5 is usually performed in a N_2 atmosphere or in vacuum. The dielectric properties of Ta_2O_5 can also be improved by O_2 or N_2O plasma treatments or O_2 annealing performed at lower temperatures of 350–550 °C at which oxidation of the ruthenium bottom electrode does not take place.

4.2. Ferroelectric Random Access Memory Electrodes

In FRAMs, the data is stored as the polarization state of the ferroelectric film. Switching between two stable polarization states is realized by applying an electric field across the ferroelectric film. FRAMs have two different types of operation principles; the data can be stored in ferroelectric capacitors or in ferroelectric transistors. Ferroelectric capacitors are otherwise similar to DRAM capacitors (Figure3) but the high-k dielectric is replaced by the ferroelectric film. In ferroelectric capacitors, an excess charge is drawn from the circuit and stored to the capacitor if the polarization state of the ferroelectric film is changed during the read-out process. Thus, the polarization state of the ferroelectric film can be determined by detecting the presence or the absence of the switching charge pulse. The magnitude of the switching charge can be increased by using three-dimensional capacitor structures [8]. The metal -ferroelectric-insulator-semiconductor (MFIS) FET shown in Figure 4 is an example of a ferroelectric transistor [7].

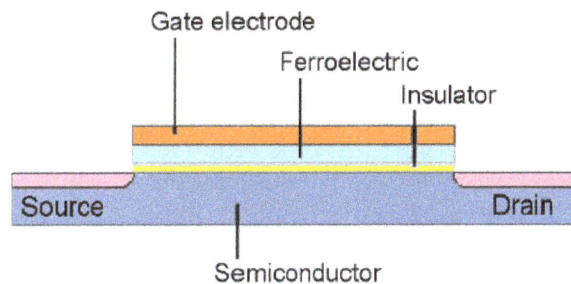

Figure 4. *A schematic of a MFIS field effect transistor.*

4.2.1. Electrodes for Pb(Zr, Ti)O₃

Iridium has been shown to be a suitable electrode material for PZT because high remnant polarization, low leakage currents, and good fatigue properties have been obtained for PZT grown on iridium. The good fatigue properties of the PZT films are most probably due to oxidation of the iridium surface during the PZT deposition at temperatures higher than 530 °C. In addition, the IrOx layer formed on the surface of the iridium electrode prevents diffusion of lead and oxygen through the electrode. Iridium can also be used as an oxygen barrier layer in multilayer electrode structures [8].

4.2.2. Electrodes for SrBi2Ta₂O₉

Iridium is a potential electrode material both in ferroelectric capacitors and in ferroelectric transistors having SBT as the ferroelectric. In ferroelectric capacitors, SBT films grown on iridium electrode have been reported to show poor ferroelectric properties, possibly because of oxidation of the iridium surface during SBT processing [8,9]. However, iridium and IrO_2 can be used as oxygen barrier layers together with platinum electrodes in multilayer electrode structures. Platinum cannot be grown directly on iridium, because platinum and metallic iridium inter diffuse at temperatures higher than 650 °C. Iridium, on the other hand, cannot be grown directly on the poly-Si plug because it forms iridium silicide. In MFMIS transistors, iridium has been used as the lower metal electrode on which the ferroelectric SBT layer is grown. However, bismuth was reported to diffuse into the iridium electrode during high temperature annealing.

5. Discussion and Conclusion

The results obtained in this work show that high quality noble metals such as iridium and iridium films can be grown by ALD. All of the processes are based on reactions of the metal precursor with oxygen, the growth temperatures being in the range of 200–450 °C. Iridium films were grown from $Ir(acac)_3$ and oxygen at a temperature range of 225–375 °C. The films were metallic iridium with a preferred (111) crystal orientation. Depending on the thickness, the films had resistivities in the range of 9–18 $\mu\Omega\cdot$cm. The films had smooth surfaces; a 65-nm film grown at 300 °C had an rms surface roughness of 1.6 nm. At 300 °C, the film growth rate saturated to 0.40 Å/cycle, and the film thickness depended linearly on the number of the growth cycles.

ALD is a suitable method for growing noble metals for the potential applications in ICs which include electrodes in DRAMs and FRAMs, gate electrodes in MOSFETs, and seed layers for copper in interconnect metallization. Noble metal ALD processes taking place in oxidizing conditions can be used when growing DRAM and FRAM top electrodes, gate electrodes for MOSFETs and ferroelectric FETs, and combined seed and barrier layers for copper in interconnect metallization. The DRAM and FRAM bottom electrodes as well as the copper seed layers grown on nitride barriers need to be grown in reducing process conditions.

References

[1] Ritala, M. and Leskelä, M., Atomic layer deposition of thin films for microelectronics, Electrochem.Soc. Proc., 2003-08 (2003) 479–490.

[2] Aaltonen, T., Ritala, M., and Leskelä, M., Atomic layer deposition of noble metals, in Advanced Metallization Conference 2004 (AMC 2004), (Eds: D. Erb, P. Ramm, K. Masu, and A. Osaki), Materials Research Society (2005) 663–667.

[3] Jylhä, O., Saarinen, R., Lindblad, M., and Krause, O., Iridium on porous supports prepared by ALD for heterogeneous catalysis applications, ALD 2004, Helsinki, August 16-18 (2004).

[4] Aaltonen, T., Ritala, M., Sammelselg, V., and Leskelä, M., Atomic layer deposition of ridium thin films, J. Electrochem. Soc., 151 (2004) G489–G492.

[5] Aaltonen, T., Ritala, M., and Leskelä, M., Atomic layer deposition of noble metals, in Advanced Metallization Conference 2004 (AMC 2004), (Eds: D. Erb, P. Ramm, K. Masu, and A. Osaki), Materials Research Society (2005) 663–667.

[6] Mandelman, J. A., Dennard, R. H., Bronner, G. B., DeBrosse, J. K., Divakaruni, R., Li, Y., and Radens, C. J., Challenges and future directions for the scaling of dynamic random-access memory (DRAM), IBM. J. Res. Develop., 46 (2002) 187–212.

[7] Iwai, H. and Ohmi, S., Silicon integrated circuit technology from past to future, Micro electron. Reliab. 42 (2002) 465–491.

[8] Ishiwara, H., Recent progress of ferroelectric memories, Int. J. High Speed Electron. Syst., 12 (2002) 315–323.

[9] Yoon, D.-S., Roh, J. S., Baik, H. K., and Lee, S.-M., Future direction for a diffusion barrier in future high-density volatile and nonvolatile memory devices, Crit. Rev. Solid State Mater. Sci., 27 (2002)143–226.

15

LabVIEW based design implementation of M-PSK transceiver using multiple Forward Error Correction coding technique for Software Defined Radio applications

Nikhil Marriwala[1], Om Prakash Sahu[2], Anil Vohra[3]

[1]Electronics & Communication Engineering Department, University Institute of Engineering and Technology, Kurukshetra University, Kurukshetra, India
[2]Electronics & Communication Engineering Department, National Institute of Technology, Kurukshetra, India
[3]Electronics & Science Department, Kurukshetra University, Kurukshetra, India

Email address:
nikhilmarriwal@gmail.com (N. Marriwala), ops_nitk@yahoo.co.in (O. P. Sahu), vohra64@gmail.com (A. Vohra)

Abstract: Software-Defined Radio (SDR) is an enabling technology which is useful in a wide range of areas within wireless systems. SDR offers a perfect solution to the problem of spectrum scarcity in wireless communication. With the significant increase in the demand for reliable, high data rate transmission these days, a different number of modulation techniques need to be adopted. The main objective of this paper is to design and analyze an SDR based M-Phase Shift Keying (PSK) transceiver using LabVIEW (Laboratory Virtual Instrumentation Engineering Workbench) and to measure the Bit Error Rate (BER) in the presence of Additive White Gaussian Noise (AWGN) introduced in the channel. Forward Error Correction (FEC) is used as a channel coding scheme in this paper. FEC codes are used where the re-transmission of the data is not feasible, thus redundant bits are added along with the message bits and transmitted through the channel. This paper describes the fundamental concept for the design & development of an SDR -based transceiver simulation model under PSK Scheme & analyses the performance of two Forward Error Correction channel coding algorithms namely the Convolution and the Turbo Codes. In this paper we have shown that how fast and effectively we can build a PSK transceiver for interactive Software Defined Radio. With the help of this design we are able to see and prove that data errors can be minimized using coding techniques, which in turn improves the Signal to noise ratio (SNR).

Keywords: Software Defined Radio, Bit Error Rate, Additive White Gaussian Noise, Phase Shift Keying, Signal-To-Noise Ratio, Forward Error Correction

1. Introduction

The term Software Defined Radio refers to reconfigurable or reprogrammable radio that shows different functionality with the same hardware. The entire functionality of the SDR can be defined in software [1]. The aim of this paper is to simulate SDR for next generation wireless communication systems by using the M-PSK modulation technique in LabVIEW.

SDR provides an alternative to systems such as the third generation (3G) and the fourth generation (4G) systems [2]. A Complete hardware based system has many limitations. SDR technology provides many benefits including increased interoperability, reduced cost, and improved life cycle for communication systems [1, 2]. SDR's can be reconfigured and can talk and listen to multiple channels at the same time. The transmitter of an SDR system converts digital signals to analog waveforms. The analog waveforms generated are then transmitted to the receiver. The received analog waveforms are then down converted, sampled, and demodulated using software on a reconfigurable baseband processor [3]. SDR systems can be used in ubiquitous network environments because of its flexibility and programmability [4, 5]. The use of digital signals reduces hardware, noise and interference problems as compared to the analogue signal in transmission, which is one of the main advantages of digital transmission [6, 7].

In this paper, the software simulator of the PSK transceiver has been designed using LabVIEW. PSK is chosen to be the modulation scheme of the designed interactive Software Defined Radio system as this modulation scheme is widely used for transmission of data for various applications over band pass channels such as paging systems and Cordless, Telephone-line modems, Caller ID, Microcomputers, Radio control etc. A PSK, SDR which is fully implemented, will have the ability to navigate over a wide range of frequencies with programmable channel bandwidth and modulation characteristics [8, 9]. The role of modulation techniques in an SDR is very crucial since modulation techniques define the core part for any wireless technology [10, 11].

The aim of this design is to implement interactive Software Defined Radio system in a shorter time and provide a cost effective solution compared to other text-based programming languages [12, 13]. With the help of this design we will be able to see and prove that data errors can be minimized using FEC coding techniques, which in turn improves the Signal to noise ratio (SNR). This interactive design of SDR will also help understand and analyze how the signal can be recovered with very less probability of error and which FEC codes are the best suited for transmission using M-PSK modulation scheme. The general Block Diagram of a generic Digital transceiver is shown in Figure 1.

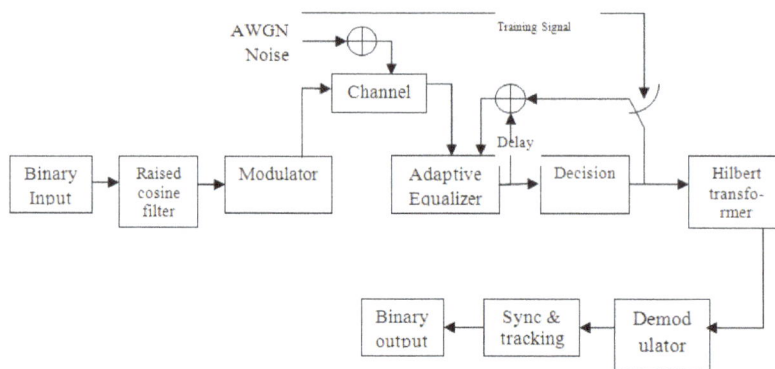

Figure 1. *Block Diagram of a Generic SDR Transceiver.*

2. Phase Shift Keying Transceiver (M-PSK)

In This paper, digital modulation scheme Phase shift Keying is used as a modulation scheme for the SDR transceiver. PSK has dynamic characteristics of the carrier signal with respect to time and this alteration results in a sine gesticulate in a divergent phase, amplitude or frequency. This results in, contrasting "states" of the sine curve are referred to as symbols which represent few digital bit ornamentation. The building blocks of the PSK transceiver

system are stated in this section. This system has two parts: transmitter and receiver. The Front Panel for a PSK transceiver with multiple Encode and Decode techniques is shown in Figure: 2.

The organization of this paper is as follows: In Section I Introduction for SDR is explained, Section II gives the implementation of M–PSK transceiver in LabVIEW, in Section III the PSK transmitter parameters are described, Section IV gives the BER Vs SNR comparison, Section V describes the simulation results of a PSK transceiver using LabVIEW and finally in Section VI we draw the Conclusions.

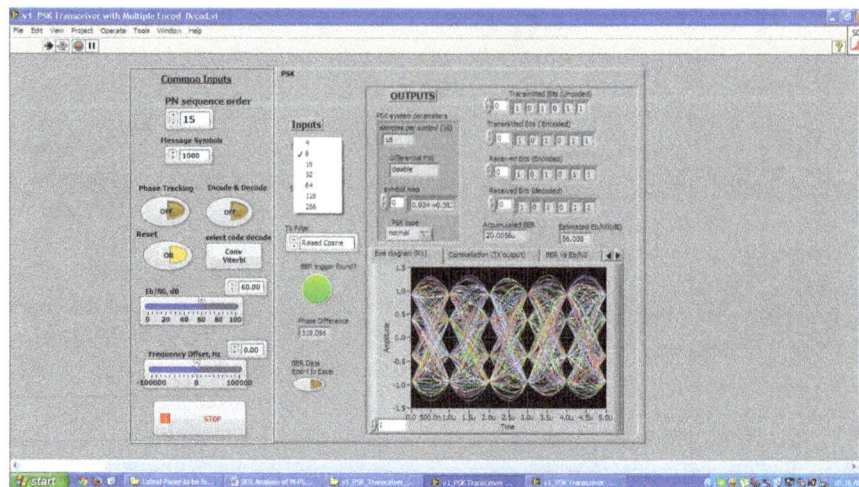

Figure 2. *PSK Transceiver with multiple Encode and Decode.*

2.1. Message Source

In this design at the transmission end, pseudo noise (PN) sequences are generated which serve as our message signal. A PN Sequence Generator block generates a sequence of pseudorandom binary numbers using a linear-feedback shift register (LFSR). The LFSR is implemented using a simple shift register generator. Here, the PN sequence is generated with a five-stage LFSR structure, whose connection polynomial is given by

$$h(D) = 1 + D^2 + D^5 \qquad (1)$$

where D denotes delay and the summations represent modulo 2 additions. The sequence generated by the above equation has a period of $31 (= 2^5 - 1)$ as shown in Figure: 3. Two PN sequence generators are used in order to create the message sequences for both the in-phase and quadrature phase components. Note that frame marker bits are inserted in front of the generated PN sequences.

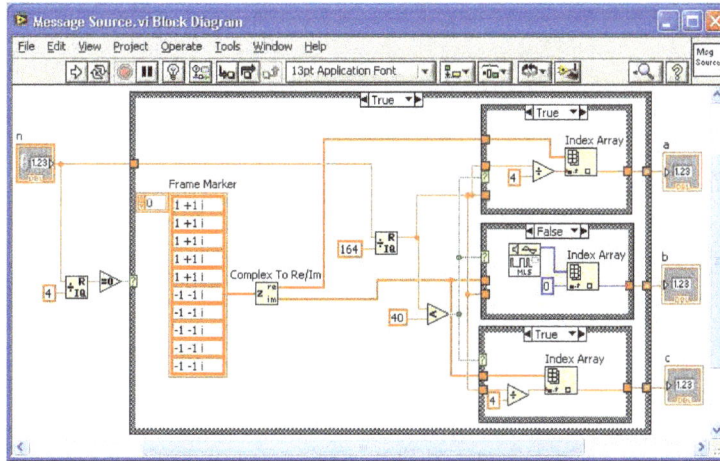

Figure 3. *Message Source VI.*

2.2. Source Encoder

The source encoder is to improve efficiency by reducing redundant bits, compressing the digital sequence into a more competent symbol for transmission.

2.3. Pulse Shaped Filter

The purpose of pulse shaping is to make the transmitted signal better suited to the communication channel by limiting the effective bandwidth of the transmission. By doing this ISI caused by the channel can be controlled before modulation [16]. The raised cosine filter is one of the pulse shaping filter as shown in Figure: 4. It is used in digital modulation due to its ability to minimize Intersymbol

interference. Its name stems from the fact that the non-zero portion of the frequency spectrum of its simplest form ($\beta = 1$) is a cosine function, 'raised' up to sit above the f (horizontal) axis.

The raised-cosine filter is an implementation of a low-pass Nyquist filter, i.e., one that has the property of vestigial symmetry. This means that its spectrum exhibits odd symmetry about $\dfrac{1}{2T}$, where T is the symbol-period of the communications system.

and characterized by two values; β, the roll-off factor, and T, the reciprocal of the symbol-rate.

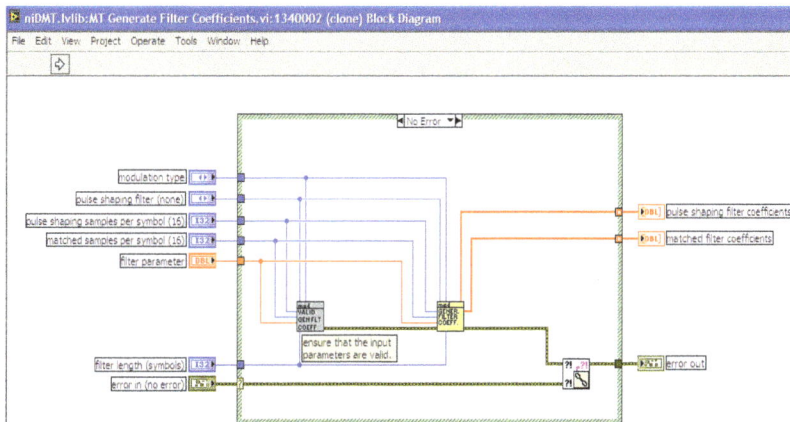

Figure 4. *Pulse Shape Filter VI.*

2.4. PSK Encoder

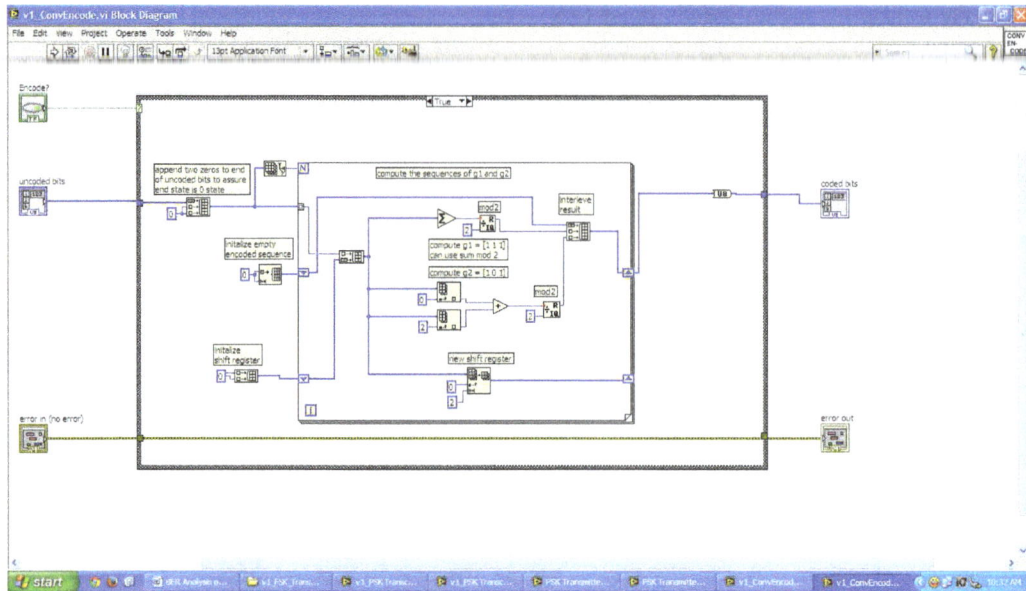

Figure 5. *Convolution Encoder VI.*

The channel encoder is to improve reliability by adding redundant bits to the compressed information in order to control the errors offered by channel impairments. Convolution Encoder is a Finite State Machine (FSM), processing information bits in a serial manner. In this case we have implemented ½ rate convolution Encoder. Shift Registers are used through which data shifts in and out linearly. They have rather good correcting capability and perform well even on very bad channels (with error probabilities of about (10^{-3}). Convolution Encoder Performs Convolution of the input stream with encoders impulse responses. Mathematically, it is written as

$$y_i^j = \sum_{k=0}^{\infty} h_k^j x_{i-k} \qquad (2)$$

where x is an input sequence, y^j is a sequence from output j and h^j is an impulse response for output j. Convolution Encoder VI is shown in Figure:5.

In this design we have used the Turbo Encoder as shown in Figure: 6 which works by using two Convolutional Encoders. One encoder receives the data to be sent and the other receives an interleaved version of the data to be sent. The convolutional encoders are identical and are rate 1. Each has 3 linear shift registers with a feedback loop. The original data, the output from encoder 1, and the output from encoder 2 are then interleaved together before being transmitted.

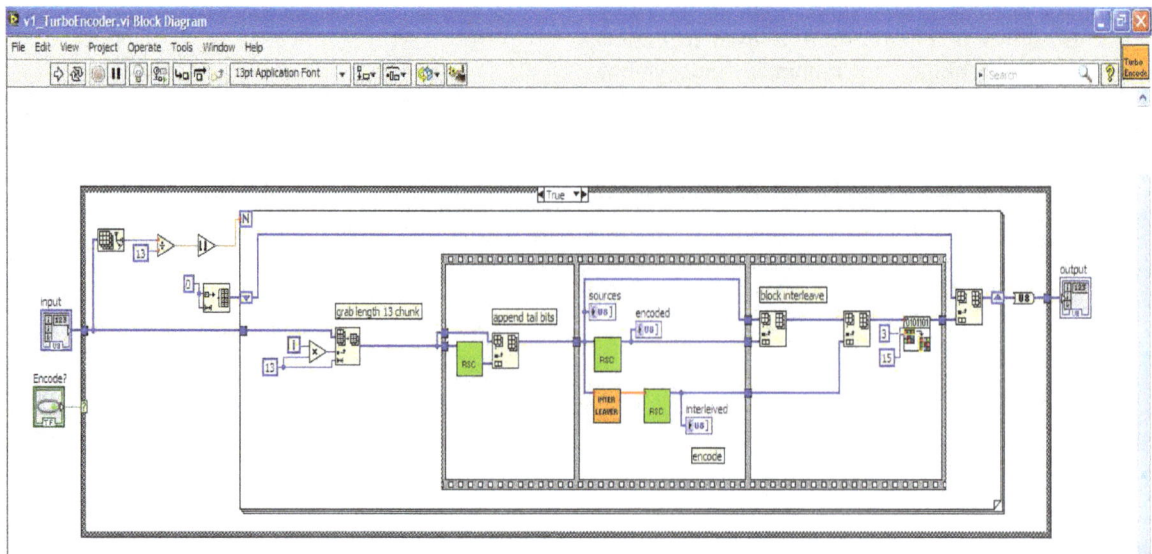

Figure 6. *Turbo Encoder VI.*

2.5. PSK Modulator

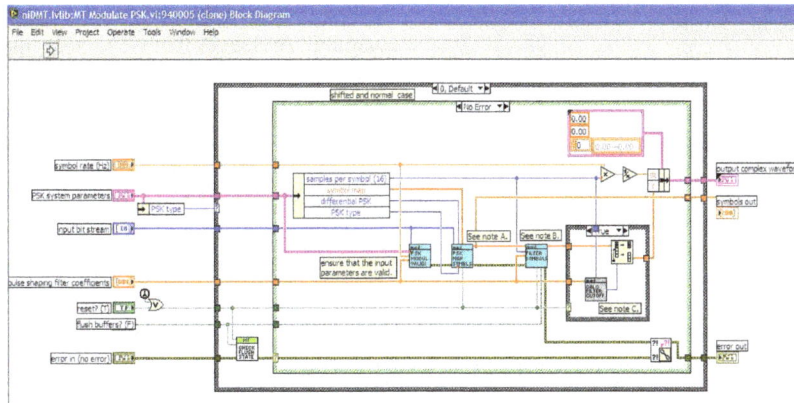

Figure 7. *PSK Modulator Vi.*

The PSK modulator converts the input bit stream into an electrical waveform suitable for transmission over the communication channel. In this design we have used the Modulator to minimize the effects of channel noise, also to match the frequency spectrum of the transmitted signal with channel characteristics, and to provide the capability to multiplex many signals as shown in Figure: 7.

The output of the raised cosine filter is then used to build a complex envelope. The data bits are transmitted by shifting the frequency of a continuous carrier in a binary manner to one or the other of two discrete frequencies. One frequency is designated as the "mark" (1) frequency and the other as the "space" (0) frequency.

2.6. Time Varying Channel

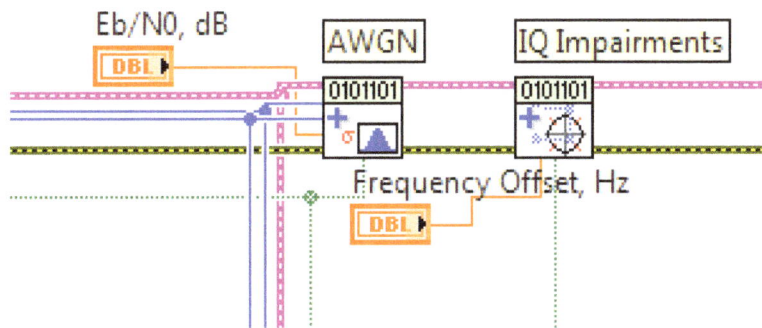

Figure 8. *AWGN & IQ Impairments.*

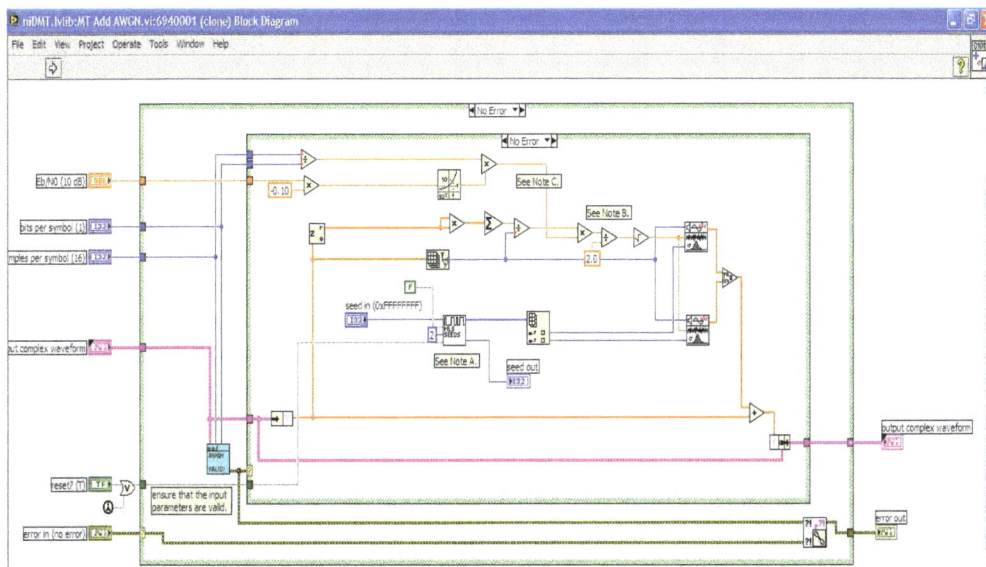

Figure 9. *Add AWGN VI.*

To be able to observe the adaptability of the system, a time-varying channel is added. The noise source, which is the Additive White Gaussian Noise (AWGN), is passed through the channel before it is added to the input signal. The channel is Gaussian in nature because its probability density function can be accurately modeled to behave like a Gaussian distribution and it is called white as it has a constant power spectral density. The characteristic of the channel is varied with time by swinging the filter pass band from 100 to 900 Hz. Figure: 8 shows the time varying channel with the AWGN noise source. Additive White Gaussian Noise is used to generate zero mean with uniform spectral density and adds it to a complex baseband modulated waveform. The VI shown in Figure:9. computes, generates, and adds the appropriate amount of AWGN to a complex-valued input signal, given a desired output Eb/N0 [10,11,12]. For true AWGN, the I and Q components of the additive noise must be uncorrelated. We accomplish this by using two separate Gaussian noise generators independently seeded. The user has the option of providing a seed in the event that they want to generate deterministic white Gaussian noise.

2.7. PSK Demodulator

PSK demodulation is the process of recovering the original message from the information bearing waveform produced by the modulation is accomplished by the demodulator. Demodulates the modulated complex baseband waveform & returns the time aligned oversampled complex waveform, demodulated bit stream. This step attempts to remove carrier & phase offset by locking to the carrier signal.

Viterbi decoding is an optimal (in a maximum-likelihood sense) algorithm for decoding of a Convolution code as this simplifies the decoding operation [17]. The decoder is a Viterbi decoder which then solves for the global optimum bit sequence. The algorithm updates a path cost as it steps through each stage of the possible output sequences. At each state, it also calculates the likelihood of entering each possible new state based on the cost of the previous state. The algorithm then needs two additional zero bits after every sequence in order to force the encoder back into the zero state and to assume that the encoder ends at the all zero state. These two tail bits represent a fractional loss rate between the coded and that of uncoded bit sequence. The Viterbi Decoder VI is shown in Figure: 10.

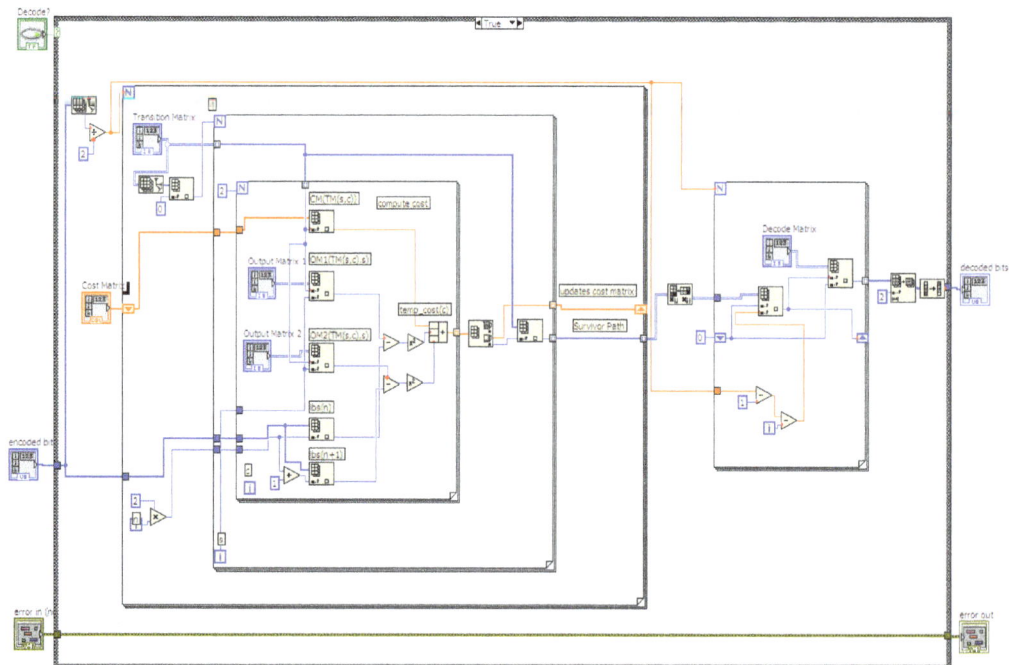

Figure 10. Viterbi Decoder VI.

Turbo Decoding used in this system as shown in Figure. 11. Works by using a set of maximum aposteriori probability (MAP) decoders. When the data is received, it is deinterleaved back into the three streams which were sent from the transmitter:

1. Original Data
2. Output from Convolutional encoder 1
3. Output from Convolutional Encoder 2.

The first MAP decoder takes as an input stream 1 and stream 2 and also the output from MAP decoder 2 (initialized to zeros for the first iteration) [16, 17]. The second MAP decoder takes in an interleaved version of stream 2 (the same interleaver used to interleave the original data before it was sent to the Convolutional Encoder), stream 3, and the output from the first MAP Decoder [18,19]. The two MAP

Decoders then work together to converge on a solution: the   most likely original bit sequence.

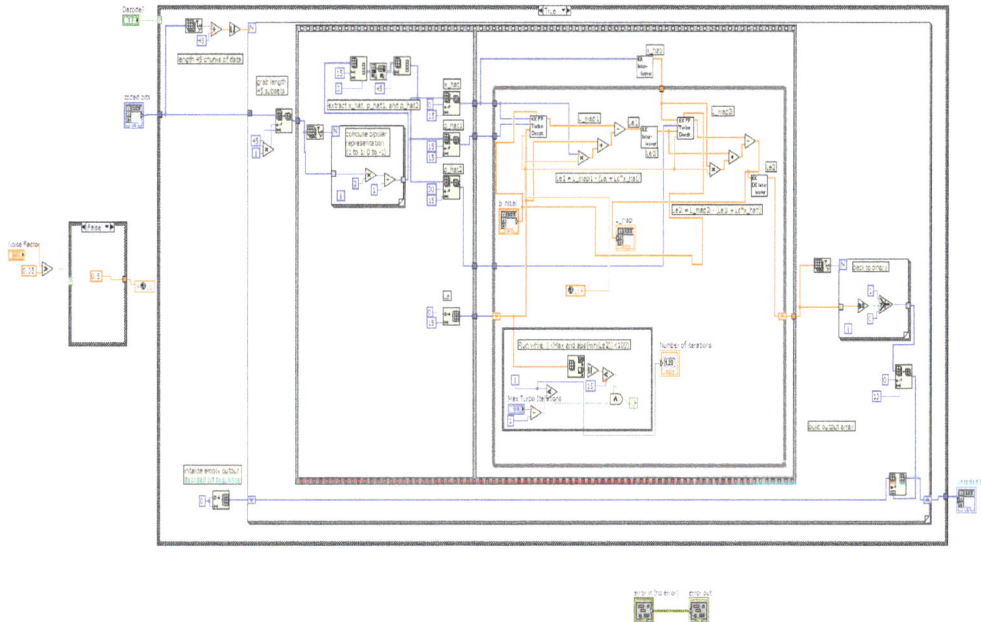

Figure 11. *Turbo Decoder VI.*

2.8. Sync & Tracking VI – Frame Synchronization Mode

Sync & Tracking VI is used for Frame synchronization and Phase/Frequency tracking. Synchronization is the act of synchronizing, i.e. concurrence of events with respect to time. In VI shown in Figure: 12 the input samples are passed through Complex Queue Pt By Pt VI, which creates a data queue of complex numbers to obtain beginning of frame. A case structure is not executed until the queue is completely filled. Extra 16 bits are added due to delays related to filtering operations in transmitter. A counter is used to count number of samples filling up the queue (as loop count VI). A Boolean (sync) is a primitive data type that can have one of two values: TRUE or FALSE. The initial value of the local variable, which is denoted by Sync, is set to true to execute the frame synchronization. Then, it is changed to false within the case structure so that it is not invoked again. The other two local variables, Initial Const and Delay Index, are used as the inputs of the phase and frequency tracking module. The queue length is chosen to be 51 in order to include the entire marker bits in the queue. This length is calculated as under: 31[(one period of MLS sequence) + 2×10 (frame marker bits)]. The sync and tracking VI for the frame synchronization mode is shown in Figure: 12.

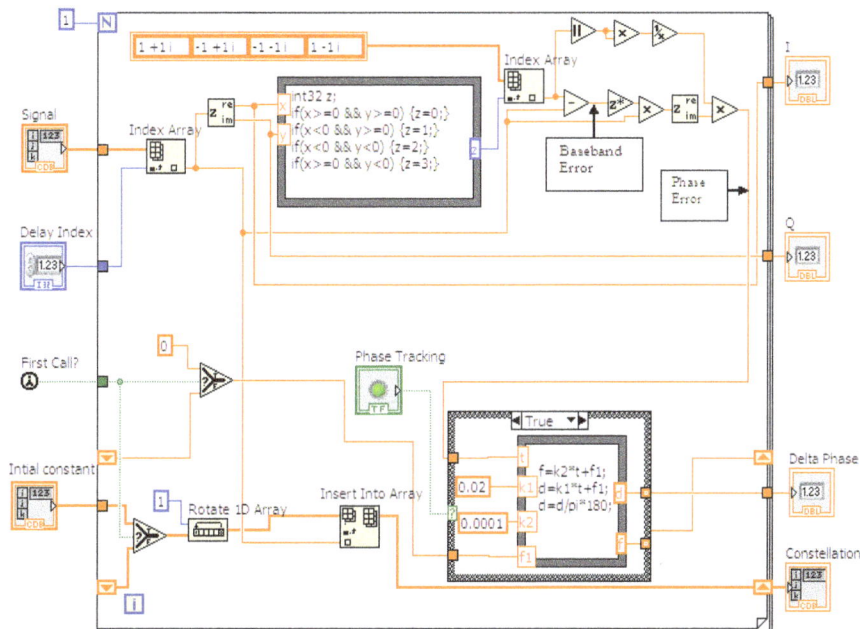

Figure 12. *Sync & Tracking VI.*

3. PSK Transceiver Parameters

Transmitter filters: Transmitter filter defines the type of band-limiting filter employed at the transmitter for pulse shaping the symbols output by the modulator. In this design the user has the option to choose any of the varieties of the filters from the given filters Raised cosine (Nyquist), Square-root raised cosine, Gaussian filters as depicted through Table: 1. Thus, this design makes it a unique SDR where the user has the option to select the required filter and see that which filter gives the minimum BER.

Table 1. Simulation Parameters.

Sl. No.	Parameters That can be Decided by the User	Values Taken by the user
1	PN sequence order	15 or any Value
2	Eb/No	80 dB or any Value
3	Message symbol	1000 or any Value
4	Transmission B.W (BT)	0.5 or any Value
5	Symbol Phase Continuity	Continuous
6	PSK frequency deviation (Hz)	25KHz or any Value
7	Filter used	Gaussian, Root Raised Cosine Filter, Root Raised Cosine Filter
8	Symbol Rate	100.00 KHz or any Value
9	Eb/No Sample	5 or any Value
10	Sample per symbol	16 or any Value
11	Modulation Index	0.5 or any Value
12	BER vs Eb / No (without filter)	None

Raised cosine Filter: The raised cosine filter is one of the most common pulse-shaping filters in communications systems. In addition, it is used to minimize inter symbol interference (ISI).

Root Raised Cosine Filter: The root raised cosine filter at low frequency produces a frequency response with unity gain and complete at higher frequencies.

Gaussian filter: The Gaussian pulse-shaping filter reduces the levels of side-lobes of the PSK & GMSK spectrum.

4. Bit Error Rate (BER) & Signal-to-Noise Ratio (SNR)

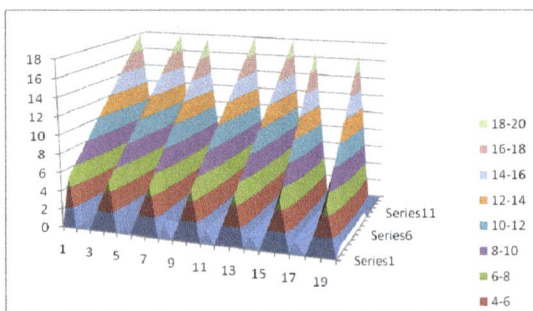

Figure 13. BER Vs $E_b/N_0(db)$ (4,8,16,32,64,128,256 bit FSK) Output Results for Convolution coding.

Bit Error Rate (BER): In this section we discuss the BER Vs the SNR achieved for different M-PSK in a noisy channel for both the Convolution and Turbo coding. The bit error rate (BER) is the number of bit errors divided by the total number of transferred bits during a considered time interval. BER is a unit less performance measure which is often expressed as a percentage (%). A pseudorandom data sequence (15) is used for the analysis in this design. The BER parameter represents the current operating BER of a specific modulation type and in this design the modulation scheme selected is M-PSK. This value depends on various channel characteristics, including the transmit power and noise level.

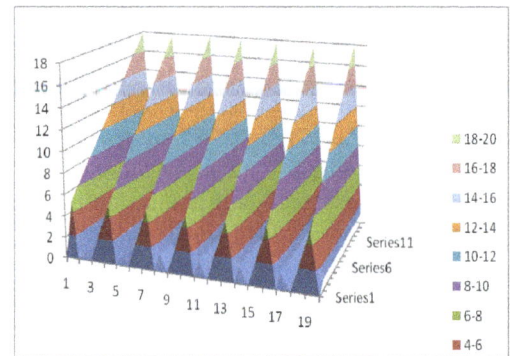

Figure 14. BER Vs $E_b/N_0(db)$ (4,8,16,32,64,128,256 bit FSK) Output Results for Turbo coding.

5. Simulation Results & of PSK Transceiver Using Labview

In this section we describe the simulation results of M-PSK transceiver system for a noisy channel. BER Vs $E_b/N_0(db)$ for (4,8,16,32,64,128,256 bit PSK) has been given in Figure 13 & Figure 14. Output Results for Convolution coding and Turbo coding has been been illustrated with the PSK parameters for Simulation being described in Table: 1. By taking a look at the output results we can very clearly say that Turbo coding gives a much improved and better minimization of the data errors than the Convolution coding. The simulation results conclude that minimum BER achieved using Turbo coding is in the range of 10^{-8} as compared to that of Convolution which is in the range (10^{-7}) at a particular value of SNR. Hence, even at larger values of

SNR, the BER achieved is extremely small. With the help of this design we can also show that how fast and effectively we can build a PSK transceiver for Software Defined Radio.

6. Conclusion

In this section we discuss the simulation results of the M-PSK transceiver VI for noisy channel. From the results it becomes clear that the wireless system designed based on PSK technique provide high data rate and SNR. This can be very clearly seen in terms of the BER Vs E_b/N_o output graph. We can also see very clearly with these results that data errors can be minimized using coding techniques, which in turn improves the Signal to noise ratio (SNR) further, we can also say looking at the results that Turbo coding gives a much improved and better minimization of the data errors that the Convolution & Viterbi coding. The performance of M-level PSK systems (4,8,16,32,64,128,256) for additive white Gaussian noise channel has been evaluated and compared on the basis of the simulations in LabVIEW as shown in Figure 13 & Figure 14. In this paper we have shown that how fast and effectively we can build a PSK transceiver for Software Defined Radio. We have used the Graphical programming language LabVIEW for building a PSK transceiver system which consists of a message source, a pulse shape filter, a modulator on the Transmitter section and demodulator, a frame synchronizer, a phase continuity and frequency deviation on the Receiver section. With the help of LabVIEW an interactive Software Defined Radio system has been built in a shorter time as compared to other text-based programming languages. With the help of this design we are able to see and prove that data errors can be minimized using coding techniques, which in turn improves the Signal to noise ratio (SNR). Also we can say by looking at the results that Turbo coding gives a much improved and better minimization of the data errors than the Convolution coding. In the end, we can say that the signal can be recovered with very less probability of error in Turbo coding than in Convolution coding with the increase in the M (number of levels) at the destination.

References

[1] J. Mitola III,"Software Radios –Survey, Critical Evaluation and Future Directions," in Proc. National Telesystems Conference, 1992, pp. 13/1513/23.

[2] Matthew N. O. Sadiku and Cajetan M. Akujuobi "Software-defined Radio: A brief Overview", IEEE Potentials Journal, October/November 2004, pg. 14-15.

[3] Wipro Technologies Innovative Solutions, Quality Leadership "Software-Defined Radio" White Paper: A Technology Overview, August 2002.

[4] Nikhil Marriwala, O. P. Sahu, Anil Vohra,: "8-QAM Software Defined Radio Based Approach for Channel Encoding and Decoding Using Forward Error Correction", Wireless Personal Communications, 1st May-2013, Springer US, 10.1007/s11277-013-1191-z.

[5] Nikhil Marriwala, O. P. Sahu, Ritu Khullar and Anil Vohra, "Software Defined Radio (SDR) 4-bit QAM Modem using LabVIEW for Gaussian Channel"CIIT International Journal of Wireless Communication". March 2011.

[6] C. Berrou, A. Glavieux, and P. Thitimajshima. Near Shannon limit error correcting coding and decoding: Turbo codes. In Proceedings of the IEEE International Conference on Communications, Geneva, Switzerland, May 2003.

[7] W. Tuttlebee, Software Defined Radio: Baseband Technologies for 3G Handsets and Base Stations, John Wiley & Sons, 2004.

[8] Friedrich K. Jondral "Software-Defined Radio—Basics and Evolution to Cognitive Radio" (EURASIP Journal on Wireless Communications and Networking 2005:3, 275–283).

[9] N. KIM, N. KEHTARNAVAZ, and M. TORLAK LabVIEW-Based Software-Defined Radio: 4-QAM Modem Proceedings of ICASSP, vol. 2, 2006, pp. 985-988.

[10] Eric Nicollet and Lee Pucker, "Standardizing Transceiver APIs for Software Defined and Cognitive Radio", www.rfdesign.com, February 2008,

[11] P. Burns, "Software Defined Radio for 3G", Artech House, 2002. ISBN 1-58053-347-7.

[12] Amanpreet Singh Saini, "The Automated Systems For Spectrum Occupancy Measurement And Channel Sounding In Ultra-Wideband, Cognitive, Communication, And Networking" Master of Science in Electrical Engineering, August 2009.

[13] RituKhullar, Sippy Kapoor, Naval Dhawan, "Modulation technique For Cognitive Radio, Applications", International Journal of Engineering Research and Applications (IJERA) ISSN: 2248-9622 www.ijera.com Vol. 2, Issue 3, May-Jun 2012, pp. 123- 125.

[14] Hiroyasu Ishikawa, "Software Defined Radio Technology for Highly Reliable Wireless Communications," *Wireless Personal Communications*, 64 (2012), 461–72 dx.doi.org/10.1007/s11277-012-0596-4.

[15] P. Prakasam and M. Madheswaran, "Intelligent Decision Making System for Digital Modulation Scheme Classification in Software Radio Using Wavelet Transform and Higher Order Statistical Moments," *Wireless Personal Communications*, 50 (2008), 509–28 ,dx.doi.org/10.1007/s11277-008-9621-z.

[16] Ying Chen and Linda M. Davis, "A Cross-Layer Adaptive Modulation and Coding Scheme for Energy Efficient Software Defined Radio," *Journal of Signal Processing Systems*, 69 (2011), 23–30,dx.doi.org/10.1007/s11265-011-0644-4.

[17] Shu-Ming Tseng, Yueh-Teng Hsu and Hong-Kung Lin, "Iterative Channel Decoding for PC-Based Software Radio DVB-T Receiver," *Wireless Personal Communications*, 69 (2012), 403–11, dx.doi.org/10.1007/s11277-012-0580-z.

16

A Compact Model of Mosfet Transistors Including Dispersion and Thermal Phenomena

Mohamed Ali Belaïd[1, 2], Ahmed Mohammad Nahhas[1], Momamed Masmoudi[3]

[1]ECED, CEL, Umm-Al-Qura University, Makkah, Saudi Arabia
[2]SAGE-ENISo, University of Sousse, Tunisia
[3]GPM-UMR CNRS, University of Rouen, Saint Etienne du Rouvray, France

Email address:
mohamedali.belaid@yahoo.fr (M. A. Belaïd)

Abstract: This paper propose a new electro-thermal model nd presents a method of studying the thermal phenomena in power MOSFET transistors, utilizing Advanced Design System techniques by a Symbolic Defined Device (SDD). The model incorporates the thermal effects and the temperature evolution in the device and captures the heat dissipation from the silicon chip to the ambient air by providing three thermal capacitances and three thermal resistances (thermal network). It enables a better estimation of the device's reliability and lifetime. Furthermore, it can be used to make a connection between the electrical parameter drifts and the existing failures types. The developed model reflects superior performance in terms of accuracy and flexibility and the results obtained indicate a good agreement with the operating conditions.

Keywords: Modelling, Electro-Thermal, Power MOSFET, Temperature, Self-Heating, Thermal Network

1. Introduction

Electrical characteristics of semiconductors are sensitive to changes in temperature, in particular power devices. The study device is used in different application areas; like telecoms (base station), radars and space activities. The continuing trend towards devices miniaturization makes the consideration of the thermal aspects very critical, both in the design and operation [1-2]. The technology evolution puts stringent requirements on performance and hence requires powerful characterization and simulation tools that can adequately assess the quality and reliability of the produced devices in different operating modes.

We present an electro-thermal model capable of evaluating the temperature effects of a power MOSFET. The temperature effects can limit the lifetime of MOSFETs and play an essential role in the failure mechanisms [3-5]. These effects have been satisfactorily modelled in previously works by Miller et al [6], based on analytical expressions, and by Angelov et al [7] Chalmers model. In this paper we present an electro-thermal model capable of capturing the thermal dependencies and yields superior performance in terms of modelling accuracy and flexibility.

This paper is organized as follows: Section II describes our electro-thermal model. The results and discussion are presented in Section III. Conclusion and perspectives are given in Section IV.

2. Electro-thermal Model

By capturing the thermal dependencies of the power MOSFET our modelling approach can provide a fuller characterization that evaluates the electro-thermal interactions during operation at different levels of the component (silicon chip to the ambient air). An empirical non-linear model is shown in Figure 1. This was implemented using the symbolic device (SDD) in the ADS software.

The power MOSFET model is composed of several passive linear and non-linear components, non-linear drain-source current generator dependent on gate-source and drain-source voltages and a thermal circuit as shown in Figure 1.

The current model of proposed channel is formulated as follows [8-9]:

$$I_{ds} = Beta \cdot \left(V_{gm}^{Vg\exp}\right) \cdot \left(1 + Lambda \cdot V_{ds}\right) \cdot \tanh \cdot \left[\frac{\left(V_{ds} \cdot Alpha\right)}{V_{gm}}\right]$$

$$\cdot \left[1 + k_1 \cdot \exp \left(V_{breff1} \right) \right] \quad (1)$$

The thermal network is a low-pass filter with the total thermal resistance R_{th} correlated with static characteristics (chip-ambient air). The total heat capacity C_{th} represents the dynamic characteristics of the heat flow inside the transistor. The static temperature of the channel may be written as below: [10]:

$$\Delta T = R_{th} \cdot P_{diss} + T_a \quad (2)$$

$$P_{diss} = I_{ds} \cdot V_{ds} \quad (3)$$

Where ΔT is the junction temperature, P_{diss} is the instantaneous total power absorbed in the transistor, and T_a the ambient temperature. R_{th_CP}- the thermal chip-package resistance (a technological characteristic specific to a transistor and given by the manufacturer), R_{th_PH}- the thermal package-heat sink resistance (referring to a conduction transfer; this resistance can be decreased by improving the contact between the package and the heat sink surface, by using silicone oil), R_{th_HA}- the thermal heat sink-ambient air resistance (depending not only on the size, form and structure of the heat sink, but also on its orientation and on the air stream flowing around it).

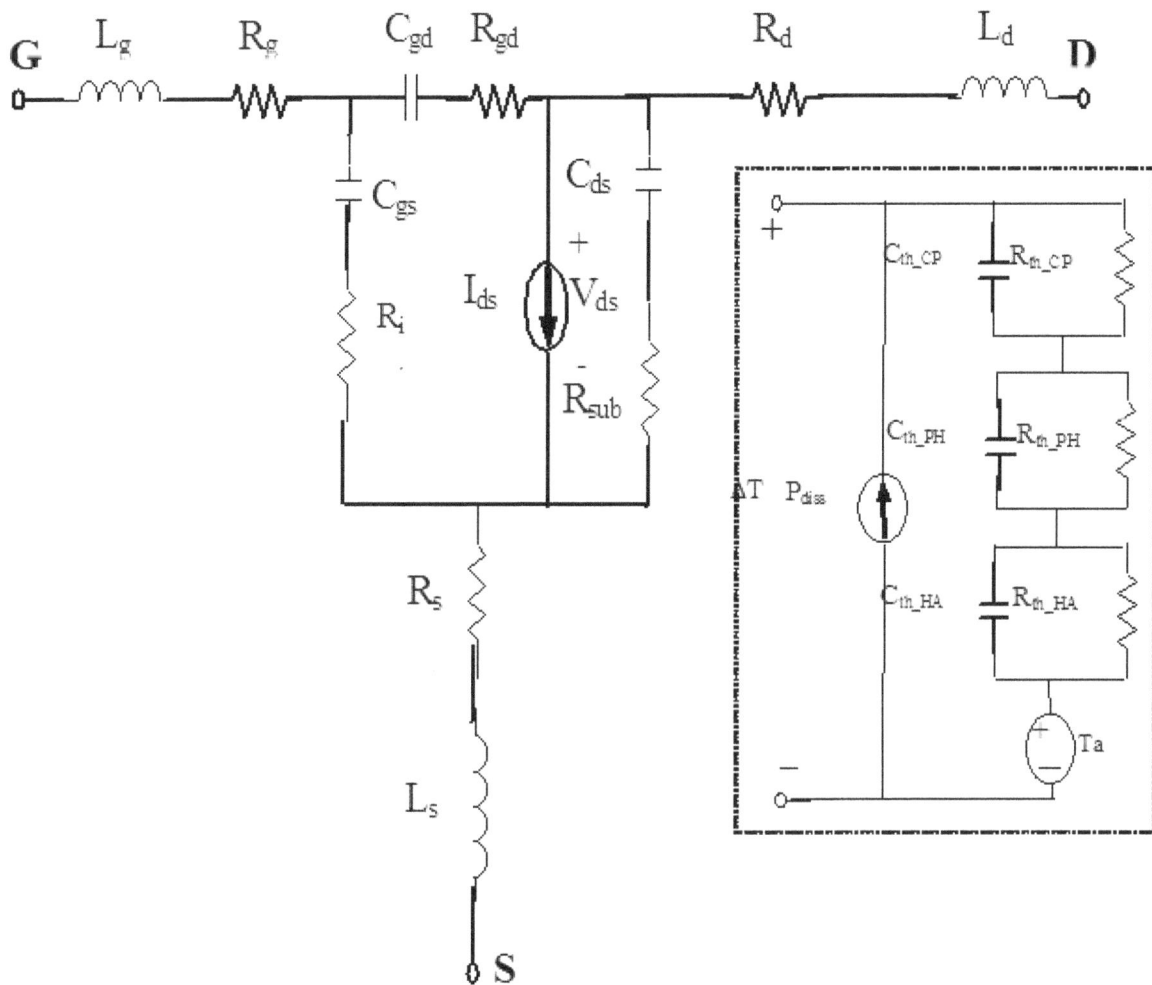

Fig. 1. *Large-signal equivalent circuit of power MOSFET model with a thermal network.*

Non-linear functions of capacity C_{gs} (gate-source), C_{gd} (gate-drain) and C_{ds} (drain-source capacitance) are used in this model with linear temperature dependence [6]:

$$C_{gd}(T) = \left[C_{gd1} + \frac{C_{gd2}}{1 + C_{gd3} \cdot \left(V_{gd} - C_{gd4} \right)^2} \right] \cdot \left(1 + C_{gdT} \cdot \Delta T \right) \quad (4)$$

$$C_{ds}(T) = \left(C_{ds1} + \frac{C_{ds2}}{1 + C_{ds3} \cdot V_{ds}^2} \right) \cdot \left(1 + C_{dsT} \cdot \Delta T \right) \quad (5)$$

The electro-thermal model will be used as a reliability tool that facilitates linking the change in the transistor electrical parameters (after accelerated aging tests at different temperatures or functioning) and the degradation phenomenon.

3. Results and Discussions

Fig. 2 (line) shows the instability and the thermal runaway which occurred due to temperature effects. To limit and decrease this problem and to improve the measurement conditions, a solution which proves to be a precious working tool would make it possible to thermal protects the component. It consists of mounting the component on a heat dissipator. Thanks to this solution, we can obtain more thermally stable measurement conditions (Fig. 2, dashed). The importance of cooling systems like the heat sink and the Peltier module could be observed.

The model of Figure 1 was implemented using the symbolic defined device (SDD) in ADS software. The results obtained are discussed next.

Fig. 3 shows the variations of the channel current from the transistor output characteristic with changing parameter values (Alpha, Beta and V_{br}). Each parameter has a unique role in describing the channel current, with a slight interdependence between them. The different parameters play on the shape in order to fit the measured curve.

The I-V characteristics of the transistor in various conditions (chip, package, heat sink) and the fitted results are in good agreement, as has been presented in Fig. 4.

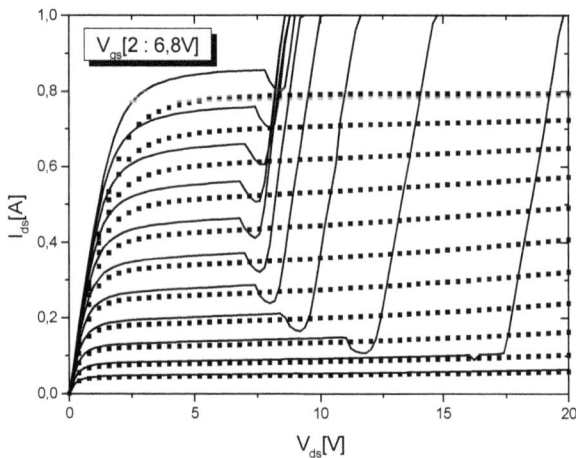

Fig. 2. Output characteristics: Thermal instability effect (line) and stable temperature conditions (dashed).

Figure (5) shows the channel current variation at different temperatures. It is evident that channel current I_{ds} vary linearly with temperature at low current levels. This is due to the threshold voltage reduction [11].

Indeed, for a MOSFET, the current relative dependence on temperature is linked to two contradicting phenomena. The current I_{ds} is proportional to the mobility μ and to the electrons concentration in the n-channel [12]. When the temperature increases, n increases while μ decreases. For low current values, the decrease effect in mobility is negligible compared to the increase in the bearer's number and thus, the current value increases [13]. On the other side, when the current value is important (high current level), the mobility effect becomes dominant, hence resulting in a decrease in current with temperature [11] [14].

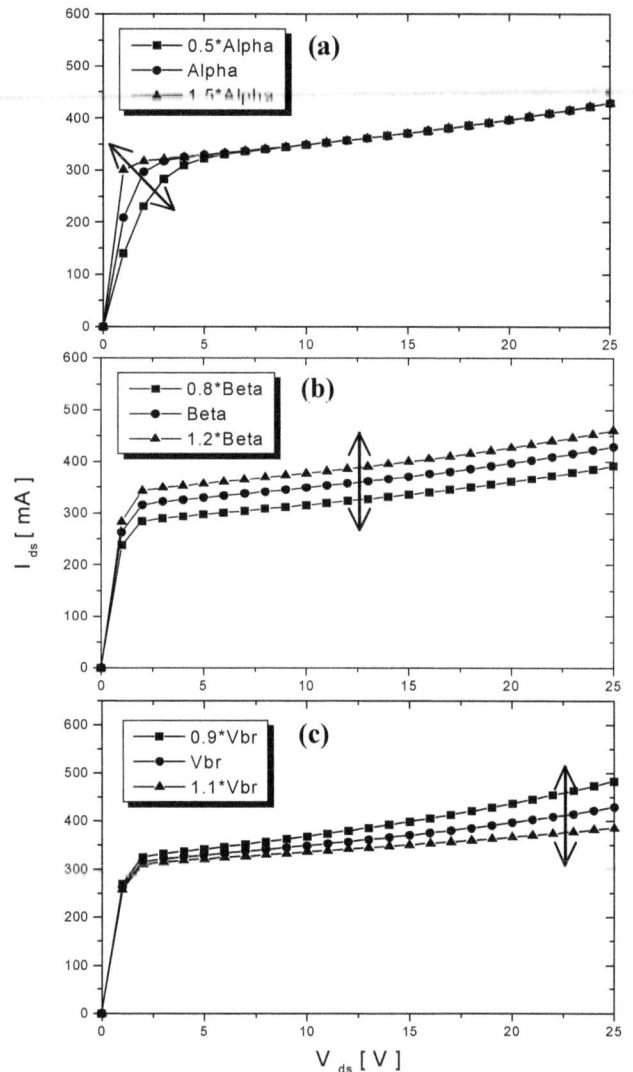

Fig. 3. Calculated channel current variation versus drain voltage with varying parameter values for: (a) Alpha, (b) Beta and (c) Vbr.

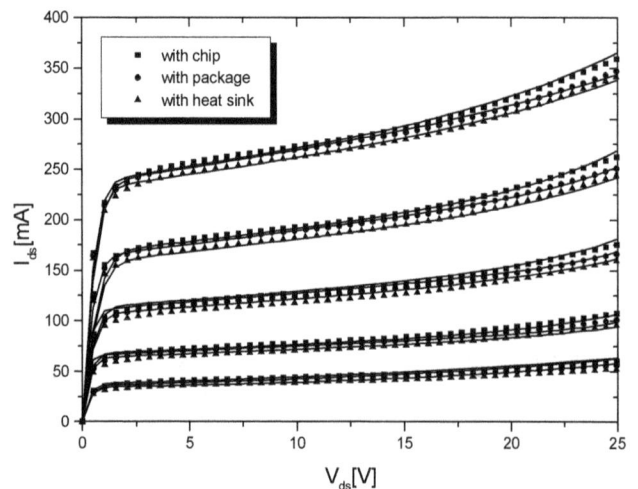

Fig. 4. Measured (dashed) and modelled (line) output characteristic with various conditions, Vgs=[3: 6V], step=600mV.

A more detailed study on the effects of the MOS physical and technologic parameters shows that the threshold voltage decreases with temperature [6]. This variation with temperature is linear and can be represented as follows [6]:

$$V_t = V_{T0} + (V_{tT} \cdot T_j) \tag{6}$$

Where T_0 is the ambient temperature, V_{T0} is the threshold voltage at T_0, V_{tT} is the coefficient of the equation and T_j is the junction temperature.

The threshold voltage is a parameter strongly dependent on temperature, as shown in Figure 6, and affects the performance of the power MOSFET. We note that the threshold voltage decreases with temperature. The change of this voltage is mainly caused by the change of Fermi potential Φ_F with temperature [12].

The temperature effect was observed and modelled, as shown in Figure 7. It occurs and increases when a voltage is applied to the device [10]. The channel current is proportional to the drain voltage and gate voltage. This illustrates the consistency of our electro-thermal model as any additional current passing in the transistor yields a significant increase in temperature.

The resistance in the conducting state, which we believe, is defined by the ratio of drain voltage to the drain current linear [12], and is inversely proportional to the later.

$$R_{ds\_on} = (V_{ds} / I_{ds})_{V_{ds \to 0}} \tag{7}$$

The expression for the instantaneous junction temperature of the transistor was developed by making use of the existing duality between heat transfer and electrical phenomena summarized in Table 1. The main significant parameters of the model are shown in Table 2.

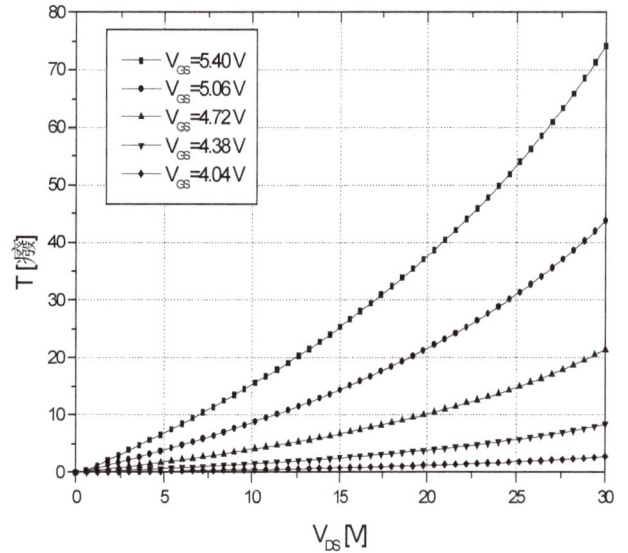

Fig. 7. Simulation of the temperature evolution inside the component.

The on-resistance decreases when the gate-source voltage increases [15]. The parameter R_{ds_on} is dependent on the distribution and dissipation of heat, which represents the heat flow in the transistor [10]. It also reflects the different values of the junction temperature. We note that R_{ds_on} decreases with temperature in the case of low polarization (low value of V_{gs}), and increases for high polarizations [15].

The C-V characteristics of the transistor in various temperatures and the fitted results form IC-CAP plot optimiser are in good agreement, as presented in Fig. 6 and 10.

The figure 10 shows the modelling feedback capacitance (C_{rss}) at different temperatures. It is deduced from the intrinsic capabilities with the following combinations [12]:

$$C_{iss} = C_{gd} + C_{gs} \tag{8}$$

$$C_{oss} = C_{gd} + C_{ds} \tag{9}$$

$$C_{rss} = C_{gd} \tag{10}$$

The final result of dynamic modelling is illustrated in Figures 8,9 and 10, where the superimposition is very good (error rate lower correlation to 2%).

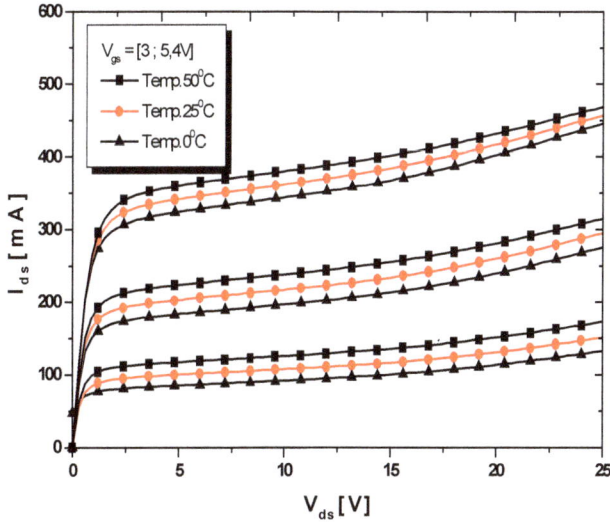

Fig. 5. Variation of low levels of channel current at different temperatures.

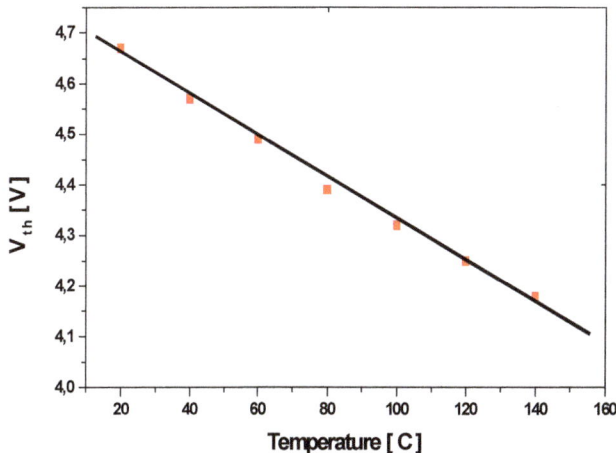

Fig. 6. V_{th} variation as a temperature function.

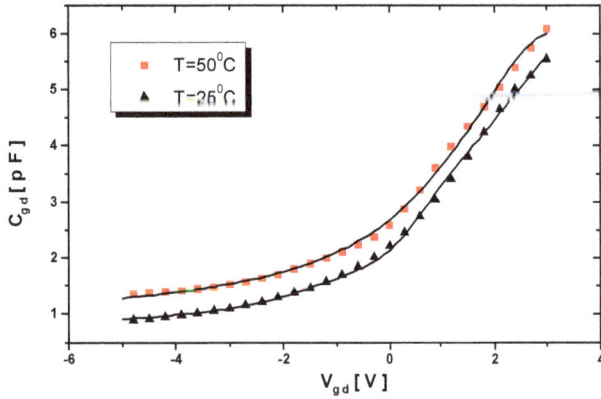

Fig. 8. *Gate-drain capacitance at different Ta, with Freq=1 MHz: Measured (dashed) and modelled (line)*

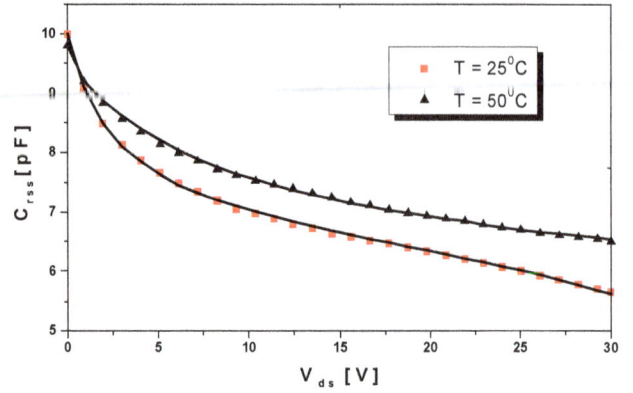

Fig. 9. *C_ISS, C_RSS and C_OSS capacitances measured (dashed) and modelled (line) for $V_{DS} = 0$ V-26 V, with Freq=1 MHz.*

The simulation results show good dynamic performance prediction of MOS device at different temperatures. Physically, the variation of that could be explained by the effect of the temperature on the Fermi level, which affects thereafter the width of the space charge zone and induces the variation of capacitance value.

Table 1. *Thermal and Electrical quantities equivalence.*

	Thermal quantity		Electrical quantity
P_{diss}	Power heat flow (W)	I	Current flow (A)
T_j	Temperature (K)	V	Voltage (V)
R_{th}	Thermal resistance (K /W)	R	Electrical Resistance (Ohm)
C_{th}	Thermal capacitance (J/W)	C	Electrical capacitance (F)

Table 2. *Model parameters.*

Parameter	Definition	Unit
Beta	Transconductance parameter	Siemens
V_t	Threshold voltage	V
Delta	Vt variation according to V_{ds}	V
V_{br}	Break down voltage	V
$K_{1/2}, M_{1/3}$	Break down parameters	--
V_K	Ids equation coefficient	V
Alpha	Linear range	$1/\Omega$
Gamma	Slope of the channel current	--
Lambda	Shapes of the Ids saturation	1/V
V_{gexp}	Term of power	--

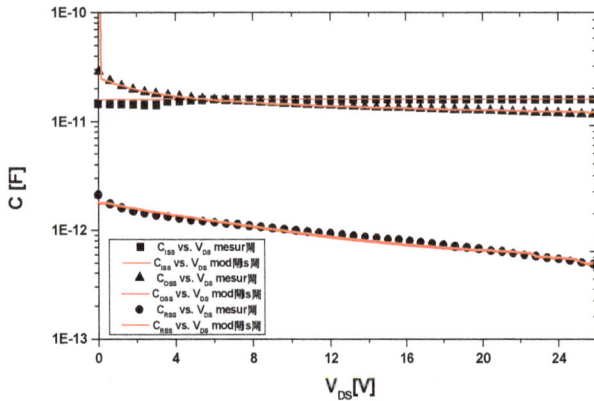

Fig. 10. *Crss capacitance at different Ta, with Vds= [0 V; 30 V] and Freq=1 MHz: Measured (dashed) and modelled (line).*

The integration of the electric field on all the thickness of the deserted zone gives the tension value of the barrier potential [16]:

$$V_b = q\left(N_A x_p^2 + N_D x_n^2\right)/2\varepsilon \qquad (11)$$

As we know that W = x_p + x_n:

$$W = \sqrt{\frac{2\varepsilon}{q}\left(\frac{1}{N_A} + \frac{1}{N_D}\right)V_b} \qquad (12)$$

$$C_j(0) = \frac{\varepsilon S}{W} = \frac{\varepsilon S}{\sqrt{\frac{2\varepsilon}{q}\left(\frac{1}{N_A} + \frac{1}{N_D}\right)V_b}} \qquad (13)$$

The deserted zone between the two conducting zones is similar to a plane condenser of S section (junction surface) such as shown in equation (13). When the operating temperature of the junction increases, E_g varies, therefore the height of the potential barrier decreases. Thereafter the capacitance increases.

Indeed, each defect type has activation energy i.e. a temperature from which it becomes active and acts on the device electric behaviour, and it has an important effect on the capacitance value [17-18].

In the last few days, we got the results of the temperature effect on the S-parameters and the evolution of electroma-gnetic interference in an application of static converters series chopper. Our aim is to make varying the junction temperature of the power transistor in the chopper circuit, while applying different temperatures to achieve a high junction temperature (T_j), which exceeds the value described by the constrictor. A test of the electromagnetic interference evolution will be determined.

4. Conclusion and Perspectives

The electric macroscopic parameters of the MOSFET (threshold voltage, channel current, transconductance R $_{ds-on}$, and capacitances) are highly sensitive to temperature variation.

The temperature increase can modify transistor behaviour and cause irreversible degradation of its performances. Therefore, the use of cooling systems is very essential to enhance measurement conditions of the device and to obtain better heat dissipation. Electro-thermal model that takes into account self-heating and the temperature effects has been developed for power MOSFETs. The model has been used to better understand the physical dynamics responsible of the degradation of performance. An increase in temperature can alter and degrade the transistor behaviour irreversibly, hence the importance of controlling the cooling systems and improving techniques for temperature measuring to better evaluate the thermal effects. Different levels of the system have been modelled successfully and the thermal dependencies have been properly captured. More heat cells can be added to better fit and reflect the heat dissipation in the device. Future work will focus on the S-parameters simulation for a complete compact model and studying the impact of temperature on electromagnetic interferences.

References

[1] Z. Radivojevic, K. Andresson, J. A. Bielen, P. J. van der Wel, J. Rantala, "Operating limits for power amplifiers at high junction temperatures," Microelectronics Reliability. 44 (2004) 963-972.

[2] Juraj Marek, Ales Chvála, Daniel Donoval, Patrik Príbytný, Marián Molnár, Miroslav Mikolášek, "Compact model of power MOSFET with temperature dependent Cauer RC network for more accurate thermal simulations," Solid-State Electronics 2014, Volume 94, pp. 44–50.

[3] M. Riccioa, A. Castellazzib, G. De Falcoa, A. Irace, "Experimental analysis of electro-thermal instability in SiC Power MOSFETs," Microelectronics Reliability 2013, volume 53, pp. 1739–1744.

[4] C. Kun-Ming, H. Guo-Wei, W. Sheng-Chun, Y. Wen-Kuan, F. Yean-Kuen, Y. Fu-Liang, Characterization and modelling of SOI varactors at various temperatures, IEEE Trans. Electron Devices. 51 (3) (2004) 415-420.

[5] M.N. Sabry, W. Fikry, K. Abdel Salam, M. M. Awad and A. E. Nasser, A lumped transient thermal model for self-heating in MOSFETs, Microelectronics Journal. 32 (2001) 847-853.

[6] M. Miller, T. Dinh, E. Shumate, "A new empirical large signal model for silicon RF LDMOSFET's," IEEE MTT-S Technol. Wireless Applicat. Dig. (1997) 19-22.

[7] I. Angelov, N. Rorsman, J. Stenarson, M. Garcia, H. Zirath, "An empirical table-based model," IEEE Trans. Microwave Theory Tech. 47 (1999) 2350-2357.

[8] Y. Yang, Y.Y. Woo, J. Yi, B. Kim, "A new empirical Large-Signal model of Si LDMOSFETs for High-power amplifier design," IEEE Trans. Microwave Theory Tech, 49 (9) (2001) 1626-1633.

[9] Motorola on line, "Motorola's Electro Thermal (MET) LDMOS Model MosnldMet", 2003.

[10] M. A. Belaïd, H. Maanane, K. Mourgues, M. Masmoudi, K. Ketata, et J. Marcon, "Characterization and Modelling of Power RF LDMOS Transistor Including Self-Heating Effects," 16th Int. Conf. Microelectronics (IEEE), Tunis, Tunisie (2004), 262-265.

[11] Y. Yang, Y. Y. Woo, J. Yi, B. Kim, "A New Empirical Large-Signal Model of Si LDMOSFETs for High-Power Amplifier Design," IEEE Trans. Microwave Theory Tech., vol. 49, pp. 1626-33, Sep. 2001.

[12] D. Moncoqut, "Propriétés physiques et modélisation du transistor de puissance LDMOS," Thèse de doctorat, Université Paul Sabatier de Toulouse, Octobre 1997.

[13] P. Aloïsi, "Les semiconducteurs de puissance de la physique du solide aux applications," Ellipses édition marketing S.A., 2001.

[14] M. A. Belaïd, K. Ketata, K. Mourgues, H. Maanane, M. Masmoudi, J. Marcon, "Comparative analysis of accelerated ageing effects on power RF LDMOS reliability," Microelectronics Reliability, vol. 45, pp. 1732–1737, 2005.

[15] M. A. Belaïd, "Contribution à l'analyse des dégradations d'origine thermique et des interactions électrothermiques dans les dispositifs LDMOS RF de puissance," Thèse de doctorat, Université de Rouen, décembre 2006.

[16] B.J. Baliga, Modern Power Devices, General Electric Company Schenectady, John Wiley & Sons, 1987.

[17] M.A. Belaïd et al, "Analysis and Simulation of Self-Heating Effects on RF LDMOS Devices", in Proc. SISPAD (IEEE), Tokyo, 2005.

[18] H. A. Jayatissa, L. Zhiyu, Effect of temperature on capacitance-voltage characteristics of SOI, Materials Science and Engineering B, (2005) 331-334.

[19] M. Tlig et al, Conducted and radiated EMI evolution of power RF N-LDMOS after accelerated ageing tests, Microelectronics Reliability, 2013, pp. 1793–97.

[20] M.A. Belaïd *, K. Ketata, M. Gares, J. Marcon, K. Mourgues, M. Masmoudi, 2-D simulation and analysis of temperature effects on electrical parameters degradation of power RF LDMOS device, Nuclear Instruments and Methods in Physics Research B, 2006, pp. 250–25.

17

Diagnosis of Water Tree Aging in XLPE Cable by the Loss Current Harmonic Component Under Variable Frequency Power

Chen Jie[1, *], Li Hongze[2], Zhou Li[1], Hu Libin[1], Li Chenying[1], Cao Jingying[1]

[1]State Grid Jiangsu Electric Power Research Institute, Nanjing, China
[2]State Grid Jiangsu Electric Power Company, Nanjing, China

Email address:
15105161377@163.com (Chen Jie)

Abstract: Water treeing is one of the main aging in XLPE cable insulation, which has the direct effect on the reliability of cable long-term operation. The diagnosis of water tree aging is a premise to ensure the safe operation of cables. The water treed XLPE insulation shows a nonlinear conductivity characteristic, which leads to a harmonic component in the corresponding lose current under the standard sinusoidal voltage. So the loss current harmonic component method can be an effective diagnosis of water tree uging. The loss current measurement system based on the high voltage current comparator bridge is used in this paper, which is designed by Harbin University of Science and Technology and Jiangsu Electric Power Company Research Institute. The feasibility of the application of variable frequency series resonant power source in the loss current measurement, which is aimed to diagnose the water treeing in XLPE insulation, is studied in the laboratory and field test. The results show that the output voltage waveform of the variable frequency series resonant power source is quite close to the standard sinusoidal voltage, although there are some high frequency sharp peaks. So the variable frequency series resonant power source can be used as the high voltage test power supply in the loss current harmonic component measurement system. The loss current harmonic component has power frequency interference during the field test, but this can be solved by differential frequency principle. The microscopic observation of water tree structures during the anatomy of the cable not only indicates the effectiveness of the results and diagnosis during the field test, but also proves the feasibility of the application of variable frequency series resonant power source in the loss current measurement. The existence and degree of the water tree can be estimated according to the test results, which can be a strong basis of the standardization of water tree aging diagnosis. Furthermore, the expensive standard sinusoidal wave generator and linear power generator is replaced by the variable frequency series resonant power source, which reduces the cost and weight of the testing system.

Keywords: Loss Current Harmonic Component Test System, Variable Frequency Resonant Power Source, Water Tree Aging Diagnosis, Field Test

1. Introduction

XLPE power cable is widely used due to its large transmission capacity, good insulation performance, and easy installation and maintenance. But the unplanned shutdown accidents of cable increase because of aging. And the water tree aging is one of the main aging in XLPE cable insulation, which has the direct effect on the reliability of cable long-term operation[1]. When the insulation material is suffered a long-term alternating electric field and humidity, the chemical component and micro structure may be changed, and finally lead to the irreversible degradation, which is called water tree aging. The water tree is a permanent degradation and can grow under a low electric field and a very few humidity. The diagnosis of water tree aging is essential to insure the safety of the cable line[2]. Therefore, the study of water tree aging diagnosis in XLPE insulation is always a hot research topic[3, 4].

The vehicular loss current harmonic component test system was developed in Japan, and the system was applied in 7 high voltage cable lines successfully, which proved the feasibility of the test method and formed the corresponding test standard. The output voltage of the system is 0~20 kV (50 Hz), and the maximum load is 1 μF. In the test system, a standard sinusoidal voltage is first produced by the signal generator, and then enlarged by the linear power amplifier to become a high power voltage signal which can drive the experimental transformer. Finally, the high voltage standard sinusoidal wave is generated by the experimental transformer, which is the voltage source for tests[5, 6]. However, the linear power amplifier is very expensive, and the equipment weight is 4 tons. These are harmful to the universal application of the test system. So, the feasibility of the application of variable frequency series resonant power source in the loss current harmonic component measurement system is studied in this paper.

2. Testing Principle and Equipment

2.1. Loss Current Harmonic Component Testing Principle

When the insulation structure is describe by the equivalent circuit, the loss current has the same phase with the applied voltage, and the capacitance current is 90 degrees ahead of the applied voltage. The relationship among the loss current, the capacitance current and the total current is shown in Figure 1.

The loss current harmonic component is the loss current harmonic of each frequency after the Fourier transform. The conductivity of the non-aging XLPE cable insulation is linear, so the loss current is not distorted and has no harmonic component when the standard sinusoidal voltage is applied, and the dissipation factor is small. But when the XLPE cable is water treed, the conductivity is nonlinear because of water and the low molecular degradation products, so the loss current is distorted and the harmonic component is detected.

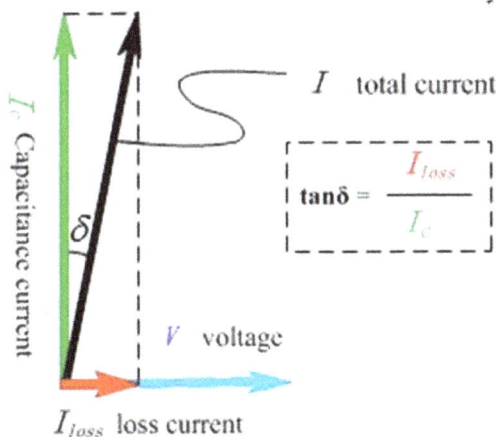

Figure 1. *Vector relationship among the loss current, the capacitance current, the total current and the applied voltage.*

The loss current harmonic component test results of water treed XLPE samples with varying degrees in the laboratory are shown in Figure 2.

Studies both at home and abroad show that: 1) when the dissipation factor increases and the loss current has no harmonic component, the dissipation factor has a certain relationship with the thermal aging degree. 2) when the dissipation factor increases and the loss current has a third harmonic component, the third harmonic component is the indication of water treeing, the corresponding amplitude is related to the quantity of water tree (the area of water trees along the cable length direction), and the phase is related to the length of water tree (the depth of water tree in the cable radial direction). Thus the measurement of the loss current harmonic component can be used to diagnose the water treeing degree in XLPE cable insulation.

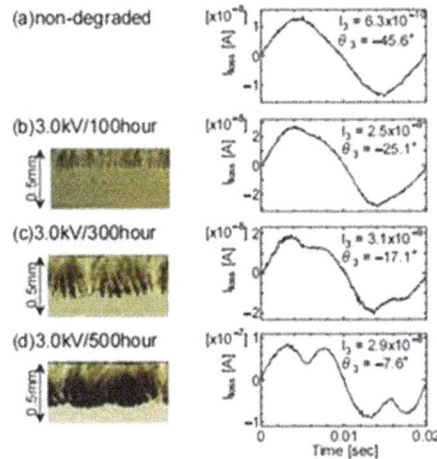

Figure 2. *Loss current harmonic component test results of different water tree aging XLPE sample.*

2.2. Loss Current Harmonic Component Testing Equipment

The loss current harmonic component testing equipment are mainly composed by the high voltage power system, the high voltage current comparator bridge, the high voltage standard capacitor, the zero current detector, the capacitance voltage divider, the data acquisition and processing circuits, the LCD displaying circuit, the filter and amplifier circuits, the oscilloscope, and the computer[7-10]. The schematic diagram of test equipment is shown in Figure 3.

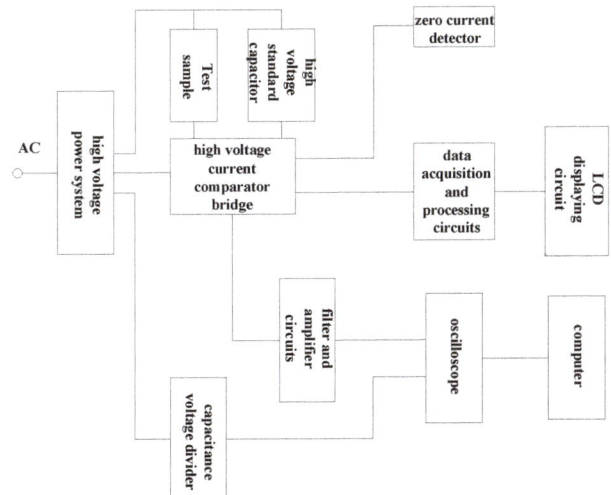

Figure 3. *Schematic diagram of the loss current harmonic component test equipment.*

The test system used in this paper is based on the comparison method, and the high voltage current comparator bridge is a core part. The loss current measurement of the test sample is implemented by the electromagnetic induction and ampere turn balance law. The high voltage standard capacitor is the comparison standard, which used to get the loss current information by comparing with the current through the test sample. The zero current detector is the basis of the balance of high voltage current comparator bridge. The output signal is captured and processed by the data acquisition and processing circuits. Then the dielectric loss tangent (tanδ), the capacitance, the test voltage and the current through the standard capacitor are displayed by the LCD displaying circuit. The loss current is filtered and amplified by the the filter and amplifier circuits, which is observed with the output voltage signal of power source by the oscilloscope. The loss current harmonic data is obtained and analyzed by the computer.

The connection mode of the test system can be forward or backward connection, which are fit for laboratory and field test separately. In the forward connection, the high testing voltage is supplied by the common high voltage testing transformer, and the loss current is imported to the current comparator from the low voltage side of the sample, which is connected with the proportional coil. The laboratory test for short cable (small capacitance sample) can be carried out by adjusting the proportional coil. In the backward connection, the high voltage power source connected with the testing cable through the current comparator by the coaxial shielded high voltage cable, and the loss current is imported to the current comparator from the high voltage side. In this situation, the current comparator is single-turn cored structure, and the field test for long cable can be carried out.

The loss current testing equipment is not only able to measure the dielectric loss tangent (tanδ), the capacitance, the test voltage, the current through the standard capacitor, the loss current and the loss current harmonic component, but also can observe the waveform of the test voltage and the loss current. The adjustable parameters during the test are mainly the test voltage U_0, the sensitivity η, the amplification ratio K_u and the resonant frequency f.

3. Feasibility Verifying of the Variable Frequency Series Resonant Power Source

3.1. Establishment of the Verification Program

The following program is established to verify the feasibility of the variable frequency series resonant power source in the diagnosis of water tree aging. First, the test system with the variable frequency series resonant power source is used in the laboratory test to see the feasibility. Then, the test system is used in the field test as the key task to diagnose the cables with water trees. Finally, the

above-mentioned cables are sliced to see water trees, which is aimed to confirm the diagnosis results. The completed verification and conclusion are obtained according to the laboratory and field tests.

3.2. Laboratory Verification Test

The capacitance of the short cable is very small, so the tested cable needs to be paralleled with compensation capacitors to get a necessary capacitance to satisfy the resonance condition, which makes the variable frequency series resonant equipment work in the required frequency range.

Two compensation capacitors (4000 pF) were paralleled with a new 35 kV XLPE cable for testing. The photo of this laboratory test is shown in Figure 4. The applied alternating test voltages U0 are 3.4 kV, 7.5 kV, and 10.0 kV in turn, the acquired waveforms of the test voltages, the loss currents and harmonic components are shown in Figure 5. Channel 1 of the oscilloscope is the signal of the test voltage, channel 2 is the signal of the loss current, and the nethermost signal is the harmonic component of loss current after Fourier transform. All the subsequent similar figures are exactly the same and will not be described again.

Figure 4. *Photo of the laboratory verification test.*

The voltage waveform of the variable frequency series resonant power source in Figure 4 is quite close to the standard sinusoidal voltage, although there are some high frequency sharp peaks. The high frequency sharp peaks superimposed on the sinusoidal waveform are produced by the switch off of the power electronic devices in the variable frequency resonant power source, and the frequency range is kHz~MHz, which is much higher than the main test frequency and the frequency of harmonic. As a result, the sharp peaks have no effect on the measurement of the loss current harmonic component. The laboratory test demonstrate the feasibility of the variable frequency series resonant power source preliminarily.

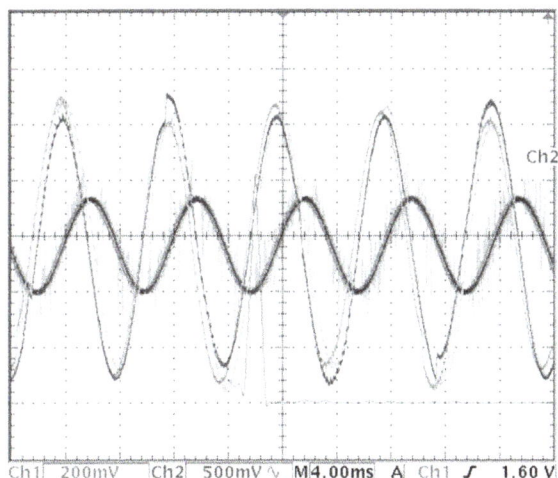

(a) U0= 3.4kV; η– 7, Ku= 10000; f=126.47Hz

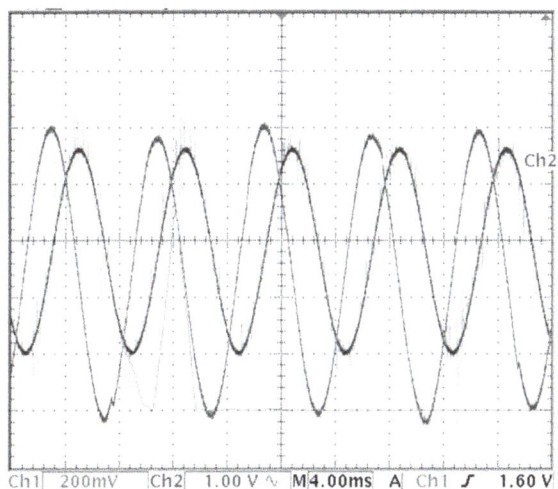

(b) U0= 7.5kV; η= 7; Ku= 10000; f=126.43Hz

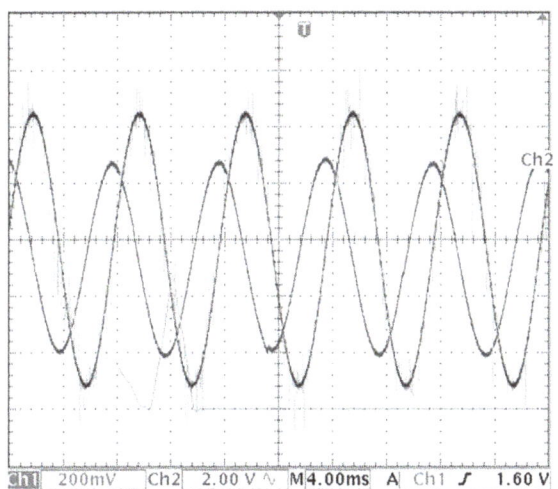

(c) U0= 10.0kV; η= 7; Ku= 10000; f=126.43Hz

Figure 5. Waveforms of the test voltages, the loss currents and harmonic components under different test voltages.

3.3. Field Verification Test and Analysis

The field verification tests are conducted in Jiangsu province. 70 cables in distribution grid are tested in the field, and 16 cables have water trees in total, of which 12 cables are slight water tree aging while 4 cables are serious aging. Two typical cable lines are taken for examples: a 35 kV cable line in Zhenjiang is marked as #1 cable, a 10 kV cable line in Changzhou is marked as #2 cable. Three phases in the two cable lines are all water treed. The photo of the field test is shown in Figure 6. The test voltage, the amplitude of the third harmonic component and the dielectric loss tangent (tanδ) of the two typical cables are listed in Table 1.

The test results of #1 cable (test voltage U_0 is 3.5kV or 8.0kV) and #2 cable (test voltage U_0 is 8.0kV or 10.0kV) are shown in Figure 7.

It is evident from Figure 7 that the loss current of both cable lines are distorted, and there are obvious third harmonic components in the loss currents.

Table 1. The test voltage, the amplitude of the third harmonic component and the dielectric loss tangent (tanδ) of the two typical cables.

#1 cable line (η= 7; K_u= 10000)			#2 cable line (η= 7; K_u= 10000)		
Test voltage/ kV	Amplitude of the third harmonic component/ mV	Tanδ ×10^{-4}	Test voltage/k V	Amplitude of the third harmonic component/ mV	Tanδ ×10^{-4}
3.5	265	1.26	4.0	80	3.82
5.0	640	2.27	6.0	145	5.02
6.5	1299	3.33	8.0	1000	10.2
8.0	2133	4.79	10.0	4100	25.1
9.5	3700	6.60	—	—	—

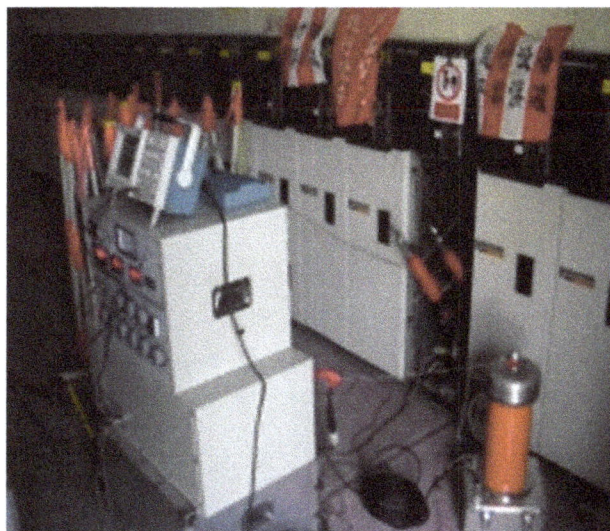

Figure 6. Photo of the field verification test.

Fourier transform results of the tested loss currents under different test voltage are compared in Figure 8. It is clear that the loss currents have the apparent third harmonic components, which means the tested cables are apparently water treed.

(a) #1 cable
U0=3.5kV, f=125.04Hz

(b) #1 cable
U0=8.0kV, f=125.07Hz

(c) #2 cable
U0=4.0kV, f=54.17Hz

(d) #2 cable
U0=8.0kV, f=54.20Hz

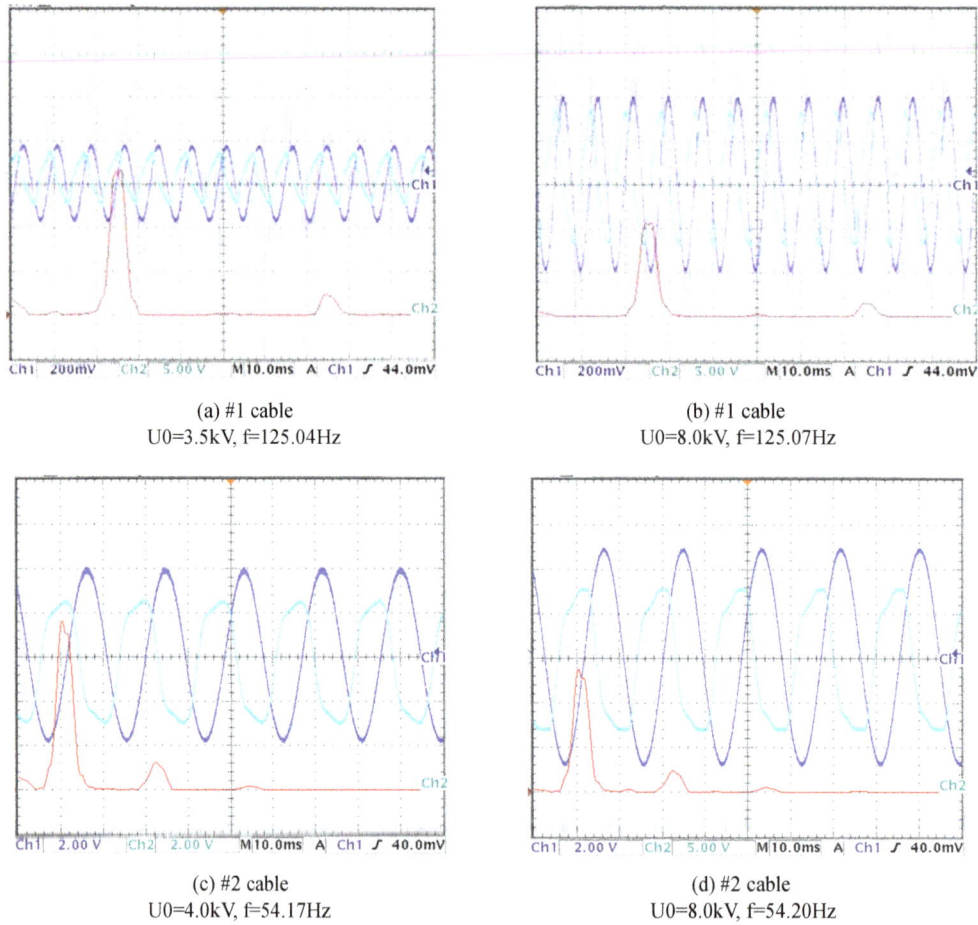

Figure 7. Test results of #1 cable and #2 cable.

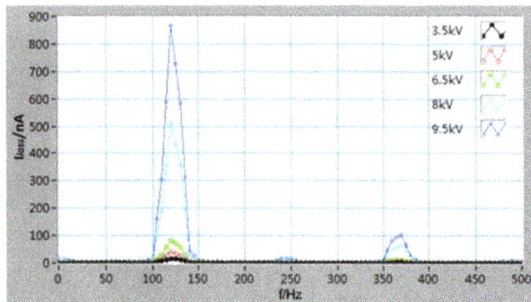

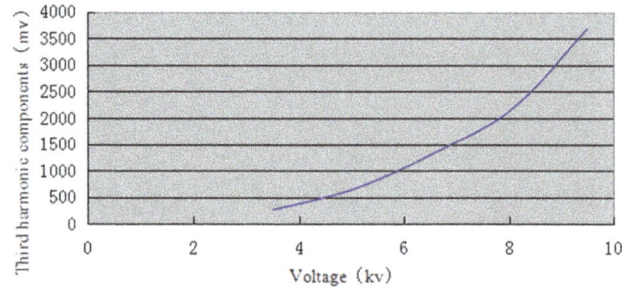

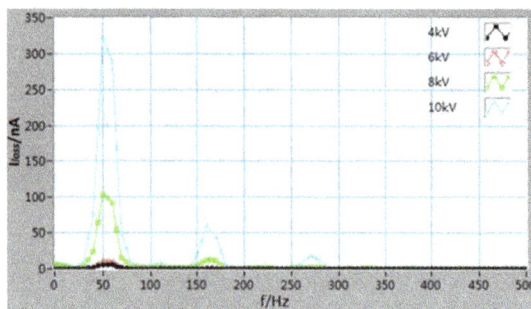

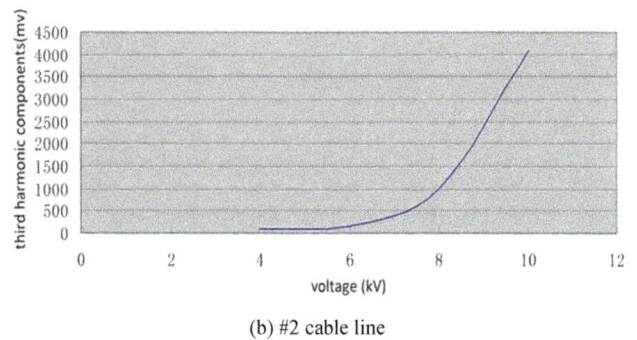

(a) #1 cable line

(a) #1 cable line

(b) #2 cable line

(b) #2 cable line

Figure 8. Fourier transform results of the tested loss currents under different test voltage.

Figure 9. Varying laws of the third harmonic components with the test voltage.

The third harmonic components change with the test voltage, and the varying laws are shown in Figure 9. It is indicated that the conductivity of the insulation material is nonlinear. The nonlinear relationship between the third harmonic components change and the test voltage also proves the existence of water tree aging in insulation.

The changes of the dielectric loss tangent (tanδ) with the test voltage are given in Figure 10, which also show the nonlinear relationship. This is another evidence of aging in cables. It can be seen that the largest tanδ of #1 cable is 6.6% and of #2 cable is 25.1%, which indicates that #1 cable is serious aging and #2 cable is slight aging.

3.4. Verification Observation of Water Tree in Cables

The results of the laboratory and field tests are verified further by the variable frequency withstand voltage tests of the 16 cables with water trees. All the 12 cables with slight water tree aging passed the withstand voltage tests, meanwhile all the 4 cables with serious aging were breakdown. The breakdown cables were sliced and observed by China Electric Power Research Institute (CEPRI) and Harbin University of Science and Technology (HUST). Figure 11 is the optical microscope observation of water trees in the breakdown cables.

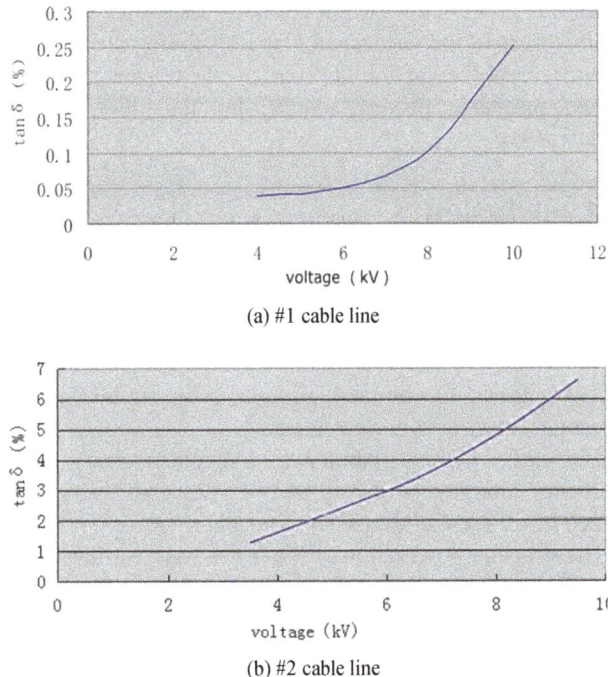

(a) #1 cable line

(b) #2 cable line

Figure 10. Varying laws of the dielectric loss tangent with the test voltage.

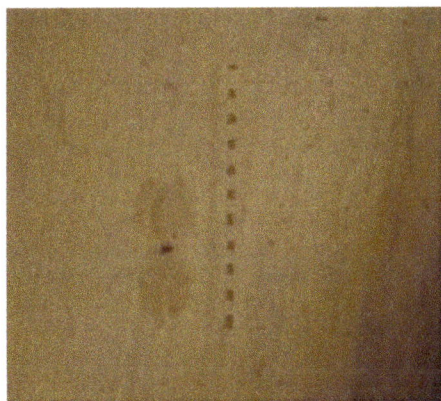

(a) observed water tree by CEPRI

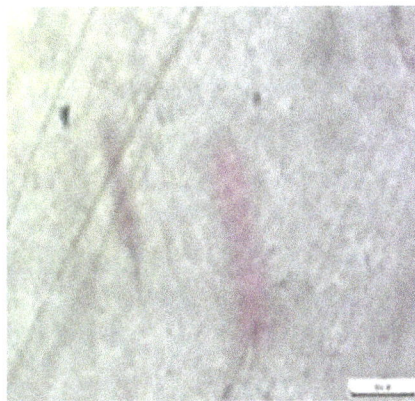

(b) observed water tree by HUST

(c) observed water tree by HUST

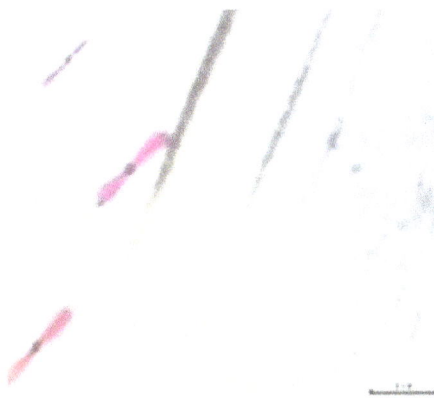

(d) observed water tree by HUST

Figure 11. Optical microscope observation of water trees in the breakdown cables.

The butterfly water trees are observed both by CEPRI and HUST. The area of water tree appears larger in the results of CEPRI, but the number of water tree is more in the results of HUST. Anyway, the structure observation of water tree have confirmed the serious water aging in cables, which prove the effectiveness of the results and diagnosis during the field test and the feasibility of the application of variable frequency series resonant power source in the loss current measurement.

4. Conclusion

The voltage waveform of the variable frequency series resonant power source is quite close to the standard sinusoidal voltage, although there are some high frequency sharp peaks. The variable frequency series resonant power source is feasible to be the high voltage for the loss current harmonic component measurement.

The loss current harmonic component has power frequency interference during the field test, but this can be solved by differential frequency principle.

The structure observation of water tree have confirmed the serious water aging in cables, which prove the effectiveness of the results and diagnosis during the field test and the feasibility of the application of variable frequency series resonant power source in the loss current measurement.

The existence and degree of the water tree can be estimated according to the test results, which can be a strong basis of the standardization of water tree aging diagnosis.

The expensive standard sinusoidal wave generator and linear power generator can be replaced by the variable frequency series resonant power source, which will reduce the cost and weight of the testing system.

References

[1]   Zheng Xiaoquan, Wang Jinfeng and Li Yanxiong. Transformation of Electrical Tree from Water Tree Degradation in XLPE [J]. Proceedings of the CSEE, 2013, 36(22): 166-174.

[2]   KENICHI HIROTSU. Development of Hot-line Diagnosis Method for XLPE Cables by Measurement of Harmonics Current[C]. Proceedings of 1994 International Joint Conference: 26th Symposium on Electrical Insulation Materials, Osaka, Japan, 1994: 455~458.

[3]   Shen Feifei, Zhang Jianliang, Lv Peiqiang and et al. Early Feature Evaluation of Distribution Network Cable Insulation Water Tree Based on Loss Current Harmonic Component [J]. East China Electric Power, 2013, 41(12): 18-24.

[4]   Gao Zhen. Measurement of the loss current harmonic component in cable insulation under variable frequency power [D]. Harbin University of Science and Technology, 2012.

[5]   Y. YAGI, H. TANAKA, H. KIMURA. Study on Diagnostic Method for XLPE Cable by Harmonics in Loss Current [J]. Electrical Insulation and Dielectric Phenomena, 1998: 653 ~656.

[6]   TANAKAATSUSHI, YAGIYUKIHIRO, TANAKA HIDEO. On-site Diagnostic Method for Water Treed XLPE Cable by Harmonics in AC Loss Current[J]. Electrical Insulation New News in Asia, 2003, (10): 29~30.

[7]   E. MOREAU, C. MAYOUX and C. LAURENT. The Structure Characteristics of Water Trees in Power Cables and Laboratory Specimens [J]. IEEE Trans. on EI, 1993, (1): 54~64.

[8]   Han Jiajia. Design of the loss current harmonic component measurement system based on FPGA and the current comparator bridge [D]. Harbin University of Science and Technology, 2013.

[9]   Zhao Q, Amagasaki M, Iida M. An automatic FPGA design and implementation framework[C]. Field Programmable Logic and Applications (FPL), 2013 23rd International Conference on. IEEE, 2013: 1-4.

[10]   Xinlao Wei, Bo Zhu, Bing Pang, Song Wang, Ruihai Li, "On-line Insulation Monitoring Method for Long Distance Three-Phase Power Cable," Proceedings of the Chinese Electrical Engineering Science, vol.35 No.8, Apr. 20, 2015.

18

Synergetic and sliding mode controls of a PMSM

Nourdine Bounasla[1, *], **Kamel Eddine Hemsas**[1], **Hacene Mellah**[2]

[1]Laboratory of Automatic, Department of Electrical Engineering, Ferhat Abbas University Setif-1, Sétif 19000, Algeria
[2]Department of Electrical Engineering, Hassiba Benbouali University Chlef, Chlef 02000, Algeria

Email address:
salemtour@gmail.com (N. Bounasla), hemsas.kamel@gmail.com (K. E. Hemsas), has.mel@gmail.com (H. Mellah)

Abstract: Permanent magnet Synchronous machines (PMSM) provide high efficiency, compact size, robustness, lightweight, and low noise; these features qualify them as the best suitable machine for medical applications. Without forgetting its simple structure, high thrust, ease of maintenance, and controller feedback, make it possible to take the place of steam catapults in the future. This paper presents the synergetic control approach for PMSM. Synergetic control theory is purely analytical and is based on nonlinear models, provide asymptotic stability. This approach allows to reduce the chattering phenomenon. To verify the performance characteristics of this approach, we compare it with sliding mode. Simulation results are presented to show the effectiveness of the proposed control method.

Keywords: PMSM, Synergetic Control, Sliding Mode Control, Asymptotic Stability

1. Introduction

In a modern industrialized country about 65% of electrical energy is consumed by electrical drives. Constant-speed, variable-speed or servo-motor drives are used almost everywhere: in industry, trade and service, house-holds, electric traction, road vehicles, ships, aircrafts, military equipment, medical equipment and agriculture [1]. Permanent magnet (PM) machines provide high efficiency, compact size, robustness, lightweight, and low noise, [2], these features qualify them as the best suitable machine for medical applications [3]. Without forgetting its simple structure, high thrust, ease of maintenance, and controller feedback, make it possible to take the place of steam catapults in the future [4]. The PM motor in an HEV power train is operated either as a motor during normal driving or as a generator during regenerative braking and power splitting as required by the vehicle operations and control strategies. PMSM with higher power densities are also now increasingly choices for aircraft, marine, naval, and space applications [2]. Permanent magnet synchronous motor (PMSM) has been attracting more and more attention in high performance electric drive applications since it has certain superiorities such as; high efficiency, high power factor, superior power density, large torque to inertia ratio and long life over other kinds of motors such as DC motors and induction motors [5]. However, precise control of a PMSM is not easy due to nonlinearities of PMSM servo systems, parameter and load torque variations. Thus the linear control schemes such as PI control cannot guarantee satisfactory performances. To get around this problem, various methods of nonlinear control methods have been developed for PMSM system, such as input-output linearization control [6], robust control [7], sliding mode control [8], back-stepping control [5], and fuzzy control [9] and so on.

Sliding mode control (SMC) [1] attracts the attention of many researchers in the field control of electrical machines has been suggested as an approach for the control of systems with nonlinearities, uncertain dynamics and bounded input disturbances. The most distinguished features of the SMC technique are: insensitivity to parameter variations, external disturbance rejection but the commutations of the control at high frequencies induce chattering problem. This problem can degrade the performance of mechanical systems because it causes excessive energy consumption and reduces the life of mechanical equipment. To remedy this problem, an asymptotic state observer is proposed to limit the chattering [11]. Another solution based on synergetic control is introduced. This command like sliding mode control is based on the basic idea that if we could force a system to a desired manifold with designer chosen dynamics using continuous control law, we should achieve similar performance as SMC without its main inconvenient: chattering phenomenon [12-13].

The aim of this paper is to give a comparison between sliding mode control (SMC) and synergetic control (SC) applied on PMSM based drive.

2. PMSG Model

The electrical model of PMSM is given by the following equations [14].

$$\begin{cases} \dfrac{di_d}{dt} = -\dfrac{R_s}{L_d} i_d + \dfrac{L_q}{L_d} p\omega + \dfrac{V_d}{L_d} \\[2mm] \dfrac{di_q}{dt} = -\dfrac{R_s}{L_q} i_q - \dfrac{L_d}{L_q} p\omega i_d - \dfrac{\phi_f}{L_q} p\omega + \dfrac{V_q}{L_q} \end{cases} \quad (1)$$

Where i_d and iq as state variables and V_d V_q are control variables.

The expression of the electromagnetic torque and the equation of motion of the rotor are given by the following equations.

$$\begin{cases} J\dfrac{d\omega_r}{dt} = T_{em} - T_L - f\,\omega \\[2mm] T_{em} = \dfrac{3}{2} p(\phi_f i_q + (L_d - L_q)i_q i_d) \end{cases} \quad (2)$$

3. Sliding Mode Control

The sliding mode control algorithm design is to determine three different stages as follow [15]:

3.1. Commutation Surface

J. Slotine proposes a general form of equation to determine the sliding surface [16].

$$s(x,t) = \left(\dfrac{d}{dt} + \lambda\right)^{r-1} e \quad (3)$$

$e(t)$: is variation of the variable to be regulated.

$$e(t) = x_{ref}(t) - x(t) \quad (4)$$

Where λ : is positive constant and r : relative degree.

3.2. Convergence Condition

The convergence condition is defined by the following Lyapunov equation.

$$s(x)\dot{s}(x) < 0 \quad (5)$$

3.3. Calculation Control

The control algorithm includes two terms, the first for the

exact linearization, and the second discontinuous one for the system stability.

$$U = U_{eq} + U_n \quad (6)$$

U_{eq}: is calculated starting from the expression

$$\dot{s}(x) = 0 \quad (7)$$

U_n: is given to guarantee the attractively of the variable to be controlled towards the commutation surface. Its simplest equation is given by:

$$U_n = k\,\mathrm{sgn}\,s(x); k > 0 \quad (8)$$

Fig. 1 shows the diagram of the sliding mode control (SMC) of a PMSM supplied by voltage source inverter.

The sliding surfaces are chosen according to the relation of J.Slotine and the output relative degree.

$$\begin{cases} s(\omega_r) = \omega_{ref} - \omega_r \\ s(i_q) = i_{qref} - i_q \\ s(i_d) = i_{dref} - i_d \end{cases} \quad (9)$$

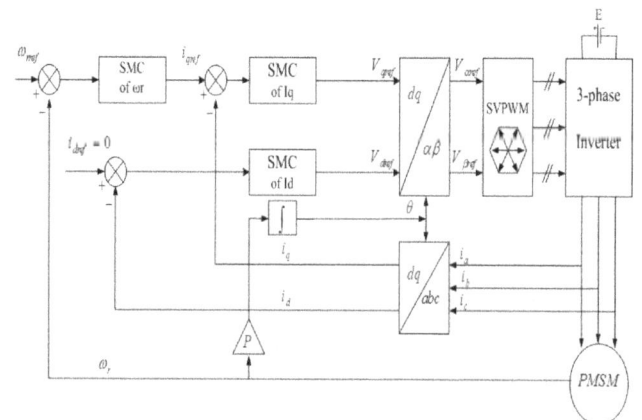

Figure 1. SMC scheme for PMSM.

4. Synergetic Control Design

Let us consider an nth order nonlinear dynamic system described by (10):

$$\dot{x} = f(x,u,t) \quad (10)$$

In which x represents the state vector, u represents the control input vector and $f(x,u,t)$ represents a nonlinear function.

The synergetic controller synthesis procedure is completely analytical, which consists of the following steps [13], [17-18]:

• Start by defining a macro-variable as a function of the state variables, for a dynamical system described by equation (10):

$$\psi = \psi(x,t) \quad (11)$$

The control will force the system to operate on the manifold $\psi = 0$. The designer can select the characteristics of this macro-variable according to the control specifications (e.g. limitation in the control output, and so on). In the trivial case, the macro-variable is a simple linear combination of the state variables.

- Repeat the same process defining as many macro-variables as control channels.
- Fix the dynamic evolution of the macro-variables according to the equation:

$$T\dot{\psi} + \psi = 0, \ T > 0 \qquad (12)$$

T: designates the designer chosen speed convergence to the desired manifold. Differentiating the macro-variable (11) along (10) leads to (13):

$$\dot{\psi} = \frac{d\psi}{dx}\dot{x} \qquad (13)$$

Combining equation (10), (12) and (13), we thus obtain:

$$T\frac{d\psi}{dx}f(x,n,t) \qquad (14)$$

- Synthesize the control law (evolution in time of the control output) according to equation (14) and the dynamic model of the system, leads to (15):

$$u = u\left(x, \psi(x,t), T, t\right) \qquad (15)$$

From (15), it can be seen that the control output depends not only on the system state variables, but also on the selected macro-variable and time constant T.

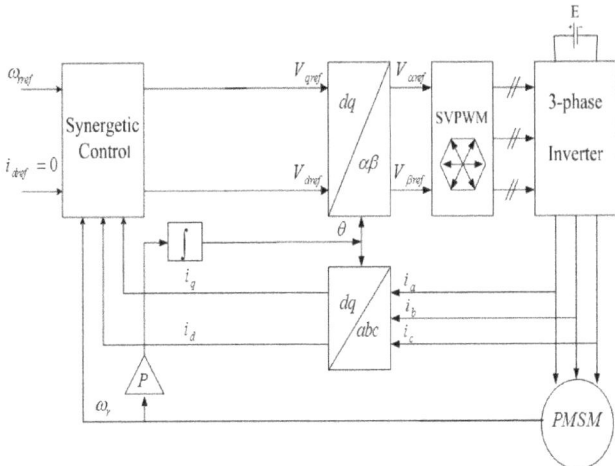

Figure 2. *SC scheme for PMSM.*

In other words, the designer can choose the characteristics of the controller by selecting a suitable macro-variable and a time constant T.

The procedure summarized above can be easily implemented as a computer program for automatic synthesis of the control law. Moreover, the synergetic control system can be global stability, parameters insensitivity and noise

suppression by suitable selection of macro-variables.

The method described in the previous paragraph requires that we define the same number of macro-variables as control channels in the system. Thus, it requires the definition of two macro-variables, which are functions of the state variables as shown in (11). We chose these two terms:

$$\begin{cases} \psi_1 = i_d - i_{dref} \\ \psi_2 = \left(\dot{\omega}_r - \dot{\omega}_{rref}\right) - k\left(\omega_r - \omega_{rref}\right) \end{cases} \qquad (16)$$

Where k is controller parameters.

Fig. 2 shows the diagram of the synergetic control (SC) of a PMSM.

5. Simulation Results

The performances of the proposed controls were tested by simulation on a 1.5kw PMSG whose parameters are given in the appendix.

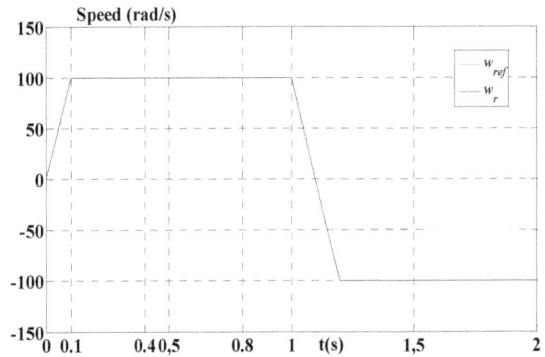

(a) sliding mode control.

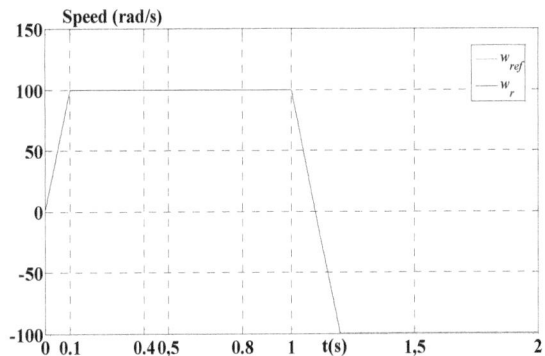

(b) synergetic control

Figure 3. *Speed responses of a PMSM controlled by sliding mode and synergetic.*

The simulation results are obtained on the Matlab\ Simulink environment.

The following figure shows the speed curve of a PMSG controlled by sliding mode and synergetic technics, in this figure we can see that the robustness tests are applied for the two controllers.

However, a moderate vibration on the case of the synergetic controller of a magnitude less than 1N.m. we think that the cause of these electromagnetic torque oscillations is the chattering phenomenon.

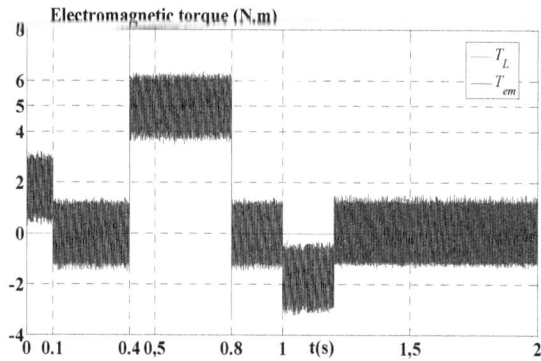

(a) sliding mode control.

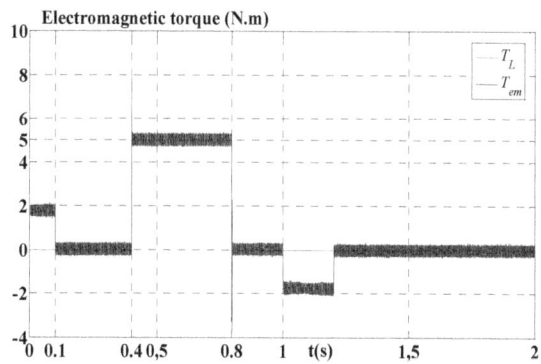

(b) synergetic control

Figure 4. *Electromagnetic torque curve of a PMSM controlled by sliding mode and synergetic.*

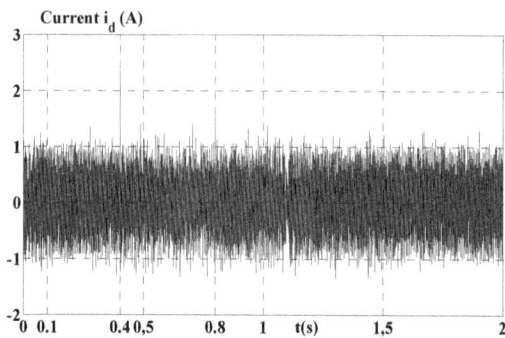

(a) sliding mode control.

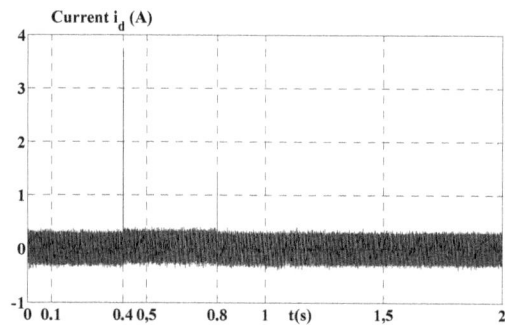

(b) synergetic control

Figure 5. *id curve of a PMSM controlled by sliding mode and synergetic.*

The *id* variations are illustrate in the following figure both for the sliding mod and synergetic controllers, we can see that the *id* oscillations at the sliding mode case are more important that the synergetic case.

The *iq* variation of a PMSG controlled by sliding mode and synergetic technics are presented on the fig.6, by comparison between the results of application of the two controllers on the iq current wave, we see an important oscillation on the case of sliding mode controller compared with synergetic controller.

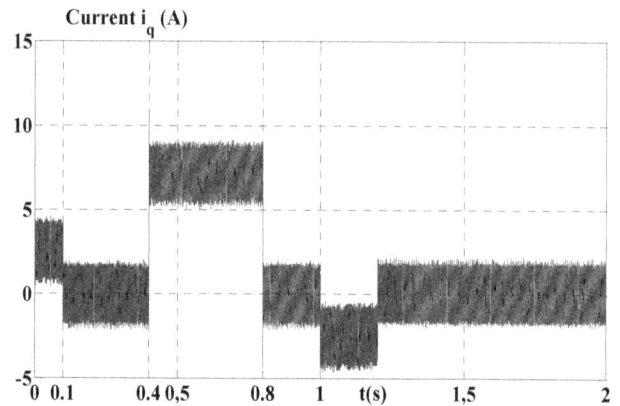

(a) sliding mode control.

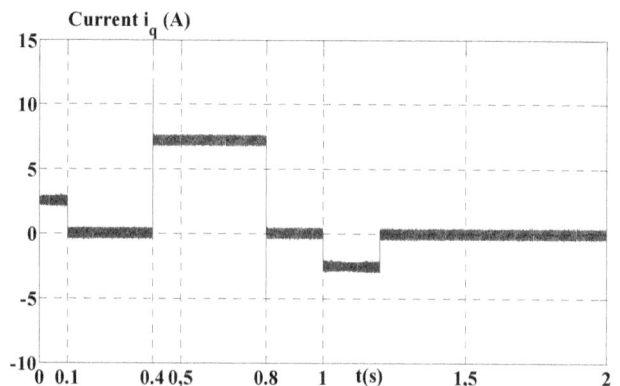

(b) synergetic control

Figure 6. *iq curve of a PMSM controlled by sliding mode and synergetic.*

6. Conclusion

In this paper two different control of permanent magnet synchronous machine (PMSG) are presented. It is a matter of sliding mode control and synergetic control. To compare their performance, many tests are performed under the same conditions.

Simulations results show that the speed and the current *id* follow perfectly their references. The response of the electromagnetic torque and the current in both cases are compared. It is clear that the synergetic control reduces the chattering better than the sliding mode control.

Simulation results show clearly the effectiveness of the synergetic control in reducing chattering problem.

Nomenclature

V_d, V_q	Direct-and quadrature-axis stator voltages.
id, iq	Direct-and quadrature-axis stator current.
L_d, L_q	Direct -and quadrature-axis inductance.
p	Number of poles.
R_s	Stator resistance.
ϕ_f	Rotor magnet flux linkage.
ω_r	Mechanical rotor speed.
J	Inertia.
f	Damping coefficient.
T_{em}	Electromagnetic torque.
T_L	Load torque.

Appendix

Table 1. PMSM Parameters

Components	Values
Rs	1.4 Ω
Ld	0.0066 H
Lq	0.0058 H
ψ_f	0.1546 Wb
f	38.818e-5 Nm/rad
J	1.76e-3 kg.m.s
p	3
TL	5 Nm

References

[1] F. JACEK, Permanent Design and Applications," Second Edition, Revised and Expanded United Technologies Research Center Hartford, Connecticut London, United Kingdom 2002.

[2] H. Mellah, K E. Hemsas, "Dynamic Design and Simulation Analysis of Permanent Magnet Motor in Different Scenario of fed Alimentation," Journal of Electrical and Control Engineering (JECE), pp. 55–61, Vol.3, No.4, 2013.

[3] A. Nasiri, Salaheddin A. Zabalawi, and Dean C. Jeutter, "A Linear Permanent Magnet Generator for Powering Implanted Electronic Devices," IEEE transactions on power electronics, pp. 192–199, vol. 26, no. 1, Jan. 2011.

[4] L. Li, H. Junjie, L. Zhang, Y. Liu, S. Yang, R. Liu, L. Xiaopeng, "Fields and Inductances of the Sectioned Permanent-Magnet Synchronous Linear Machine Used in the EMALS," IEEE Transactions on Plasma Science, pp. 87-93, Vol. 39, 2011.

[5] M. Karabacak , H.I. Eskikurt," Design, modeling and simulation of a new nonlinear and full adaptive Backstepping speed tracking controller for uncertain PMSM," Applied Mathematical Modelling, vol. 36, no. 11, pp 5199-5213, 2012.

[6] B. Grcar, P. Cafuta, M.Znidaric, and F.Gausch, "Nonlinear control of synchronous servo drive," IEEE Trans. on Control Systems and Technology, vol. 4, no. 2, pp. 177-184, 1996.

[7] T.L. Hsien, Y.Y. Sun and M.C. Tai, "H1 control for a sensorless PMSM drive," IEE Proc. of Electric Power Applications, vol. 144, no. 3, pp. 173-181, 1997.

[8] R.J. Wai, "Total sliding-mode controller for PM synchronous servo motor drive using recurrent fuzzy Neural network," IEEE Trans. on Industrial Electronics, vol. 8, no. 5, pp. 926-944, 2001.

[9] Y.S. Kung and M.H.T sai, "FPGA-based speed control IC for PMSM drive with adaptive fuzzy control," IEEE Trans. on Power Electronics, vol. 22, no. 6, pp. 2476-2486, 2007.

[10] F. Benchabane, "Robust position and speed estimation algorithms for permanent magnet synchronous drives," European Journal of Scientific Research, vol. 57 no. 1, pp. 6-14, 2011.

[11] A.G. Bondarev, S.A. Bondarev, N.Y. Kostyerva, and V.I. Utkin, "Sliding modes in systems with asymptotic state observers," Automation and Remote Control, vol. 46, no. 6, pp. 679-684, 1985.

[12] L. Medjbeur and M.N. Harmas, "Adaptive Fuzzy Terminal Synergetic Control," IEEE Communications, Computing and Control Applications (CCCA), pp. 1–16, 2011.

[13] Z. Bouchama, and M.N. Harmas, "Optimal robust adaptive fuzzy synergetic power system stabilizer design," Elsevier, Electric Power Systems Research, vol. 88, pp. 9–15, 2012.

[14] G. Sturtzer, and E. Smigiel, "Modélisation et commande des moteurs triphasés," Edition Ellipes, 2000.

[15] A. Massoum, M.K. Fellah ,A. Meroufel , P. Wira, and B. Bellabes, "Sliding mode control of a permanent magnet synchronous machine fed by a three levels inverter using a singular perturbation decoupling," Journal of Electrical and Electronic Engineering, vol. 5, no. 2, pp. 1427–1433, 2005.

[16] J. J. Slotine, and W. Li, "Applied non linear control," Prentice-Hall, 1991.

[17] L.E. Santi, A. Monti, D. Li, K. Proddutur, and R. Dougal "Synergetic Control for DC-DC Boost Converter: Implementation options", IEEE Transactions on Industry Applications, vol. 39, no. 6, pp. 1803-1813, 2003.

[18] A. Kolesnikov, G. Veselov, A. Kolesnikov, A. Monti, F. Ponci, E.Santi, and R. Dougal, "Synergetic Synthesis of DC-DC Boost Converter Controllers: Theory and Experimental Analysis," Proc. IEEE APEC, vol. 1, pp. 409-415, 2002.

19

Considerations on Developing a Chainsaw Intrusion Detection and Localization System for Preventing Unauthorized Logging

Valentin Andrei, Horia Cucu, Lucian Petrică

Speech and Dialogue Research Laboratory, University "Politehnica" of Bucharest, Bucharest, Romania

Email address:

horia.cucu@upb.ro (H. Cucu)

Abstract: This work presents a system designed to prevent unauthorized logging by detecting and locating chainsaw sound sources. We analyze the specifics of chainsaw related sounds and discuss about the possible approaches for classifying the input sounds. The work also highlights several approaches for sound source localization that can be used in wireless sensor network architecture for tracking the assumed intruders. Finally we describe the architecture of the system and discuss on how our approach is designed to be scalable, fail-safe and cost effective.

Keywords: Chainsaw Detection, Sound Source Localization, Sound Recognition

1. Introduction

Unauthorized logging is an issue that can have serious consequences over the surrounding environment and over the human settlements nearby an affected location. It has disastrous effects ranging from destroying the habitat of valuable wildlife to serious landslides that can result in human casualties. It is an issue that needs to be treated with maximum strictness. Often unauthorized logging is very difficult to prevent due to the lack of funding or inadequate surveillance methods. For example, using classical patrols is often no longer an option because the intruders that are not authorized to cut trees are very difficult to be tracked, especially in wide domains.

Over the years various automated surveillance solutions mostly based on Wireless Sensor Networks (WSNs) have been implemented to aid the authorities to detect intrusions in a forest domain and to track the intruders. The most efficient techniques rely on identifying a sound that is likely to be produced by a chainsaw and using microphone arrays for locating the source. For example, in [1] a light-weight approach to chainsaw identification is proposed, based on the auto-correlation of 256-sample blocks of audio signal. The authors refer to this method as Lightweight Acoustic Detection (LAC). Signal energy is utilized, as well as pitch

and pitch stability measures computed from the autocorrelation of the signal. These measures are chosen because chainsaws are much noisier than forest ambient sounds, have distinctive pitch signatures and have good pitch stability. LAC executes in real-time on a WSN sensor node based on the ATMega128RFA1 microcontroller. The reported accuracy of detection is up to 85%.

In [2] and [3] an algorithm called Normalized Peak Domination Ratio (NPDR) is utilized. Based solely on the signal spectrum and noise energy, NPDR looks for an overlap between the signal and pre-computed reference peaks in the spectrum, and also for a sufficiently high concentration of signal energy in the spectral vicinity of the reference peaks. Evaluation of NPDR on 1024-sample blocks of signal indicates over 99% accuracy in quiet room conditions on a range of sounds. NPDR runs on a PC and the authors indicate (but do not demonstrate) the possibility of executing NPDR on smartphones. Between the two algorithms, NPDR has better apparent accuracy and LAC a slightly simpler structure which may result in a more efficient implementation.

Other automatic surveillance solutions are strictly based on satellite image processing [4, 5] and local video surveillance [6, 7]. These solutions are focused and efficient on detecting forest fires or estimating fire risk, but they have important short-comings when it comes to (illegal) logging. The satellite signal based methods do not support real-time operation in general, and also the spatial resolution is very limited. Conventional video surveillance suffers from the following:

- If the viewing angle of the lens is wide, then the resolution is low; otherwise the monitored area is too small.
- The analysis of data requires high complexity solutions.
- The required bandwidth for transmitting images is relatively high, while the energy consumption of such devices is also relatively large.

In the current work we describe implementation details of a system designed to detect and locate chainsaws in a forest domain under surveillance. The following section concentrates on discussing the specificities of the sounds produced by chainsaws and comments on the feature extraction approaches. Section 3 is dedicated to reviewing several sound localization techniques that can be used in the sensor elements of the proposed system. The architecture of the system is proposed in section 4.We discuss about the key features of the proposed system and about strategies to ensure a high level of quality of service. Finally the last section is dedicated to conclusions and future work.

2. Detection of Chainsaw Related Sounds

The first challenge to consider in developing the proposed system is the ability to recognize from a multitude of sounds, the specific ones produced by a chainsaw. In a forest there are numerous sound sources such as the noise caused by the tree leafs in windy conditions, the sounds caused by birds or animals, walking related sound, car related sounds, sounds produced by tourists who have authorized presence and plenty of other sources. The system should not fire the alarm for example when a car passes by. This would generate high usage costs and in the end would prove to be an unreliable solution.

Detecting a sound produced by a chainsaw is a task slightly more flexible than, for example, recognizing spoken words or speakers. A much greater level of detail needs to be captured by the sound analysis algorithms in order to recognize speech or speakers, because each voice is unique and each word produced by a certain voice can be considered an unrepeatable sample. We state that detecting a chainsaw is a flexible task because *we want to determine that a captured sound belongs to a certain class* – in our case the class of sounds produced by chainsaws. With the sounds being produced by a mechanical system the task becomes relatively easier. This is because mechanical systems like engines have an extremely good periodicity rate and therefore are likely to produce extremely repeatable sounds. Let's analyze how a signal produced by a chainsaw looks like. In Figure 1 we plotted the power spectral density computed using the Welch method for 4 signals captured from 4 different chainsaw types in different recording conditions. We want to observe if we detect a similarity pattern between the captured samples. If the pattern is detected, even in different recording conditions, we are more confident that detecting chainsaw presence based on the sound footprint is feasible.

We can observe that in the figure the spectrum shape obtained for the four types of chainsaws is extremely similar. In all quadrants the power per frequency step is higher towards the lower frequencies and decreases almost linearly towards the higher frequencies. This is an easily detectable pattern. We also speculate that using classical approaches for detecting if a sound belongs to a specific class can *fade the similarity* observed in Figure 1.

In [11], the authors collected a database of 10 chainsaw produced sounds. The results we obtained are in good correlation with data presented in [11].

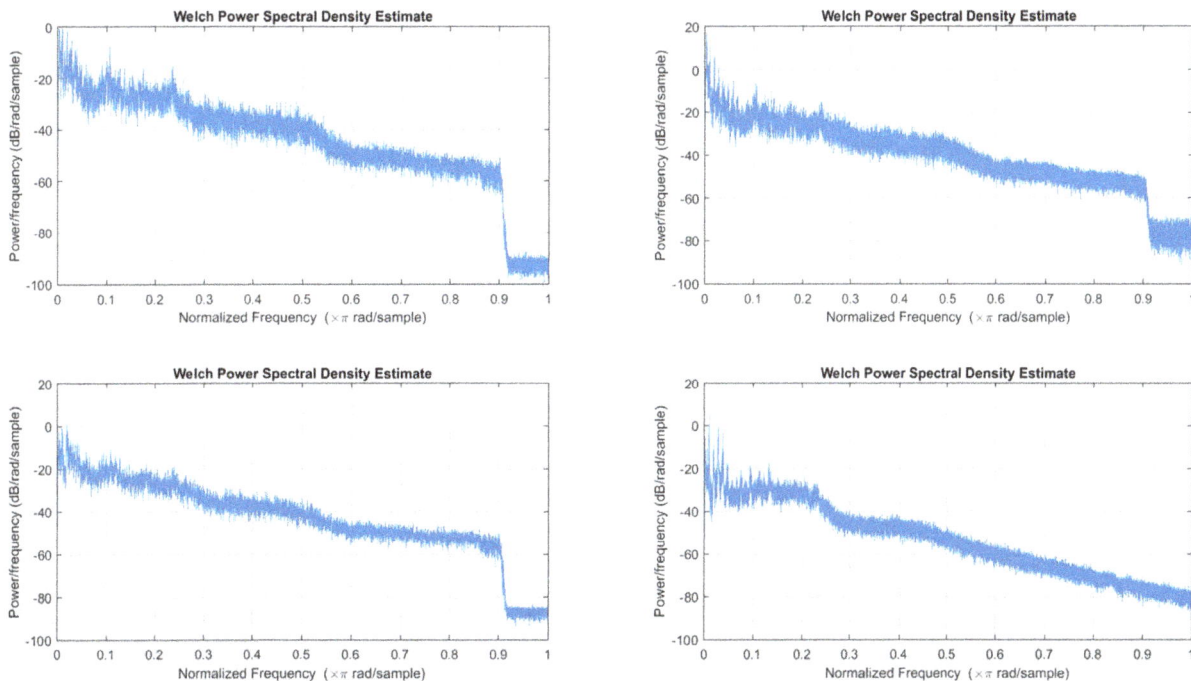

Figure 1. Example of PSD for chainsaw produced sounds.

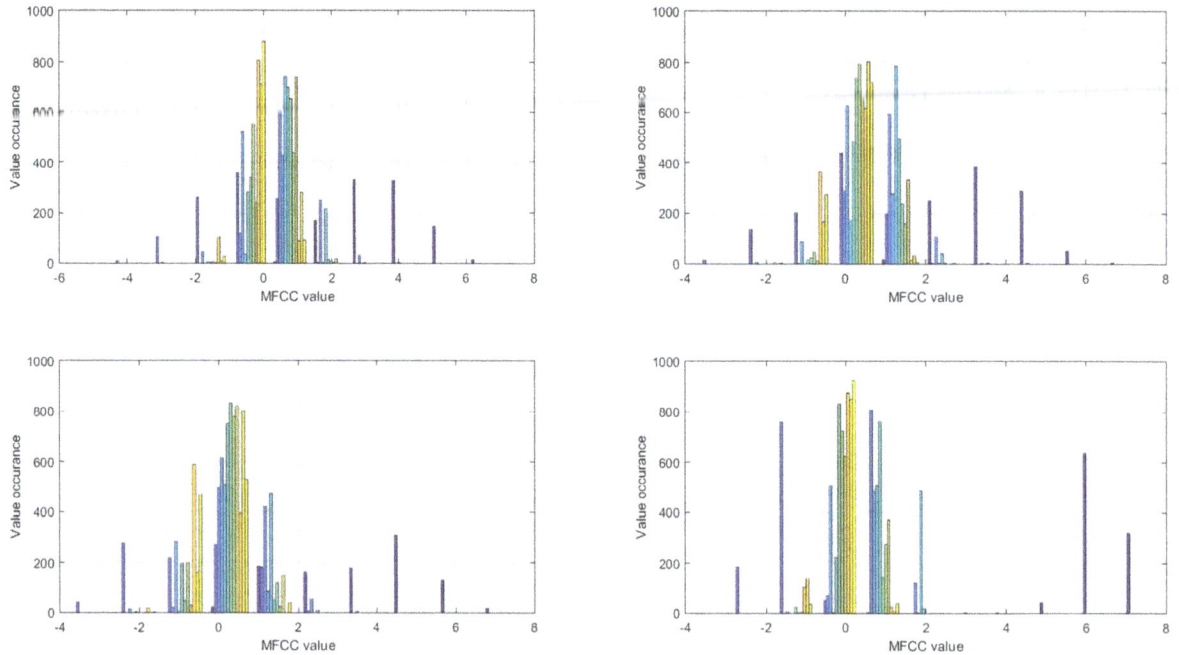

Figure 2. The usage of MFC coefficients for detecting chainsaw related sounds.

In Figure 2 we computed the MFC coefficients for the four signals produced by chainsaws, also displayed in Figure 1. In order to verify if we can observe similar trends between the four sounds, when analyzing MFC coefficients, we computed for each sound a histogram that creates value bins and counts the occurrence frequencies for each bin. We can observe that while the values seem centered around zero for all the recordings, the similarity between the histograms is not as obvious as when analyzing the signals in frequency domain. Therefore we speculate that using complex techniques for extracting features from chainsaw produced sounds may not always yield promising results.

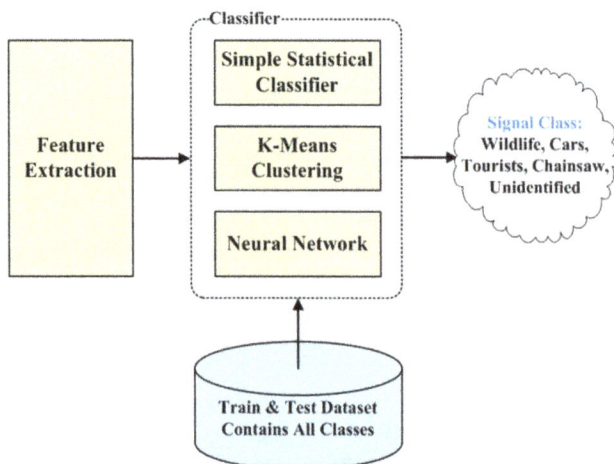

Figure 3. Chainsaw detection approach.

We used MFCC for this experiment as it is an extremely widely used technique adopted in systems for recognizing various sounds. Our results are somewhat in contrast with the work described by [8] and [9] which uses MFCC. However we firmly state that this approach is not necessarily suitable for mechanical produced sounds as it was designed for speech related sounds. For example, in [10] the same authors propose a methodology of feature extraction, based on TESPAR which extracts simple metrics from the waveform and is stated to produce better classification results.

Considering the comments stated above, the general diagram for classifying sounds into classes and marking the ones that are likely to be produced by chainsaws, is presented in Figure 3. As stated we consider that the feature extraction algorithms should be fairly simple due to the specifics of the chainsaw produced sounds (e.g. compute a linear regression on the points that describe the spectrum and compute the correlation error as input parameter). A simple threshold base statistical classifier can be used to separate input sounds into classes or a more complex approach can be designed, using K-Means clustering. We recommend using a neural network only in conditions were chainsaw sounds are demonstrated as being rather complex therefore implying the use of feature extraction methods with wide feature sets. In this case a neural network is far better at learning the links between the feature sets than a simple statistical classifier and a K-Means Clustering algorithm. We estimate that the chainsaw detection approach should tag the recorded signals using a set of commonly met classes of signals for the desired use case. In a forest we can consider the following classes: wildlife, tourists, passing cars (in the case where there is a paved road through the forest) and finally we can consider a generic class: "unidentified".

3. Sound Source Localization Challenges

Sound source localization in wireless sensor networks has been achieved for single-microphone sensor nodes in previous work through the use of distributed time difference of arrival (D-TDOA). In D-TDOA, nodes synchronize their internal clocks and listen for a reference sound at a pre-determined moment in time. Because it has already been established through synchronization that the sampling is simultaneous, any difference in audio signal phase at the sensing nodes (SNs) is caused by differences in relative position of the sound source to the respective SNs. If the geometry of the WSN is known a priori, then the individual measurements are aggregated to determine the position of the sound source relative to the WSN. D-TDOA requires that time re-synchronization, hence radio communication, must occur before each SSL event to compensate for drift in SN timers. Following signal acquisition, the data must be aggregated for the SSL to be computed. The WSN incurs an energy consumption penalty for these communication events.

A different approach to sound source localization has been explored in [12] utilizing microphone arrays. We refer to this method as Array TDOA (A-TDOA). In A-TDOA, each SN is equipped with an array of microphones arranged in a fixed geometry. To obtain positioning in a 2D plane, a planar symmetrical geometry is best suited, such as placing the microphones on a circle. Delay-and-Sum (DS) [13] is the algorithm utilized in [12] for sound direction estimation, although more sophisticated algorithms exist [14]. The multi-microphone data gathering and processing in A-TDOA is above the capabilities of a microprocessor and needs either a

DSP or a FPGA to be added to the system.

Two or more A-TDOA direction measurements by separate WSN nodes may be aggregated as in [12] in order to increase the localization accuracy if needed. A single SN can only determine through A-TDOA the probability that a sound source is located in a certain direction relative to the SN. Adjacent SNs with overlapping sensing ranges may superimpose their localization, resulting in a more accurate localization. The cost of the increased localization accuracy is energy expended for SN communication.

It must be noted that A-TDOA can only determine the direction of a sound source relative to the SNs microphone array. Therefore, for A-TDOA there must also be some way for the SN to determine the absolute spatial orientation of the microphone array.

Localization by A-TDOA avoids the need for WSN time synchronization. Conversely, on-node computation is required for A-TDOA to determine the sound direction. Both A- and D-TDOA require data aggregation between nodes to determine the sound source accurately, however A-TDOA is capable of obtaining a certain measure of localization without SN communication. In some cases, knowing the general direction of the sound source is sufficient to discriminate between legal and illegal deforestation activity (e.g., at the edges of a forest or natural reserve).

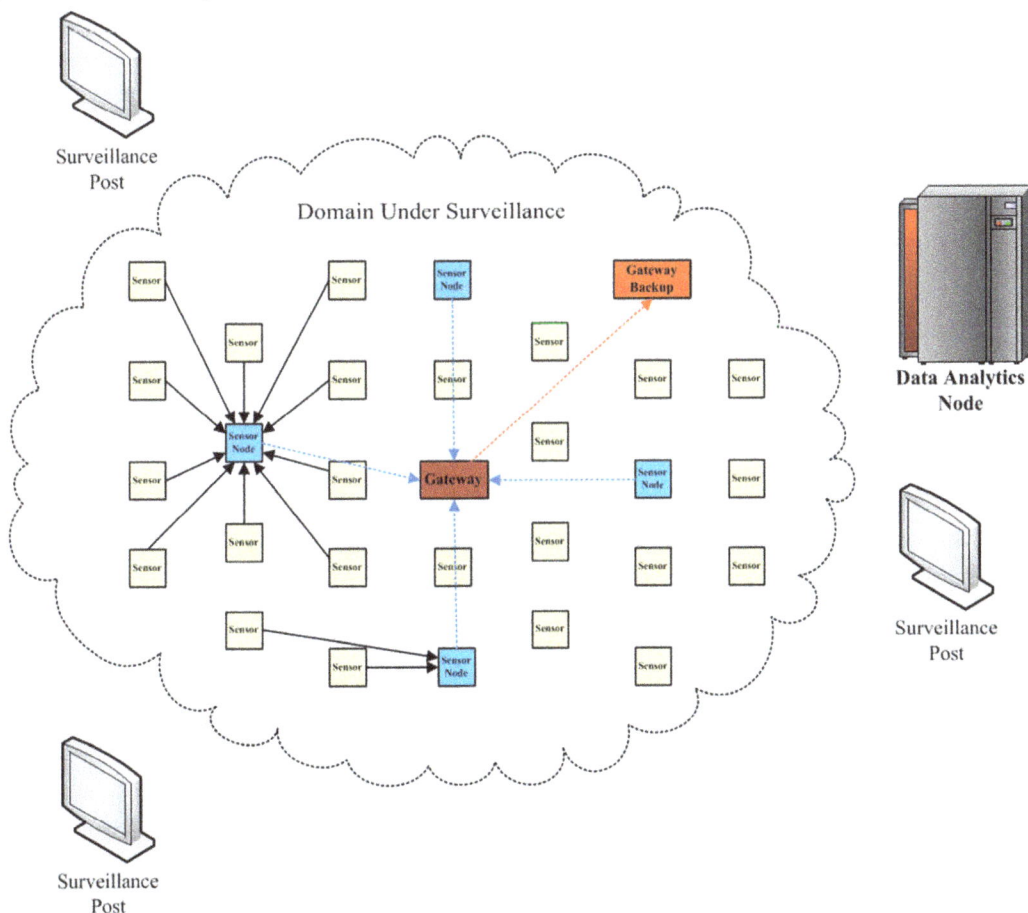

Figure 4. Intrusion localization system diagram.

4. System Design and Reliability Considerations

We propose a flexible design for detecting chainsaw related sounds in a monitored area. We consider that such a system needs to have the below features:

- *Scalability* – Extending the monitored domain needs to be done without involving redesign costs.
- *Fail-Safe* – Because it contains multiple components it should expose redundancy to ensure that in the case one component fails, the system will not fail completely.
- *Cost-Effective* – We state that the investment needed for such a system needs to have a reasonable return. Technically this involves using generally available hardware with low maintenance and support costs. Additionally the amount of energy consumed by the system needs to be kept at a reasonably low level.

The diagram of the proposed system is presented in Figure 4. It can be viewed as a distributed architecture, similar with the architecture of a wide area cellular network. Let's describe the features of each component.

4.1. Sensors

The basic element of the design is the sensor. However this component is more complex than a plain sensor. In our case, the sensor can have the logic for capturing the audio signal, for estimating the location of the source and for performing basic processing over the collected input. In Figure 3 we illustrated the architecture of the sound recognition system. We estimate that basic recognition using either a simple set of statistical thresholds or even a k-means clustering approach can be performed at the sensor level.Of course even more processing can be done, but that would greatly increase the energy footprint of the proposed system.

4.2. Sensor Node

The sensor node acts like a hub, connecting multiple simple sensors. Its main functionality is to route the acquisition data towards the gateway. Additionally it can be designed to contribute to the processing especially in the case where the simple sensors do not contain complex processing logic. Such enhancements could be related to the preprocessing of the captured sound signal, like filtering or even feature extraction.

4.3. The Gateways

The gateways have a central role in the architecture of the proposed system. First, we propose at least 2 gateways. If one should fail there should always exist a backup. The gateway number can increase as the domain under surveillance grows larger.

At the gateway level we can consider more complex processing, for example even adding a neural network for labeling sounds. Unlike the sensors and the sensors nodes, the gateway needs to have a more powerful processor because it is responsible for processing the information from a high number of sensors. The computational power demand is expected to grow seriously if the lower level components do not contain complex processing.

4.4. Data Analytics and Surveillance Posts

The final level is represented by the data-analytics node. This component will provide the ultimate decision to the surveillance post. For example the lower levels like the sensor, the sensor node or the gateway can predict the occurrence of an intruder, but the data analytics node has the final decision of raising the intrusion alarm. This can be accomplished by storing information related to the terrain where the chainsaw sound source was located. If the data analytics node has a detailed map of the domain under surveillance it can assign intrusion likelihoods to each spatial division of the domain and therefore if the intrusion is estimated to be in a highly inaccessible area (without having detected „unknown" sounds a priori) the alarm will be raised with a lower probability. The data analytics level needs to be powered by a server system.

5. Conclusions

We discussed in this paper about several aspects that need to be considered when designing a WSN solution for chainsaw intrusion detection. We analyzed that the spectrum of a chainsaw produced sound has a shape that can be extracted with fairly simple feature extraction techniques and we also illustrated that there is a high degree of similarity between spectra associated to different types of chainsaws. Nevertheless because we are talking about a mechanically produced sound, it has a high degree of periodicity. We also illustrated that the usage of complex feature extractions, like MFCC can visually lower the similarity between feature sets. We also mentioned several sound localization techniques, like A-TDOA and D-TDOA that can be used in our setup. Finally we proposed a WSN architecture for chainsaw intrusion detection that can be scaled and is equipped with fail safe mechanisms. We also discussed about techniques to distribute the processing load in order to minimize the energy footprint of the system.

Acknowledgement

This work was supported in part by the Sectoral Operational Programme "Human Resources Development" 2007-2013 of the Ministry of European Funds through the Financial Agreement POSDRU /159/1.5/S/132397 and in part by the PN II Programme "Partnerships in priority areas" of MEN - UEFISCDI, through project no. 32/2014.

References

[1] L. Czúni, P. Varga, "Lightweight acoustic detection of logging in wireless sensor networks", in Proc. DINWC 2014, pp. 120-125, 2014.

[2] M. Babis, M. Duricek, V. Harvanova, M. Vojtko, "Forest guardian – monitoring system for detecting logging activities based on sound recognition", Proc. of IIT. SRC, pp. 1-6, 2011.

[3] I. C. Yoo, D. Yook, "Automatic Sound Recognition for the Hearing Impaired", IEEE Transactions on Consumer Electronics, vol. 54, no. 4, pp. 2029-2036, 2008.

[4] Q. Guangmeng, Z. Mei, "Using MODIS land surface temperature to evaluate forest fire risk of northeast China", IEEE Geoscience and Remote Sensing Letters, vol. 1, issue 2, pp. 98-100, 2004.

[5] K. Kyzirakos, M. Karpathiotakis, G. Garbis, C. Nikolaou, K. Bereta, I. Papoutsis, C. Kontoes, "Wildfire monitoring using satellite images, ontologies and linked geospatial data", Web Semantics: Science, Services and Agents on the World Wide Web.

[6] Z. Xiong, R. Caballero, H. Wang, A. M. Finn, M. A. Lelic, P. Y. Peng, "Video-based smoke detection: possibilities, techniques, and challenges", in IFPA, fire suppression and detection research and applications - a technical working conference (SUPDET), Orlando, USA.

[7] Bo Lei; Zhijie Zhang; Song Yue and Chenzhong Wang, "Forest smoke detection in video images based on constant speed rotating platform," in Proc. ISPDI 2013 - Fifth International Symposium on Photoelectronic Detection and Imaging, 2013.

[8] G. Oltean, L. Grama, L. Ivanciu, C. Rusu, "Alarming events detection based on audio signals recognition", Proc. of 8th Conference on Speech Technology and Human Computer Dialogue, pp. 83-90, 2015.

[9] M. Ghiurcau, C. Rusu, R. C. Bilcu, J. Astola, "Audio based solutions for detecting intruders in wild areas", Journal of Signal Processing, vol. 92, Elsevier, pp. 829-840, 2012.

[10] L. Grama, C. Rusu, G. Oltean, L. Ivanciu, "Quantization effects on audio signals for detecting intruders in wild areas using TESPAR S-Matrix and Artificial Neural Networks", Proc. of 8th Conference on Speech Technology and Human Computer Dialogue, pp. 1-6, 2015.

[11] L. Tudor, V. Zoicas, L. Grama, C. Rusu, "The SPG (Signal Processing Group) Sound Database", Novice Insights in Electronics, Communications and Information Technology, Issue 15, pp. 62-65, 2014.

[12] J. Tiete, F. Dominguez, et. al. "Sound Compass: a distributed MEMS microphone array-based sensor for sound source localization", MDPI Sensors, vol. 14, no. 2, pp. 1918-1949, 2014.

[13] A. Greensted, "Delay Sum Beamforming", The Lab Book Pages, October 2012.

[14] H. Do, H. Silverman, Y. Yu, "A real-time SPR-PHAT source location implementation using stochastic region contraction (src) on a large-aperture microphone array", in Proc. ICASSP, pp. 121-124, 2007.

20

Studying the effect of existing technologies in a smart grid for reducing the distribution networks losses

Zahra Alizadeh

Sepehr Industrial Group Corporation, Ahwas, Iran

Email address:

Sg.sansgroup@gmail.com

Abstract: Nowadays, population growth and the electrical load demand on the one hand and rising cost of energy generation on the other hand have led to raises reducing the losses in distribution networks as a serious challenge. According to studies, smart grid is posed as the most important applicable solution to eliminate these problems. The purpose of this paper is to examine the impact of communication and information technologies and new control capabilities in smart grids for reducing distribution network dissipations. In this study, a new formulation and an optimal load flow algorithm are used to reduce losses in the smart distribution network. Optimal load flow for a 69-bus IEEE network as well as the installation of the smart network facilities are implemented in the MATLAB and GAMS software and the results show that distribution network losses are greatly reduced.

Keywords: Smart Network, Control Centers, Losses Reduction, Optimal Load Flow

1. Introduction

Increasing needs of electric power supply has made the transmission and distribution lines capacity insufficient in responding to the needs. Since the construction of new power plants and new transmission and distribution lines are not economically viable, power industry has been forced to cope with their problems by adding power, telecommunication and information technologies and new control features to networks and or in other words smart distribution transmission networks [1].

The different components of a distribution grid are shown in Figure 1. As it can be seen, most important infrastructures of distribution network include: distribution substations, control and protection devices, distributed energy resources, smart sensors and measurement devices and control center. Control center will be able to use telecommunications networks and interface equipment which has capabilities such as control, monitoring, protection, etc. receive information at each instant from various network components and issue desired control commands to them [2-9].

Also, the internal network components may also have some connections to control devices, regulating equipment, coordination between the substations, protection and control of sensors between themselves. Control centers of transmission and distribution networks have ability to exchange information between each other.

As it can be seen, by advancing power network structure and evolution in its nature, the necessity of control methods and new tools for network designing and analysis is revealed more than ever.

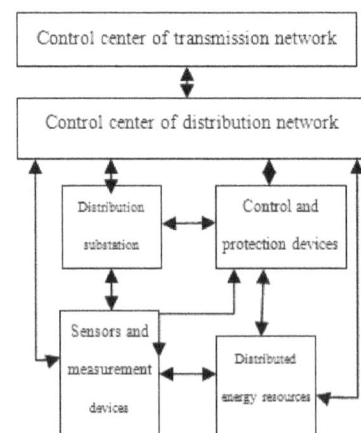

Figure 1. Different components of a smart distribution network[1]

Several research studies [10-16], optimal load flow algorithms for distribution networks are presented in order to reduce power losses and improve the voltage profile where the presence of physical network components such as dispersed generations and controllers based on power electronics devices are addressed.

In this paper, a new algorithm is presented for optimal load flow that can determine the favorable conditions to exploit a smart network with all its new features, including the presence of ICT systems. In this algorithm, role of distribution and transmission control centers is very significant. In this formulation, the active and reactive powers of transmission network loads, that can be smart distribution networks themselves and exchange power with transmission network bi-directionally, are considered as control and state variables which determine their values control center. In fact, a series for the new control and state variables for formulations related to optimal load flow is introduced that determines their values control centers immediately.

The presented algorithm is implemented for an IEEE 69-bus network in which smart network equipment are installed which its results show reduction in the network power losses.

2. The Optimal Load Flow Formulation in a Smart Grid

The optimal load flow formulation in a smart network will be according to equations 1 through 3:

$$Min_{u,s} cobj(x,u,x_s,u_s) \tag{1}$$

$$f(x,u,x_s,u_s) = 0 \tag{2}$$

$$g(x,u,x_s,u_s) \prec 0 \tag{3}$$

Where Cobj is the objective function, x is the state variable vector in a traditional network, xs state variable vector in the smart network, u the control variable vector in the traditional network and us control variable vector in the traditional network and smart network. In fact, the control and state variables are divided into two categories, first are control variables and traditional network state (x,u) whose values are determined according to the conventional characteristics of the power networks. The second are the control and state variables related to the smart network (us, xs) whose values are determined by smart control centers through real-time received data from operators immediately.

2.1. Introducing the Objective Functions

The objective function we have considered in the distribution network is minimizing the total lines losses.

$$c_{obj} = \min(\sum_{l=1}^{nl} p_l) \tag{4}$$

So that Pl is active power losses in Lth line and L is number of lines. But the active power losses in each distribution lines are derived according to Equation 5.

If we consider a part of the distribution network like

Figure 2, that power flowing from bus i to the line is (Si = Pi + jQi) and received power at bus j equals (Sj=Pj+jQji) and amount and angle of voltages in source and destination buses, respectively, from δ_i و vi و δ_j و vj, We have:

$$Plosses_{(i,j)} = \frac{r_{i-j}(P_j^2 + Q_j^2)}{V_j^2} \tag{5}$$

The active power losses between buses i and j, r (i, j) is the resistance between buses i and j, Vj is voltage at bus j and Pj and Qj are active and reactive powers at bus j respectively.

Equality and inequality constraints on the transmission line connected to the distribution network according to reference [17] and relationships related to the amount and angle of the reference voltage have been determined according to [18].

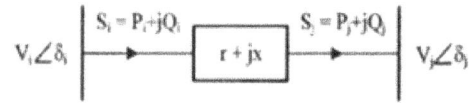

Figure 2. A part of radial distribution network.

3. Optimal Load Flow Algorithm in a Smart Distribution Network

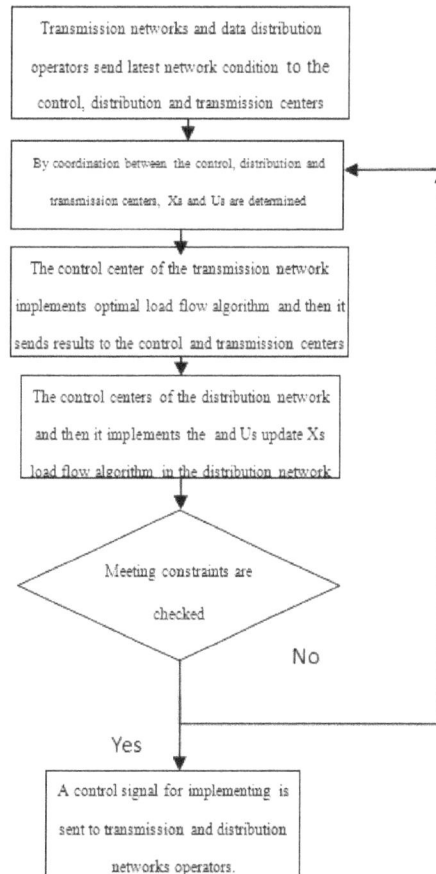

Figure 3. The optimal load flow algorithm in a smart network.

In this study, backward and forward sweep method for optimal load flow in a smart distribution network is used.

Because this method has high operation speed and will be appropriate in the smart network whose control center will be able to receive real-time data and do optimal load flow in a short time to do. The distribution control center determines Xs and Us based on received data, again and implements optimal load flow in the distribution network.

If at this step, all the objective functions and constraints are met, the results of load flow are sent to the transmission and distribution networks operators. Otherwise, Us and Xs should be determined by coordination between distributed and transmission control centers, again.

Transmission networks and data distribution operators send latest network condition to the control, distribution and transmission centers

By coordination between the control, distribution and transmission centers, Xs and Us are determined.

The control center of the transmission network implements optimal load flow algorithm and then it sends results to the control and transmission centers

The control centers of the distribution network update Xs and Us and then it implements the load flow algorithm in the distribution network.

4. The Studied Network

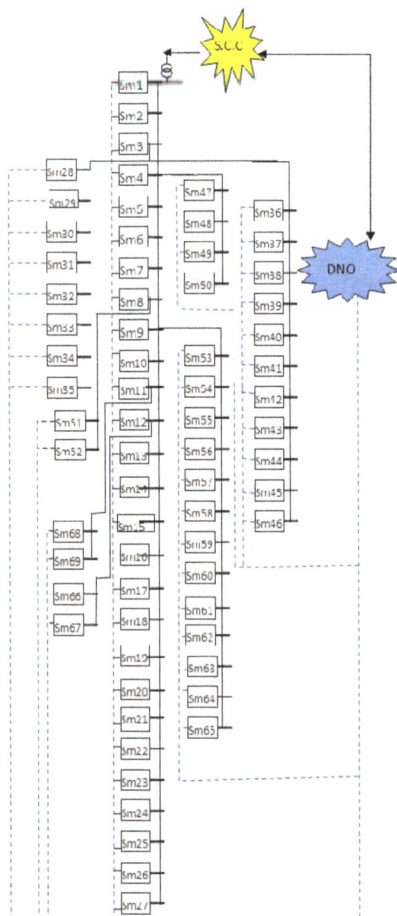

Figure 4. *69 buses IEEE radial distribution network that convert to smart distribution grid & smart control center could manage system through DNO(SM is abbrevate of smart meter).*

The studied network is an IEEE standard 69-bus radial distribution network (Figure 4) in which the smart network equipment are installed. Including the equipment, smart meter (sm) can be noted as well as the load centers.

Operator of the distribution network will be able to receive total real-time data through telecommunication lines which are shown by dashed lines and send them to the control center for analysis and implement commands issued by the control center in the entire network. Also, this distribution network has possibility to exchange the power with the transmission network in two directions.

5. Simulation Results

In this section, the results of the optimal load flow have been studied in each of the active distribution networks, in four different modes. In the first case, we assume that the distribution network is not a distribution network originally and can only receive energy from the transmission network. In the second case, dispersed generations are installed in a distribution networks and the control center allows the operator to feed the network as an island. In the third case, not only dispersed generations are installed in the distribution network, but the distribution network operator can exchange energy with the transmission network. The fourth case is similar to the third, but the difference is that in the fourth case, exchanged power between the transmission network and the distribution network is set by the presence of existing information and communication technologies in the smart networks and coordination between operators and the control centers of distribution and transmission networks such that The active power losses become minimal in the distribution network. The DG capacity installed at bus no.14 is 750 KW and the DG capacity installed at bus no.65 is 1.65 MW. Optimization in fourth case is possible by virtue of the connection between MATLAB and GAMS software.

MATLAB software transfers loads of flow data to the GAMS software and GAMS sets imported energy from transmission lines to the distribution network in a way that the distribution lines losses will be minimal, according to Equation 4.

GAMS software employs a non-linear programming method with discontinuous derivatives in optimization.

The voltage profile is shown in Figures 5 to 8. As shown in Figures 9 and 10, the minimum voltage is in each case as follows respectively: 0/913,0/943, 0/95. and 0/945 Pu. And the active power losses are 200 kW, 160 kW, 230 kW and 190 kW respectively. As you can see in the fourth case, in presence of the information and communications technology systems that have possibility of establishing a bi-directional connection between the control and operations centers than the third case that because the lack of required technology, there was no possibility of the flexibility in the exchanged power between the transmission and distribution, distribution lines losses has fallen by about 20 percent.

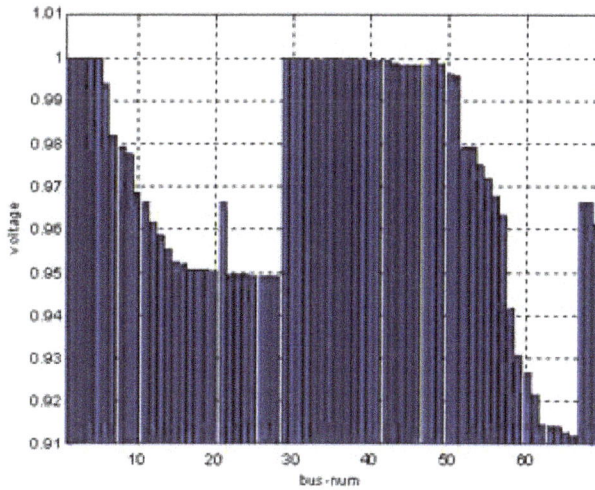

Figure 5. *The voltage profile of 69-buses IEEE radial distribution system (first state)*

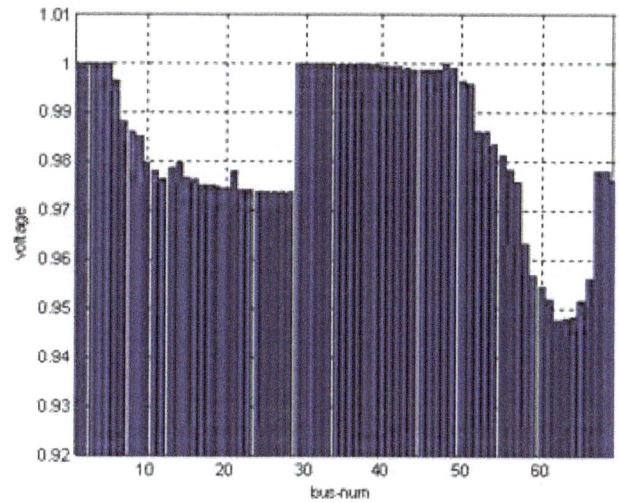

Figure 6. *The voltage profile of 69-buses IEEE radial distribution system (second state)*

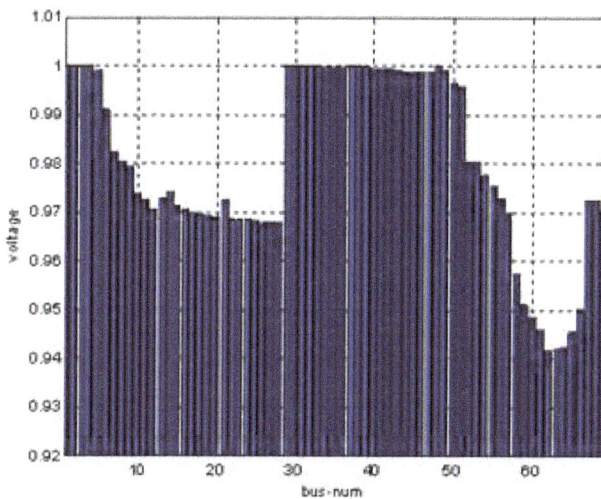

Figure 7. *The voltage profile of 69-buses IEEE radial distribution system (second state)*

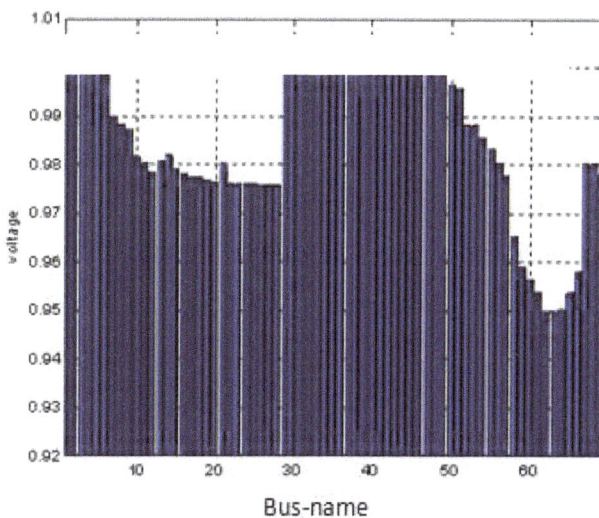

Figure 8. *The voltage profile of 69-buses IEEE radial distribution system (fourth state)*

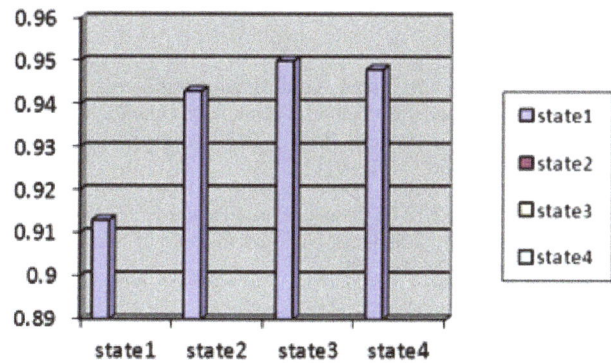

Figure 9. *Comparison of the minimum voltage at 4 different cases*

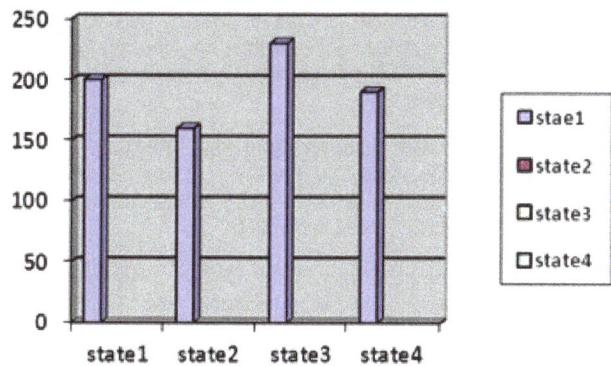

Figure 10. *Comparison of active power losses in the network at 4 cases in kW*

6. Discussion and Conclusion

Due to the reason that the conventional electricity network structure does not meet the increasing requirement for electric energy, the smart network, as an economical solution is raised to solve this problem. By advancing network and adding infrastructures such as dispersed generation resources, smart control centers, information and communication technologies and smart meters, more advanced computational algorithms are required for load flow control in the network to be consistent with the new infrastructures.

In this paper, an attempt has been done to show impact of control centers of smart transmission and distribution in the variables (xs.us), in the optimal load flow formulation. Furthermore, an optimal load flow algorithm was presented which can contain communication between the smart control centers together and with the network operators. In the algorithms that have been mentioned in previous sections, the presence of the physical components of the smart networks such as dispersed generations were considered to lower losses, but here, in addition to the physical components, the special attention has been made to the presence of the information and communication technology systems of the smart networks and special. Then the algorithm for a of 69-bus IEEE standard network was implemented in which all smarting network equipment are installed.

The results showed that the distribution network losses have been reduced about 20%.

References

[1]   IEEE Guide for Smart Grid Interoperability of Energy Technology and Information Technology Operation with the Electric Power System (EPS), End-Use Applications, and Loads. IEEE Std 2030, 2011

[2]   Second of the NIST Framework and Roadmap for Smart Grid Interoperability Standards, National Institute of Standards and Technology, October, 2011

[3]   Telvent utilities group, "the advanced distribution management system" The indispensable tool of the Smart Grid era, August 2010, www.telvent.com/smartgrid | 1.866.537.1091

[4]   Pei Zhang, Fangxing Li, Navin Bhatt "Next-Generation Monitoring, Analysis, and Control for the Future Smart Control Center" IEEE Transactions on smart grid, VOL. 1, NO. 2, September 2010

[5]   Fangxing Li, Wei Qiao, Hongbin Sun, Hui Wan , Jianhui Wang , Yan Xia , Zhao Xu, Pei Zhang " Smart Transmission Grid: Vision and Framework" IEEE Transactions on smart grid, VOL. 1, NO. 2, September 2010.

[6]   James A. Momoh "Smart Grid Design for Efficient and Flexible Power Networks Operation and Control" pp. 1-8,(March 2009),

[7]   Anjan Bose, "Smart transmission grid applications and their supporting infrastructure" This paper appears in: Smart Grid, IEEE Transactions on, Volume: 1, Issue: 1, Page(s): 11 – 19, June 2010

[8]   Chun-Hao Lo, Nirwan Ansari "The Progressive Smart Grid

[9]   M.E. Baran, F.F. Wu, "Optimal sizing of capacitor placed on radial distribution systems", IEEE Transactions on Power Delivery, vol. 4, num. 1, pp. 735-743, 1989

System from Both Power and Communications Aspects" This paper appears in: Communications Surveys & Tutorials, IEEE Third Quarter Volume: 14, Issue: 3 Page(s): 799 – 821, 2012.

[10]   A. Rost, B. Venkatesh and C. P. Diduch" Distribution System with Distributed Generation Load Flow"   Department of Electrical and Computer Engineering New Brunswick, NB, Canada, 2008

[11]   A. Milo, A. Martinez, M. Rodríguez, A. Goikoetxea" Dynamic power flow tool development for low voltage networks analysis with high penetration level of distributed generation" Renewable Energy & Power Quality Journal, March 2008.

[12]   K. Vinoth Kumar, M.P. Selvan, "A Simplified Approach for Load Flow Analysis of Radial Distribution Network" International Journal of Computer and Information Engineering 2:4, 2008.

[13]   M.Kowsalya, K.K.Ray, Udai Shipurkar, and Saranathan" Voltage Stability Enhancement by Optimal Placement of UPFC" 310 Journal of Electrical Engineering & Technology Vol. 4, No. 3, pp. 310~314, 2009.

[14]   M.Padma lalitha,N.Sinarami reddyv,C. Veera reddy " Optimal dg placement for maximum loss reduction in radial distribution system using abc algorithm" International journal of reviews in computing , 2009-2010

[15]   M. Ravichandra Bubu, D. Mary" Power Loss Minimization for Power Quality Radial Distribution System" European Journal of Scientific Research Vol.78, pp.507-521, (2012)

[16]   A. Lashkar Ara, A. Kazemi, S. A. Nabavi Niaki, Multiobjective"Optimal Location of FACTS Shunt-Series Controllers for Power System Operation Planning" IEEE Transactions on power delivery, Page(s): 481 – 490,2012

[17]   Stephan Koch, Göran Andersson," Implementation and evaluation of a distribution load flow algorithm for networks with distributed generation" Semester Work, Vanco Janev, EEH - Power Systems Laboratory, Swiss Federal Institute of Technology, Zurich, in March 2009

[18]   Jiaqi Liang, Ronald G. Harley, Ganesh K. Venayagamoorthy" Adaptive Critic Design based Dynamic Optimal Power Flow Controller for a Smart Grid" Computational Intelligence Applications In Smart Grid (CIASG), Page(s): 1 –8 , IEEE Symposium on   11-15 April 2011

[19]   Sumit Paudyal, Claudio A. Ca~nizares, Kankar Bhattacharya" Optimal Operation of Distribution Feeders in Smart Grids"Smart Grid, IEEE Transactions on, Volume: 3 Page(s): 59 – 69, March 2012

21

Proper communicative protocols in building management system

Fereshteh Kheirabadi[1], Seyyed Reza Talebiyan[2]

[1]Islamic Azad University, Science and Research Branch of Khorasan Razavi, Neyshabour, Iran
[2]Imam Reza International University, Computer and Electronics School, Mashhad, Iran

Email address:
fe.kheirabadi@gmail.com (F. Kheirabadi), talebiyan@gmail.com (S. R. Talebiyan)

Abstract: Increasing energy consumption, finite energy resources, negative effects of excessive use of energy on environment, and increasing energy price forces administrators and consumers to look for different ways of saving and accurate consumption. Implementing smart systems and managing energy consumption in buildings are examples of applying modern technology. Implementing this technology not only reduces energy consumption, but also provides appropriate situation and increase the welfare of residents. This research studies the concept of Building Management System (BMS) in construction industry, especially reviews some communicative protocols in building management system, and after comparing and analyzing them, proposes X-10 protocol in home automation from cost, time consuming and simplicity of installation aspects.

Keywords: Building Management System (BMS), Energy Management, Communicative Protocols

1. Introduction

Due to the increasing development of implementing modern technologies in the field of construction, competition in producing more modern buildings, and some issues including reducing energy consumption, saving in strategic expenses of building, and increasing the level of residents' welfare, the implementation of a combination of various technologies under the title of "Building Automation", "Smart Building", and/or "Building Management System (BMS)" are increased [1]. From a different dimension BMS can be explained as: "Energy + Information = less energy" [2]. In designing the high buildings, engineers and managers look at BMS as a real need, but not as a luxurious technology. Constant increase in the price of energy and the need for having access to accurate and integrated information justify using BMS by building investors. Comparing with the old standard controlling system, the advantage of BMS is that optimum integrated managerial and technical control is possible. In economic calculation for implementing this system, one must pay attention to this point that about 80% of equipment price in heating and cooling controlling systems which are used jointly, are also expense in common

and old systems. Another important point in choosing this system is the permanent supporting system by producer or its local representative. Hardware and software capabilities of BMS is nearly the same among the international producers, but the important point is that the art of using this capabilities is in the process of designing, programming, installing and especially setting up the system. The above mentioned advantages for this system, specially optimizing energy consumption, reducing maintenance expense, managing heating-cooling and electric facilities, and increasing the level of residents' comfort directly depend on the quality and design of programming and maintaining it, and also it has a direct relation with the scientific level of BMS operator engineers in information analysis [3].

This Article introduces the concept of using building management system, and the advantages of its implementation. The components of BMS and especially various types of communicative protocols are introduced. At first, all communicative protocols and their technical features are introduced. Afterwards, a comprehensive comparison is done. On this basis, this study proposes a road map for choosing an appropriate communicative protocol in building smart management system. Finally, it reviews the impact of BMS on energy consumption, fund return, as well as environment.

2. The Concept of Applying Building Management System (BMS)

Since monitoring the accuracy of any controlling system in industrial, commercial, office and residential environment separately requires time, energy and presence of manpower in the place, it is necessary to implement an integrated management system which is able to display information and arrange all smart controlling systems in short time. BMS as a unique and modern way to fulfill this need have been applied in most of developed countries and it has proven its abilities in the field of managing all kinds of smart controlling systems applying in industrial and non-industrial environments.

Smart controlling systems are a collection of equipment including controllers, I/Os, and field devices which are connected with various building's mechanical and electrical plants (air generators, air conditioners, chillers, cooling towers, boilers, pumps, heat exchangers, high voltage and low voltage switchgears, lightings, alarm systems, CCTV, lifts and escalators, etc.). It also carries out a set of performance and controlling scenarios (temperature, humidity, pressure, liquid level, monitoring values, status and pictures, register alarms, issuing commands to open/close, active/inactive, and on/off for mechanical, electrical, audio, and video equipment and lighting lines, defining schedule, compensating change in external temperature, summer/winter function, etc.) to make efficient the consumption of energy and provide safety and security of the buildings stand alone [4].

The building management system actually is a network of local control systems connected to a central (server) and lateral (user stations) computer which aims to aggregate, monitor, and integrate information and change data values or status of remote smart controlling systems. This system eliminates the users' need to refer to the local controlling system scattered in different parts of building. Implementing building management systems leads to saving in time and being aware, quickly and timely, of the latest changes in data and performance of local controlling systems and plants, and also ordering the requirements when necessary.

All energy consumption control scenarios are merely implemented in the local smart controlling systems and these systems actually are a series of building management systems which will continue their operations if any error occurs and BMS grid gets out of circuit [3].

Nowadays, building management systems are developed based on web; its best feature is taking advantage of World Wide Web and controlling building through common communication systems in the world. For this purpose, after constructing a site for the building and entering the user name and password, you can control your building from anywhere. In these buildings, it is possible to install an electronic display panel in especial places to display various information from the building controlling system; it will be joyful for the residents.

The most important purposes of applying BMS in buildings are using the economic advantages, reducing energy consumption, and creating a secure and comfortable environment inside them. The general advantages of applying BMS are as follows:

- Creating a secure and comfortable environment for the residents;
- Optimal use of equipment and increasing their useful life;
- Providing controlling system capable to schedule the operation;
- Reducing maintenance expenses;
- Optimizing and energy saving;
- Not needing a permanent building contractor;
- Providing the facility to monitor and control all under control points through a PC, cell phone or internet;
- Eliminating the possibility of interference, by integrating installation and system management;
- Setting energy consumption priorities smartly, in the case of necessity;
- Removing operator's errors;
- Providing statistic reports from facilities and their operation to optimize consumption and operation [3].

A smart building has a dynamic and economic environment by integrating four key elements, i.e. systems, structure, services and management.

A smart building provides its facilities through smart control systems. These systems are as follows:

- Gas leaking alarm system;
- Fire alarm system;
- Theft alarm system;
- Earthquake waning system;
- lighting control system;
- calling telephone system;
- control via remote control, digital key, inside home phone, cell phone and internet;
- scenario description;
- Integrated controlling system [5].

3. Elements of Building Smart Management

Like the other controlling systems, BMS is consisted of three parts namely: Sensors, Controllers, and Actuators. These three parts are connected through a connection mechanism which is consisted of two important parts a) Conductor such as wire, fiber optics, radio waves and b) Communication protocol or spoken language of components.

In fact sensors, controllers and actuators are connected through conductors, according to spoken language or communicative protocol [6]. Here we explain the protocols.

3.1. Different Kinds of Protocols

Protocol is a series of common language rules to communicate signals. The important advantage of applying BMS according a standard protocol is increasing the rate of compatibility among different elements of system's control equipment. Protocols have different types; each has its own advantages and disadvantages, such as:

3.1.1. KNX

KNX is an international building standard (ISO/IEC 14543). Applying information technology, KNX connects different equipment such as sensors, operations, controllers and terminals, and supports different net mediators, including TP, PL and wireless solution which are called KNX_RF. This protocol is created by combining three European technologies –EHS, BatiBUS, and EIB- to control a building or a house [7].

Advantages:

- Its protocol is open (it is possible to communicate with other supplier of BMS);
- Its life cycle in mechanical relay is up to 3 million times and other electrical equipment's is up to 1 million operations;
- The size of keys are standard;
- The network is secure (network cable can pass by power cable);
- It is controllable by computer graphic software, in which all places are shown graphically;
- It is highly safe to use it (there is just 29V electricity in the place of controlling keys);
- There is no noise taking and noise putting, because it uses separated and shielded cable;
- There is a microprocessor for each sensor (so system is flexible when a sensor is replaced another one);
- It is possible to control the weather condition, humidity, electric current consumption, lighting keys, and heating-cooling system [7].

3.1.2. X-10

X-10 is one of the most efficient open standards which is designed and produced for wiring communication in home automation. The main feature of X-10 protocol is sending controlling commands to appliances through the city's power supply network, in this way there is no need to rewiring or manipulating the situation of home electricity. To design a smart home by this protocol, you must connect your appliance by a medium to the outlet available at home. In his way, you are able to turn the appliance on and off by a sender adaptable with this protocol, according the set code.

Advantages and disadvantages:

- It is cheap to develop it completely;
- It is adaptable with all brands;
- X-10 remotes are active by RF protocols;
- The maximum number of appliances controllable with X-10 is 256.

This protocol is noise putting for some systems like TV and radio, and it is noise taking of systems which have electric engine; to reduce the noise several filter must be used.

3.1.3. LON Works (Local Operating Networks)

LON Works is an open protocol on the basis of LON Talk protocol. The Advantages of this protocol are as follow: Its standard is based on ISO-14908; Installation of this protocol's controlling grid is more simple than normal grids; It has high capacity for cooperation; It reduces the installation expenses (comparing with X-10, it is more expensive); One can choose the connection interfaces freely; Radio waves, optical fibers, coaxial, infrared, etc. can be placed in the using physical layer [8].

3.1.4. ZigBee

This standard uses the series of protocols for wireless communication in short distances. This protocol is two sided which connects the electrical devices with low energy and low expense. By this technology, all parts of home can be controlled, such as adjusting temperature, humidity level, etc.

Advantages:

- This standard is under IEEE 802.15.4;
- Home and industrial automation and any kind which needs a controlling grid can use this technology;
- Its board, expense, electricity consumption and data transfer rate is low;
- It supports more than 65635 devices in each network;
- It transfers data confidently, due to solid and self-constructed combinational gridding [9].

3.1.5. BACnet (Building Automation Control Network)

BACnet is an open protocol of building automation control network. The distinguishing point of this protocol is its exclusive structure for building automation. As it is highly sensitive, much attention is paid for applying in fire alarm system (using in large networks) [3].

3.1.6. S-BUS

S-BUS is a closed protocol which has no valid international standard. The advantages and disadvantages of this protocol include:

- It has just one producer (consumer is depended on producer and its products);
- The life cycle of devices is 300000 times (10 times less than KNX);
- There are limited options.

3.1.7. Z-Wave

Z-Wave is a two sided wireless protocol which is designed for limited power and band width. It is mostly efficient in home automation. The advantages of Z-Wave are:

- It is a good choice for temperature sensors and controlling devices, as its energy consumption is low, its expenses is low, its transmission is two sided, it has MESH technology and it is supported via battery to battery.
- The hardware size is really small, so it is suitable to combine with other devices;
- As it works on a specific frequency, it has no interference with the other wireless devices [10].

3.1.8. Bluetooth

Bluetooth technology is entirely determined as a short distance wireless communication. The main idea of this technology is to minimize the interfaces incompatibility of the different devices. A range of the Bluetooth devices is about from ten to hundred meters with line of sight contact between devices. At used indoors the range decreases depending on the surrounding factors. Bluetooth operates in the unlicensed frequency band 2.4 GHz. Because this frequency band is also used by another wireless systems (IEEE 802.11), it's necessary to prevent the interactions in-band.

A carrier frequency is changed according to the selection scheme and it's determined by a Bluetooth address and clock of a control device. The data is transferred using time division duplex method (TDD) in packets inside of a short time interval (time slot). A packet format is exactly defined. The communication proceeds in certain kind of a personal local network called piconet. The maximum number of the active Bluetooth devices can be 8 in the piconet. The maximum data rate doesn't exceed 723 kbit per second.

Advantages of Bluetooth are as follows:

* It removes the interfaces incompatibility of different types of the Bluetooth devices;
* Between devices, there isn't required straight visibility (advantage against IrDA);
* Mutual communication of different types of the Bluetooth devices (PC, mobile phone, PDA, notebook, headset, printer, modem, ...);
* It is easy to establish connection;

* Relative high range (up to 200 meters) beyond a low transmitting powers (it depends on a class device);
* Resistance in the face of interference (a system with spread spectrum);
* Small size of a radio chip and small size of a Bluetooth module;
* Technology can be implemented into optional equipment [11].

3.2. Comparing Protocols

In table 1, some valid protocols for making buildings smart are briefly compared.

KNX is the most prominent building (residential, commercial, office, industrial, hotel) automation protocol. In addition to the above mentioned important points in the table, this protocol has different communicative media (RF, PL, TP) and a wide range of choices in configuration modes (S, E, A) [7].

As they are highly safe and adaptable with other products, Lon Works and BACnet protocols are frequently used in building automation. Applying these protocols, building owners are able to choose the best technology and services provided by each company, without concerning the adaptability of selected systems with previous one [8].

X-10 is the most popular protocol in home automation, as it does not need rewiring in building, so installation cost and time are minimum.

In wireless applications ZigBee and Z-Wave are used. Z-Wave protocol is mostly used in Home Automation and Sensor Network.

Table 1. Comparing some protocols.

Protocol	Standard	Application	Adaptability with other products	Security	Price	Simplicity of Implementation	Information transfer
KNX	ISO/IEC 14543	Building automation	Is	Very high	Start from average	Average	Wire PLC Wireless (not common)
X-10		Home automation	Is	moderate	Low	Simple	Wire
Lon Works	ISO-14908	Home and industrial automation	Is	moderate	Average	Average	Wire PLC Wireless (not common)
ZigBee	IEEE802.15.4	Home and industrial automation	Is	Very high	Low	Simple	Wireless
BACnet	ISO 16484-5	Building automation	Is	More than moderate	Start from average	Average	Wire PLC ZigBee wireless
Z-Wave		Home automation	Is	More than moderate	Low	Simple	Wireless

4. Other Important Points in BMS

4.1. Accessing and Controlling Facilities in Smart Building

To access and control facilities in smart building, there are different ways, including:

* Central control panel;
* Radio wave control (controlling via RF technology, as it doesn't need direct vision, it is efficient in long distances);
* Remote control (via SMS and internet)
* Scenario (team work by pressing one key)
* Automation (controlling repetitive and preset tasks

automatically) [5].

5. BMS System's Strategies to Reduce Energy Consumption

The most common strategies applied by BMS designers are as follows:

- Turning the equipment on and off according the function schedule;
- Looking out equipment if necessary;
- Utilizing minimum of allowed capacity in utilizing the equipment;
- Limiting demand which causes electricity cut off in equipment if loading is more than limited rate;
- Monitoring equipment situation by trained operators and utilizing data to solve their problems and review their efficient functions [3].

5.1. Fund Returning

Expenses and executive levels of making a building smart depend on the type of utilization (residential, office, commercial, etc.), area, dimensions of spaces, the number of stories, and the number of units. The main purpose of implementing BMS system in a building is to save energy and true and optimum consumption of facilities. So the fund invested for implementation of BMS will return.

5.2. Building and Environment

As a main consumer of electricity, buildings consume about 70% of electricity consumption (in Europe) [12]. For example, in the US building sector consumes 40% of energy and it is responsible for producing 45% of greenhouse gas. For efficient management of building, some mechanisms must be applied to reduce the level of energy consumption and greenhouse gas production, without any change in comfortable living standards [13]. For this purpose we propose green buildings [14].

5.3. Failed BMS Project

The advantages of building management system are achieved, when the system is installed and maintained appropriately. Otherwise, 4 problems lead the projects to failure:

- Neglecting the duties
- Neglecting some effective factors
- Hiding information
- Neglecting the real world.

6. Discussion and Conclusion

As the technology moves toward creating a smart network, the need for smart buildings and the related systems will become important. During last decades, important changes have been occurred in building designing, controlling, and implementing and their related systems [2].

By emergence of microprocessors, computerized controlling systems have the vital role in the most societies and building industries. When systems operate appropriately and save energy, their reliability will increase.

This paper investigated different protocols and finally proposed that KNX, LonWork and BACnet protocols are appropriate for building automation, because of their high safety, adaptability with other products, rational prices and simple installation.

Internet, in this process, has a major role, and now is the best time to be ready for a future with efficient communication.

References

[1] Jan Bozorgi. A., Ghannad. Z., Building Smart System, Kaison Quarterly, 43, winter 2009.

[2] Lawrence. T., Thomas M. Lawrence, Richard T. Watson, Marie-C. Boudreau, Kyle Johnsen, Jason Perry, Lan Ding, A new paradigm for the design and management of building systems, Elsevier, 51, 56-63, 2012.

[3] Sinopoli. J., Smart building systems for Architects, Owners, and Builders, Butterworth-Heinemann: USA, 2010.

[4] Yin. Hang, Building Management System to Support Building Renovation, Department of Civil and Environmental Engineering, UCC, Snapshots of Doctoral Research, University College Cork, 2010.

[5] Sripan. Meensika; Xuanxia Lin; Ponchan Petchlorlean; Mahasak Ketcham, Research and Thinking of Smart Home Technology, International Conference on Systems and Electronic Engineering (ICSEE'2012) Phuket (Thailand), 2012.

[6] Abiodun. Iwayemi; Wanggen Wan; Chi Zhou, Energy Management for Intelligent Buildings, Energy Management Systems, Dr Giridhar Kini (Ed.), ISBN: 978-953-307-579-2, InTech, 20011.

[7] KNX Association, KNX handbook for home and building control, ZVEI & ZVEH: 2013.

[8] Merz, Hermann, James Backer, Viktoriya Moser, Thomas Hansemann, LeenaGreefe, ChristofHübner, Building Automation: Communication systems with EIB/KNX, LON and BACnet, Berlin: Springer, 2009.

[9] Pankaj. Jadhav; Amit Chaudhari; Swapnil Vavale, Home Automation using ZigBee Protocol, Department of Computer Engineering, University of Pune, India, 2014.

[10] John Robles. Rosslin; Tai-hoon Kim, Applications, Systems and Methods in Smart Home Technology: A Review, International Journal of Advanced Science and Technology, Vol. 15, February, 2010.

[11] Mikeska, Zdenek, Radio Specification of the Bluetooth System, Institute of Radio Electronics, Faculty of Electrical Engineering and Communication, Brno University of Technology, 2009.

[12] Agarwal. Yuraj, Bharathan Balaji, Rajesh Gupta, Jacob Lyles, Michael Wei, Thomas Weng, Occupancy-Driven Energy Management for Smart Building Automation, BuildSys, November 2010.

[13] Ko-Yang Wang; Lin, G.; Chou, P.; Chou, A. Leverage smart system services technology for smart green building management, Institute for Information Industry, IEEE, Berlin, 2012.

[14] Anastasi, Giuseppe; Francesco Corucci; Francesco Marcelloni. An intelligent system for electrical energy management in buildings, International Conference on Intelligent Systems Design and Applications (ISDA), IEEE, 2011.

22

A Detailed Comparison Between FOC and DTC Methods of a Permanent Magnet Synchronous Motor Drive

Navid Maleki[1, *], Mohammad Reza Alizadeh Pahlavani[2], Iman Soltani[2]

[1]Department of Electrical Engineering, Saveh Branch, Islamic Azad University, Saveh, Iran
[2]Faculty of Electrical Engineering, Malek-Ashtar University of Technology (MUT), Tehran, Iran

Email address:
NavidMaleki86@yahoo.ca (N. Maleki), MR_Alizadehp@iust.ac.ir (M. R. Alizadeh), i_soltani@ikiu.ac.ir (I. Soltani)

Abstract: This paper focuses on detail comparison between two common control methods including Field Oriented Control (FOC) and Direct Torque Control (DTC) of the Permanent Magnet Synchronous (PMS) motor. The main characteristics of the motor such as torque, flux and speed under different operation conditions are studied and the advantages of FOC and DTC are obtained. It can be concluded that although both the DTC and FOC methods have different structures but the motor has the same behavior on the control methods. Thus, it is concluded that both the methods can be implemented as the Direct Flux Ccontrol (DFC) and can be applied as an optimized method of PMS motor control in the industry applications.

Keywords: PMSM, Vector Control (VC), FOC, DTC, Optimized Control Method

1. Introduction

In the recent years, domain of PMSM's applications in different industries such as electricity, gas, oil, transportation, military and etc. have been developed because of their two natural specifications. In the most of variable speed industrial applications, high power density, higher rigidity, lower volume, low inertia and high efficiency compared to DC and induction motors, are superiority indices of these motors. Discovery of new magnets have been a great help in development of designing and producing process of these kinds of motors in the world of industry. These kinds of magnetic materials have very high energy and resistance against demagnetizing property and this is resulted that new generation motors have lower volume compared to past common motors which took more spaces because they used Ferrite magnets or Aluminum Nickel Cobalt (AlNiCo) [1].

The optimized control of these motors to promote effectiveness and efficiency in the industry is something necessary and inevitable. In this way in his paper by studying control methods of synchronous permanent magnet motors such as FOC and DTC [4-9], and optimized control method is proposed to control a PMSM with capability of using in the industry.

This paper focuses on detail comparison between two common control methods including Field Oriented Control (FOC) and Direct Torque Control (DTC) of the Permanent Magnet Synchronous (PMS) motor. The main characteristics of the motor such as torque, flux and speed under different operation conditions studied and the advantages of FOC and DTC are obtained. It can be concluded that although both the DTC and FOC methods have different structures but the motor has the same behavior on the control methods. Thus, it is deduced that both the methods can be implemented as the Direct Flux Control (DFC) and can be applied as an optimized method of PMS motor control in the industry applications.

2. Dynamic Modeling of Motor

Dynamic model of PMSM in the rotational two phase system d-q in the steady state is achieved:

$$v_d = R_s i_d + \frac{d\Psi_d}{dt} - \omega_e \Psi_q \tag{1}$$

$$v_q = R_s i_q + \frac{d\Psi_q}{dt} + \omega_e \Psi_d \tag{2}$$

$$\Psi_d = L_d i_d + L_{md} i_D + \Psi_f \tag{3}$$

$$\Psi_q = L_q i_q + L_{mq} i_Q \tag{4}$$

$$T_e = \frac{3}{2}\frac{p}{2}[\Psi_f i_q + (L_d - L_q)i_d i_q] \qquad (5)$$

$$\frac{p}{2}(T_e - T_L) = J\frac{d\omega_r}{dt} \qquad (6)$$

Presentation of flux vector of machine in respect with reference frame of rotor is showed in "Fig. 1".

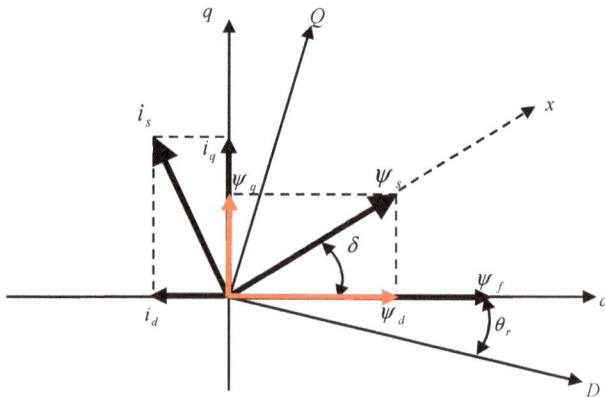

Figure 1. Flux vector in static D-Q system and rotational d-q system [2]

3. Vector Control Based on Stator Flux Reference Frame

In the method, real axe of x-y system is along with stator flux vector and control is accomplished along with field "Fig. 2". Current variables, based on (7) and (8) can be transformed from rotational rotor system to reference stator flux system and vice versa and electromagnetic torque relation in the x-y stator system is calculated by (9).

$$\begin{bmatrix} i_x \\ i_y \end{bmatrix} = \begin{bmatrix} \cos\delta & \sin\delta \\ -\sin\delta & \cos\delta \end{bmatrix}\begin{bmatrix} i_d \\ i_q \end{bmatrix} \qquad (7)$$

$$\begin{bmatrix} i_d \\ i_q \end{bmatrix} = \begin{bmatrix} \cos\delta & -\sin\delta \\ \sin\delta & \cos\delta \end{bmatrix}\begin{bmatrix} i_x \\ i_y \end{bmatrix} \qquad (8)$$

$$T_e = \frac{3}{2}\frac{p}{2}|\Psi_s| i_y \qquad (9)$$

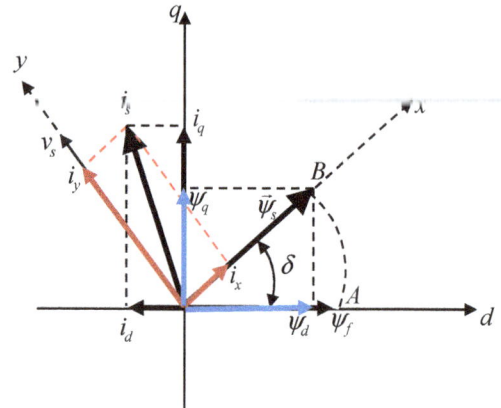

Figure 2. Phase diagram of PMSM in different systems [3]

Based on (9) relation, if flux linkage amount of stator is hold constant, torque will have a direct relation with y components of stator flux and will be controlled with this quantity. As Ψ_s is along with x axe and this vector has no component along with y axe ($\Psi_s = 0$), therefore by controlling i_x, we can easily control the stator flux. To achieve optimized operation PM motor in this method, by controlling stator flux amount, directed state or along with orator flux vector in transient state will be calculated.

In non-salient pole motors (lack of salience), because $L_q = L_d$, machine is not able to produce reluctance torque so based on relation (5), current component id has no role in torque production and for achievement to MTPA (maximum torque per Ampere), it is necessary that current equal zero. This cause simplicity of practicing of this control method in these kinds of motors, But in salient pole motor which is $L_q \neq L_d$, motor will produce both reluctance and electro mechanic torque and using this method will be a little different. In this mode, torque as a function of current in d-q system, is not prolonged to achieve MTPA, it is necessary to calculate minimum distance from in needed torque curve. "Fig. 3" shows a diagram of vector control of PMSM in x - y system. [1, 4, 5]

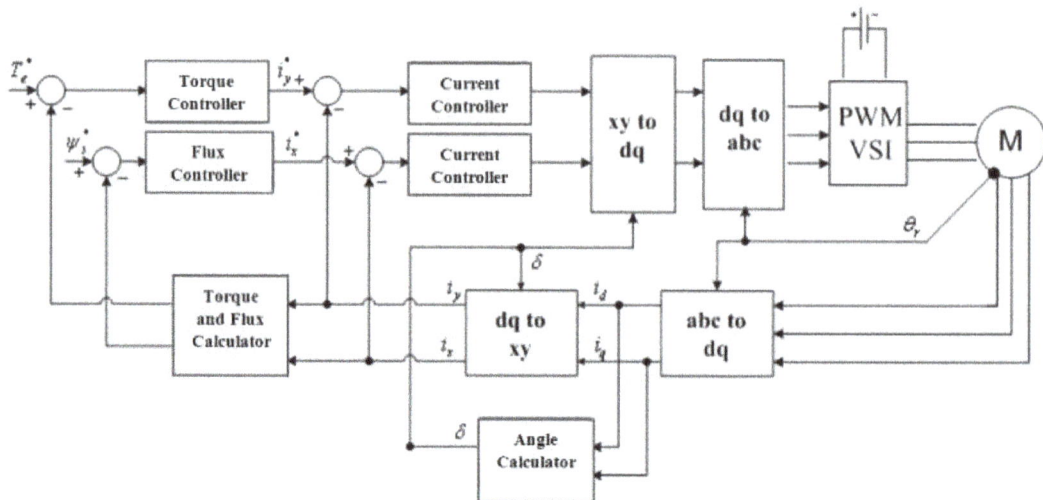

Figure 3. Block diagram of Field-Oriented Control

4. Direct Torque Control

Direct torque control method is one of modern methods to control different kinds of AC motors which is a never method compared to directed flux control method. Basic relation of torque which is a base for practicing this method is defined next.

$$T_e = \frac{3P}{4L_d L_q} |\Psi_s| [2\Psi_f L_q \sin\delta - |\Psi_s|(L_q - L_d)\sin 2\delta] \quad (10)$$

In this relation, first component of torque is resulted from excitation field and is produced by permanent magnet and second component of torque which is named reluctance torque is because of machine configuration.

By taking a derivative from (10) relation compared to time in the instant t = o (i.e. instant of load variations application), we have:

$$\left.\frac{dT_e}{dt}\right|_{t=o} = \frac{3P}{2L_d L_q} |\Psi_s| [\Psi_f L_q \dot\delta - |\Psi_s|(L_q - L_d)\dot\delta] \quad (11)$$

Which in this relation δ is angular speed of stator flux linkage with related to permanent magnet flux, round so to achieve a stable torque, following relation should be confirmed:

$$|\Psi_s| < \frac{L_q}{L_q - L_d}\Psi_f \quad (12)$$

So to have fast dynamic response and fast variation of torque, amount of stator flux linkage should be calculated by using (12) [1].

Usually appoint control command needed to correctly control of flux or torque, hysteresis comparators is used. Hysteresis comparator compares difference between needed amount and evaluated amount and then registers following data for flux and torque vectors.

Torque comparator works in all three levels while flux comparator only works in two levels, because stator flux cannot be held constant during Permanent Magnet Synchronous operation of motor "Fig. 4".

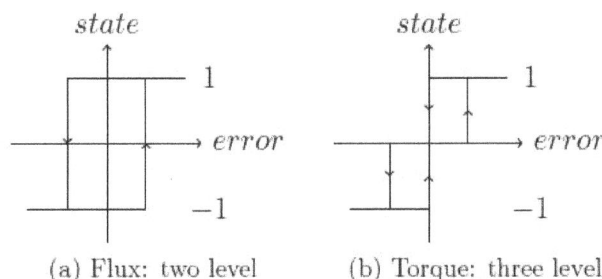

(a) Flux: two level   (b) Torque: three level

Figure 4. *Hysteresis comparators [6]*

By referencing to appointed areas and also by using amounts from table (1), proper voltage vector will be chosen and in this way different states of hysteresis *comparator (φ, τ)* is achieved.

Table 1. *Basic table for DTC switching [7]*

Flux φ	Torque τ	θ - Section (stator flux linkage position)					
		θ₁	θ₂	θ₃	θ₄	θ₅	θ₆
	τ=1	V₂(110)	V₃(010)	V₄(011)	V₅(001)	V₆(101)	V₁(100)
φ=1	τ=0	V₇(111)	V₀(111)	V₇(111)	V₀(111)	V₇(111)	V₀(111)
	τ=-1	V₆(101)	V₁(100)	V₂(110)	V₃(010)	V₄(011)	V₅(001)
	τ=1	V₃(010)	V₄(011)	V₅(001)	V₆(101)	V₁(100)	V₂(110)
φ=0	τ=0	V₀(111)	V₇(111)	V₀(111)	V₇(111)	V₀(111)	V₇(111)
	τ=-1	V₅(001)	V₆(101)	V₁(100)	V₂(110)	V₃(010)	V₄(011)

Hysteresis controllers of flux and torque give proper output based on difference between evaluated amount and command amount of flux and torque.

If φ =1, this means that real amount of stator flux linkage is lower than reference amount of flux. If φ =o, so real amount of stator flux linkage is higher than reference amount of flux. If τ=1, real amount of electromagnetic torque is lower than command torque and so torque should have a raise, therefore by application of proper voltage vector and by raising angle δ, torque will increase. If τ=-1, so real amount torque is higher than reference torque amount.

To decrease ripple of torque, instead of two levels of torque decrease and increase, state of "no change" can be added which means τ=o. This can be done by application of zero voltage vectors. This is shown in table (1) [2, 3, 8]

Six switches three phase voltage source inverter in this control method can be seen in "Fig. 5". This inverter creates six non-zero voltage vectors and two zero voltage vectors (totally eight vector V_0 , V_1 , V_2, V_3 , V_4, V_5 , V_6 ,V_7)

In switching time, voltage vector will be constant and stator resistance will also be supposed constant. In relation (13), $\Psi_s|_{t=o}$ is preliminary amount of stator flux linkage in the beginning of switching. By ignoring stator resistance, relation between stator flux linkage and voltage is described as following:

$$\Psi_s = v_s t + \Psi_s|_{t=o} \Rightarrow \Delta\Psi_s = v_s \quad (13)$$

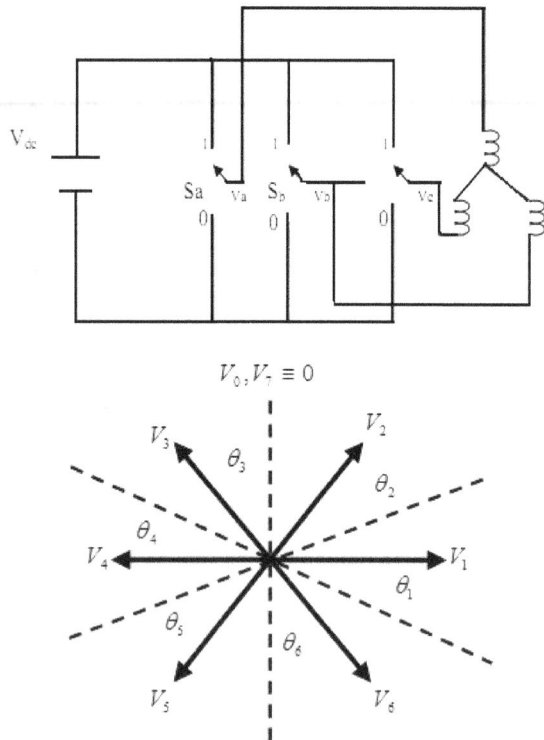

Figure 5. *Schematic of inverter and voltage vectors presentation in DTC [9]*

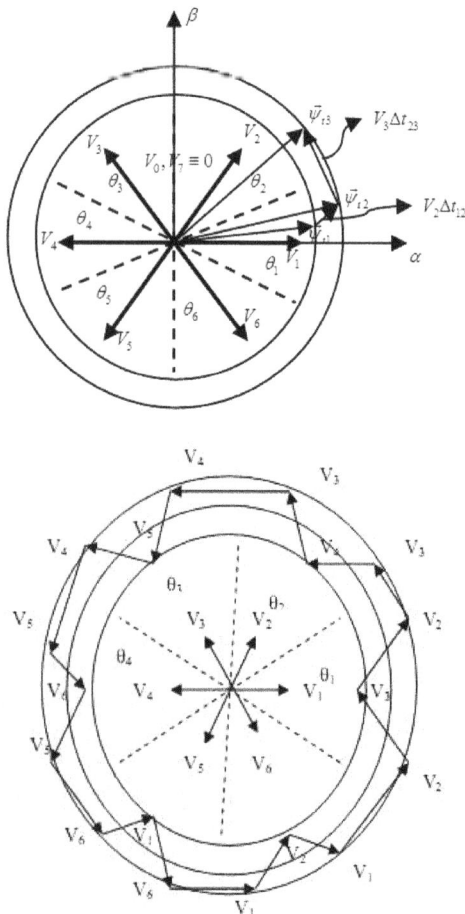

Figure 6. *Rotation of stator flux linkage vector by voltage vector [9]*

Based on relation (13), movement of stator flux linkage is coordinated with application voltage to stator "Fig. 6". So amount, movement way and movement speed of stator flux linkage can be controlled by choosing proper voltage vector.

In each six- fold area, to decrease or increase or increase stator flux amount, two voltage vectors will be chosen which result in lower switching frequency. Therefore if zero voltage is applicator to machine, vector Ψs will stay in its right place. In permanent magnet synchronous motor, stator flux linkage will be achieved by mixture stator voltage and rotor magnetic flux vectors. As permanent magnet flux is rotating all the time. Even if zero voltage is used, there is stator flux- round so using zero voltage just decreases the torque.

Therefore inverse voltage vector is usually used to decrease torque fast and zero voltage vectors are not used to control stator flux- round so stator flux vector Ψs is always moving related to rotor flux linkage.

As relation (10) shows, electromagnetic torque can be effectively controlled by controlling stator flux linkage amount and its rotation speed. For anti-clockwise operation, if real torque is lower than reference torque, voltage vector are chosen so as cause rotation of Ψs in the same direction (triangular).

In this state, angle δ changes fast and real torque of machine will increase. If motor torque is higher than reference amount, voltage vector will be chosen so as cause rotation of Ψs in inverse direction and so δ and torque decrease.

Table (1) is for controlling amount and rotation direction of vector Ψs which is used for both operation ways. In this table, τ and ϕ are outputs of hysteresis flux and torque controllers respectively, and θ_1-θ_6 show area of space vector page in which stator flux- round vector is placed. This table shows DTC control strategy for IPM motor [9, 10]

5. Study and Comparison FOC and DTC

In table (2), two control methods for FOC and DTC are compared with 5 basic indices. So with respect to aforementioned table, practicing direct control of torque is very simpler than directed flux control therefore it is preferred. From above comparison on I (we) conclude, in spite of that both these methods are in control vector group, but they have very external differences. So it seems that these two methods basically have no common point. Simulation results show this very well.

Table 2. Comparison of FOC and DTC

Method Index	FOC	DTC
System Transformation	need	no need
Voltage Modulation	need	no need
Calculation Value	high	low
Sensivity To Parameter Changes	yes	no
Sensor Position	need	no need

6. Simulation Results

In this section, system of Field-Oriented Control and Direct

Torque Control in permanent magnet synchronous motor is simulated. Simulation parameters of studied motor have been shown in table (3).

Output specifications of two control system should be compared with each other in the same operational conditions. In the following, results of two systems simulation, creator factors and finally similarities and differences of these two systems will be studied.

Table 3. PMS motor parameters

Parameter	value
P	6
R	4.1 Ω
Ld	.0511 H
Lq	.066 H
φf	.154 Wb
ωb	1000 rpm
V	350 V

Figure 7 and 8 show machine torque variations in terms of time for two control methods. At first starting torque increases, after arriving a specified speed, torque profile will have stable swings around a zero. Based on achieved results, we can conclude that machine response to load torque variations in DTC method is a little less than FOC method.

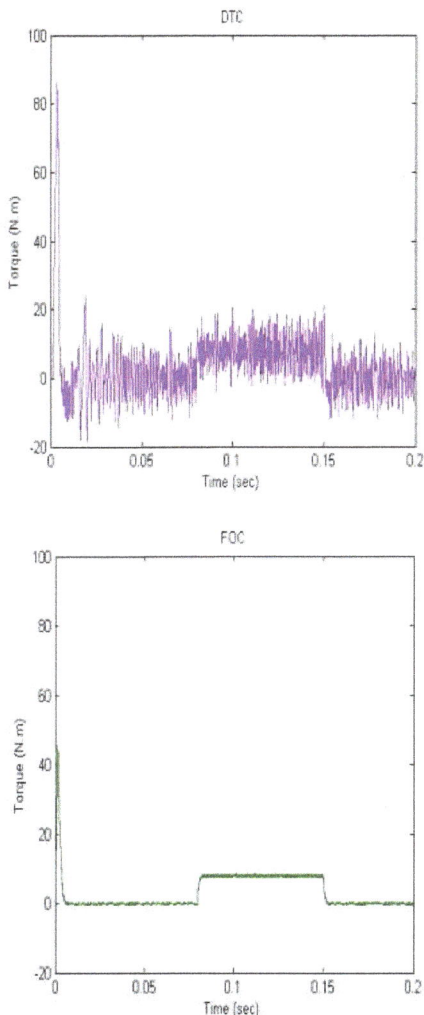

Figure 7. Torque variations in rated speed

It is seen that torque swig in DTC method is much more compared to FOC method. This is because of torque hysteresis controller in DTC method and its direct effect on machine torque.

"Fig. 8" shows torque variation diagram presented in "Fig. 7"in a smaller time range. This waveform shows high torque ripple in DTC method compared to its low swing in FOC method.

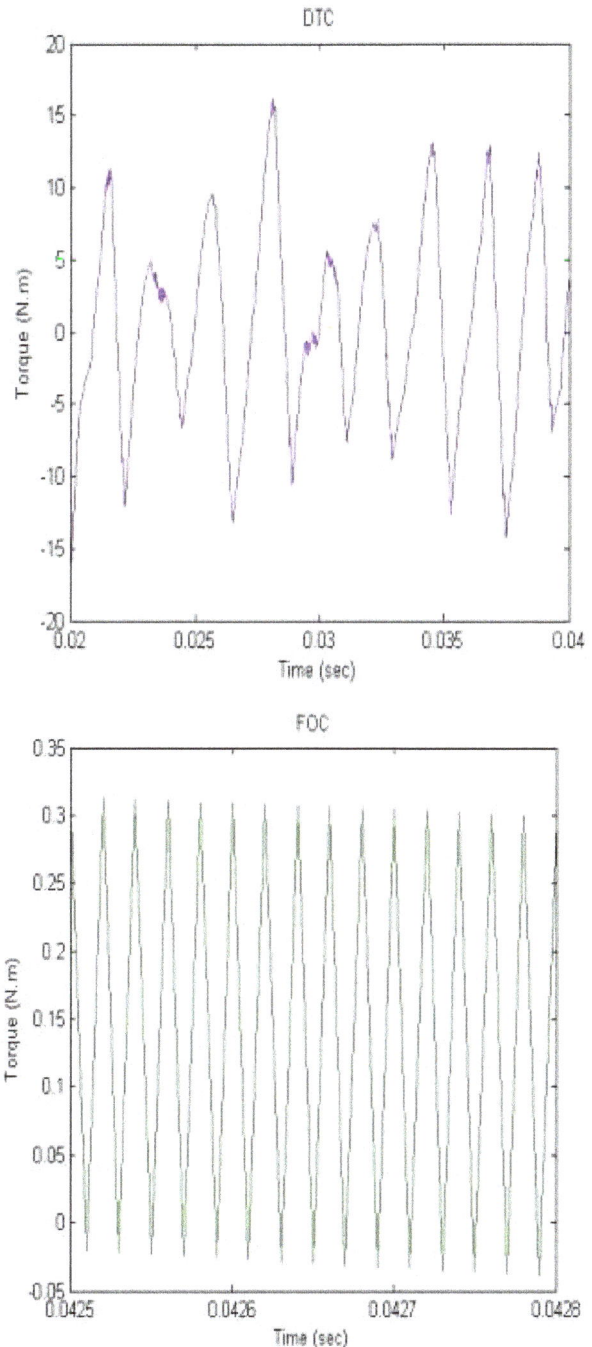

Figure 8. Fine torque variations

Figure 9 shows a motor three phases current diagram in rated speed when full load is applied to machine, current waveform in DTC method have more disorder than FOC method, which is because of hysteresis controllers is in DTC method.

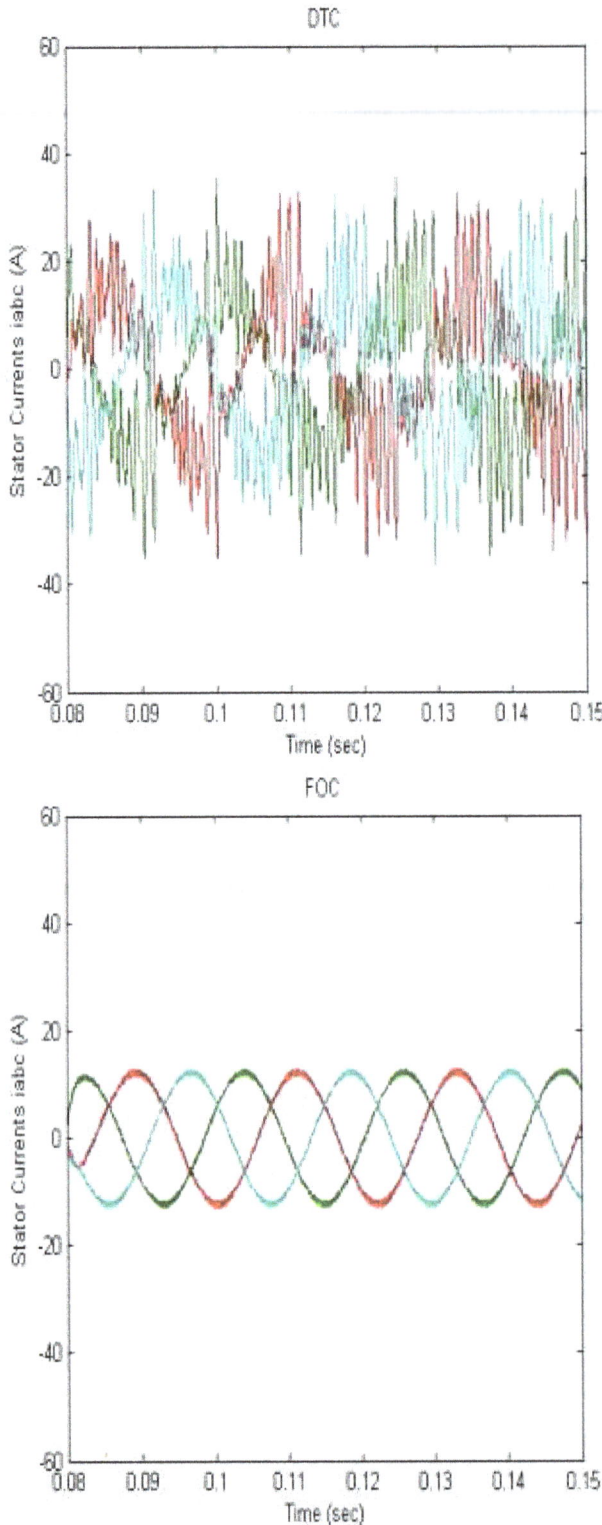

Figure 9. *Stator three phase's current diagram in full*

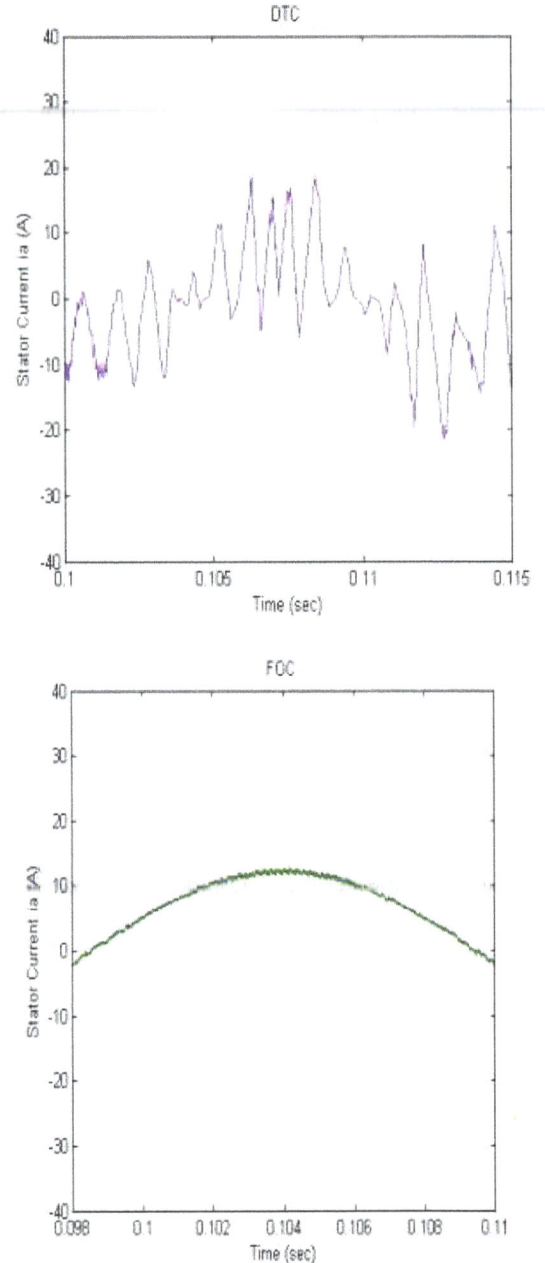

Figure 10. *Pick of phase current in full load in both methods*

Furthermore, Figure 10 shows pick of phase current in a small time range. Current swing in DTC method is more than FOC method and this swing even become more in low load. Existence of many swings in current waveform in DTC method is one the great disadvantages of this method compared to FOC method.

Figure 11 and 12 shows stator flux variations in terms of time in rated speed from no load to full load. In both method control method, amount of command flux will be considered 0.29 Weber (almost two times of permanent magnetic flux of rotor).

A desired control system is a system that its flux stays constant by applied load torque to machine and is not affected by distortion. Figure 11 shows that in both control methods it is satisfied and by applying load torque, motor flux has stayed constant.

As it is seen in Figure 12, in DTC method, because of existence of flux hysteresis controller and its effects on flux, created ripple is much more than FOC method. In DTC method, this ripple amount is equal to width of flux hysteresis band.

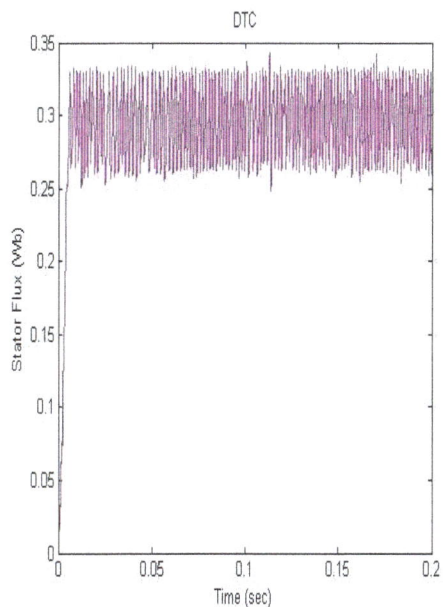

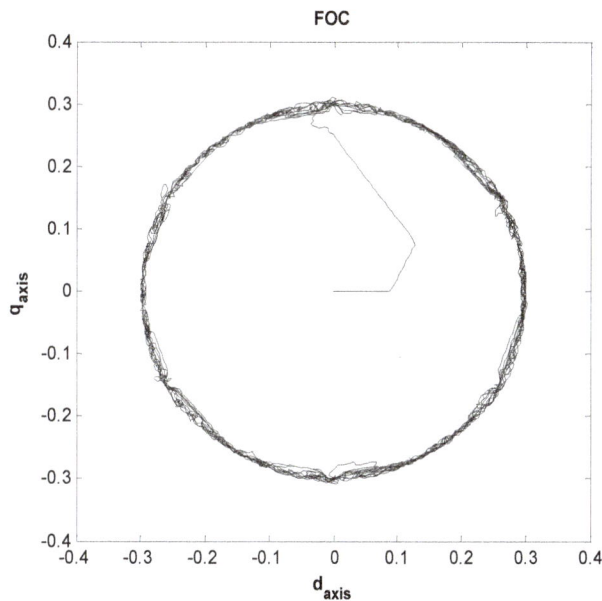

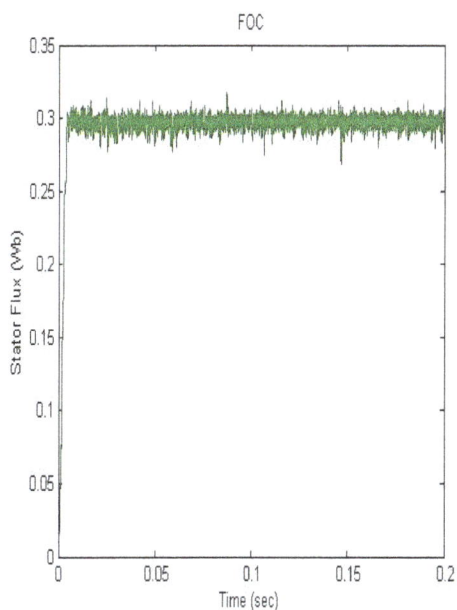

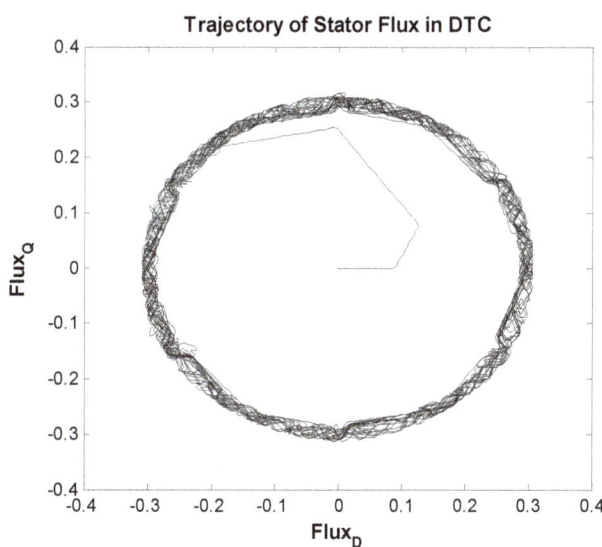

Figure 11. Stator flux variations diagram

Figure 12. Flux variation waveform in both case control methods.

(a)

(b)

Figure 13. Stator flux vector direction presentation diagram

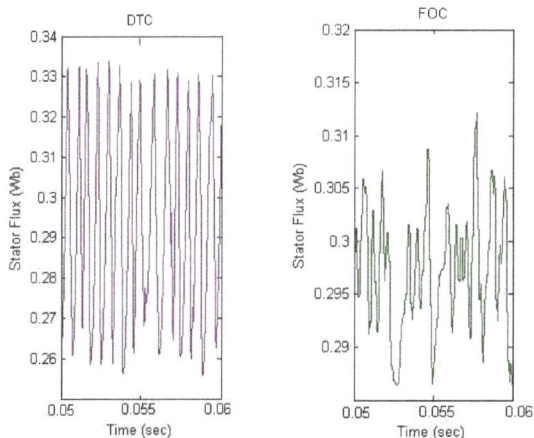

Figure 13.a and 13.b show movement direction of stator flux vector in DTC and FOC methods. These two diagrams are almost same, by this difference that flux diagram in FOC method get its desired amount a little faster, but flux ripple in DTC method is more. Reason of faster flux in FOC method can be considered flux direction finding along with stator flux. As before mentioned, reason of more ripples in DTC is existence of hysteresis flux controller.

Figure 14 shows Speed variations diagram for two control method. Observation, in DTC method we achieve early as possible to the response.

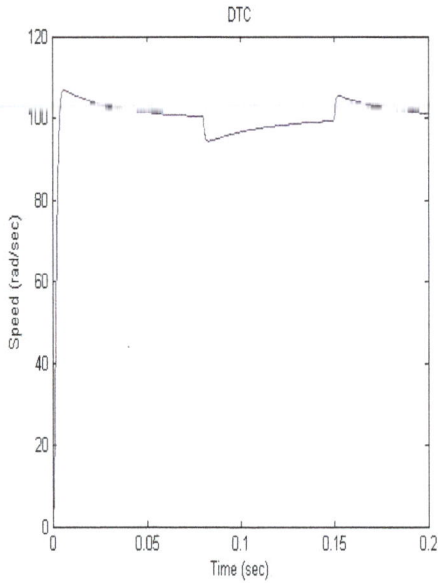

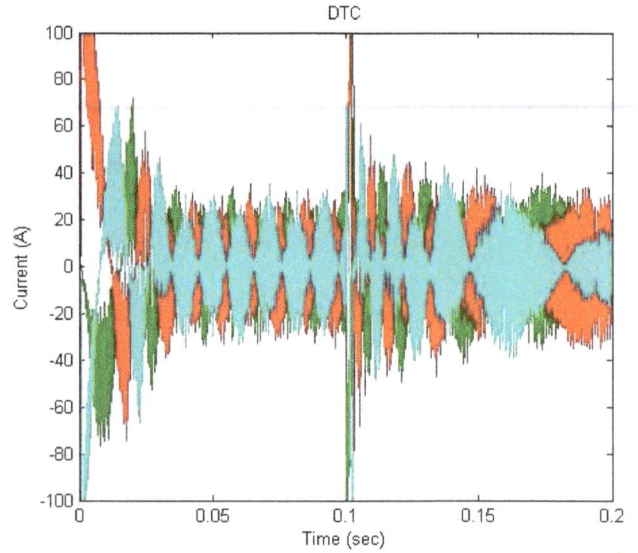

Figure 16. *Three-phase currents DTC*

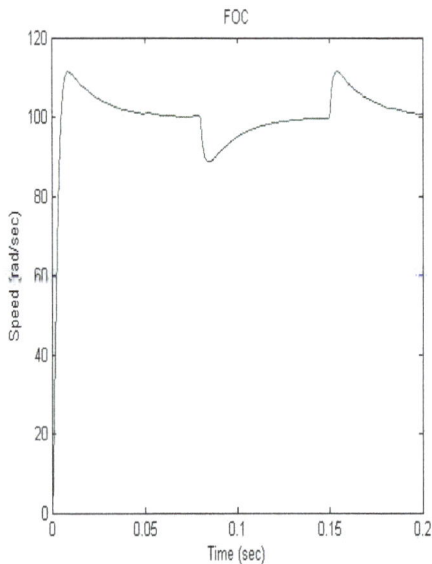

Figure 14. *Speed variations diagram*

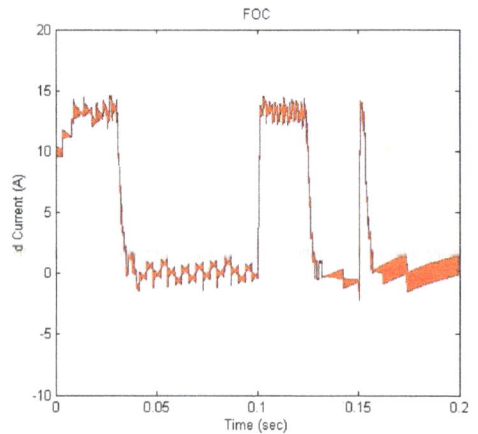

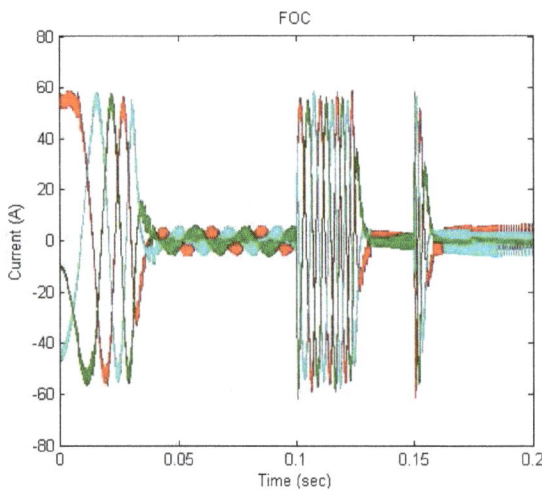

Figure 17. *The longitudinal and transverse axes (d-q) in the FOC approach*

Figure 15 and 16, the phase currents for the two methods show FOC and DTC are approximately the difference between the two methods can be observed. Flow chart of the method of phase FOC as its torque curve method is formed around zero, while the method DTC phase currents are relatively higher than zero. The reason for this behavior is that the DTC method. Unlike methods that FOC, the direct control of the d axis currents and q does not exist.

Figure 15. *Three-phase currents FOC*

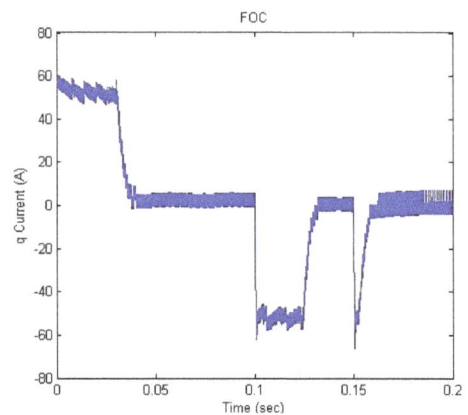

These issue forms Figure 17 and 18 can be seen. Also Three-phase current waveforms DTC compared to the FOC is having ripple and high volatility, which hysteresis in DTC method is used because it controls.

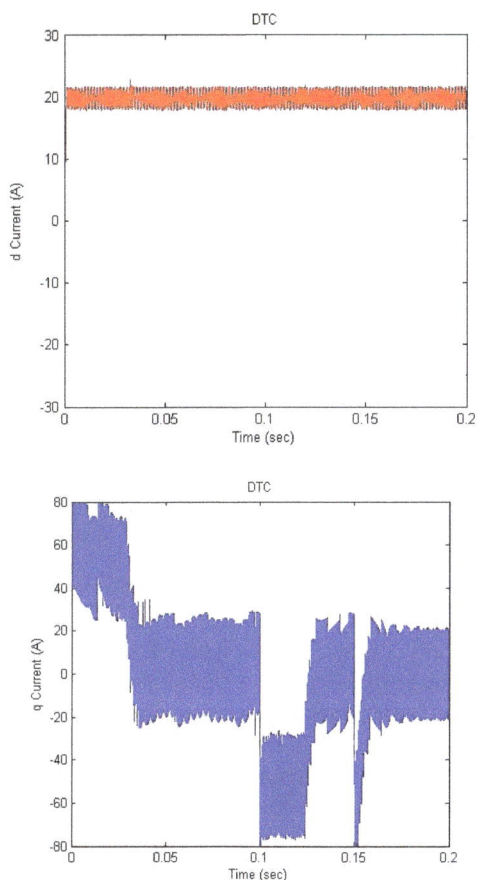

Figure 18. The longitudinal and transverse axes (d-q) DTC

7. Discussion and Conclusion

In this paper, two common control methods including FOC and DTC of the PMS motor is discussed in detail. The main characteristics of the motor such as torque, flux and speed under different operation conditions studied and the advantages of FOC and DTC are obtained. It can be concluded that although both the DTC and FOC methods have different implementation structures but the motor has the same behavior on the control methods. Therefore, it is deduced that both the methods can be named as direct flux control and can be applied as an optimized method to substitute custom control methods of these kinds of motors in the industry applications.

References

[1] Juha Pyrhonen. 2011, ELECTRIC-DRIVE-BASED CONTROL AND ELECTRIC ENERGY REGENERATION IN A HYDRAULIC SYSTEM.Book..

[2] P. Pragasan, and R. Krishnan, 1998, Modeling of permanent magnet motor drives, IEEE Trans. Industrial electronics, vol. 35, no.4, nov. 1988\

[3] Anders Kornberg, 2012, Design and Simulation of Field Oriented Control and Direct Torque Control for a Permanent Magnet Synchronous Motor with Positive Saliency.Thesis For M.S. degree ,uppsala university.

[4] Thomas M. Jahns, Gerald B. Kliman, and Thomas W. Neumann. Interior permanent-magnet synchronous motors for adjustable-speed drives. IEEE Transaction on industry applications, IA-22(4):738-747, 1986.

[5] R.D. Doncker, D.W.J. Pulle, and A. Veltman. Advanced Electrical Drives: Analysis, Modeling, Control. Power Systems. Springer, 2011. ISBN 9789400701793. URLhttp://books.google.se/books?id=_sEDx2lAKboC.

[6] Yi Wang and Jianguo Zhu. Modeling and implementation of an improved dsvm scheme for pmsm dtc. International Conference on Electrical Machines and Systems, pages 1042-1046, 2008.

[7] H. L. Huy, 1999, Comparison of Field-Oriented Control and Direct Torque Control for Induction Motor Drives, 0-7803-5589-X/99/10.00 © 1999 IEEE.

[8] [8] A. Daghigh, M.B.B. Sharifian, M. Farasat, A Modified Direct Torque Control of IPM Synchronous Machine Drive With Constant Switching Frequency and Low Ripple in Torque, in Conf. 18th Iranian conference on Electrical Engineering, ICEE, Isfahan-Iran, 2010.

[9] L. Zhong, M. F. Rahman, 1997, Analysis of Direct Torque Control in Permanent Magnet Synchronous Motor Drives, IEEE TRANSACTIONS ON POWER ELECTRONICS, VOL. 12, NO. 3, MAY 1997

[10] L. Zhong, M. F. Rahman, and K. W. Lim, 1995, A strategy for torque control for high-performance interior permanent magnet synchronous motor drives, in Proc. Australasian Univ. Power Engineering Conf., 1995, pp. 259–264.

23

A Study of Reactive Power Requirements of Traditional and Restructured Power System

Laleh Haddadi[1, *], Abolfazl Pirayesh Neghab[2], Seid Babak Mozafari[1], Morteza Ghasem Salamroodi[3]

[1]Department of Power Engineering, Science and Research Branch, Islamic Azad University of Tehran, Tehran, Iran
[2]Department of Power Engineering, Shahid Beheshti University, Tehran, Iran
[4]Power Distribution Company of West of Mazandaran, Noshahr, Iran

Email address:

l_haddadi90@yahoo.com (L. Haddadi), ab_pirayesh@yahoo.com (A. P. Neghab), mozafari_babak@yahoo.com (S. B. Mozafari), msalamroudi@yahoo.com (M. G. Salamroodi)

Abstract: Due to importance of requirements of reactive power in a power system, by employing probabilistic load flow with Monte Carlo method we can gain the exact value of requirement reactive power of system in a traditional and a restructured power system. In this paper the main object is the examination of effect of random changes in load, generation and price on all requirement reactive power of system as well as the amount of reactive power of each bus. Effective parameters in reactive power distribution are determined and an effective probable model for each of parameters is presented. The simulation results are used by network management for reactive power and voltage control service contracts. The results of analysis in two environments (traditional and restructured power system) are shown. The Monte Carlo method and MATLAB software is used, and studies are done on 30 buses IEEE system that is modeled by MATPOWER software.

Keywords: Reactive Power, Restructured Power System, Monte Carlo Simulation

1. Introduction

Reactive power supply is essential for reliably operating the electric transmission system. Inadequate reactive power has led to voltage collapses and has been a major cause of several recent major power outages worldwide.

The rules for procuring reactive power can affect whether adequate reactive power supply is available, as well as whether the supply is procured efficiently from the most reliable and lowest- cost sources.

Not only is reactive power necessary to operate the transmission system reliably, but it can also substantially improve the efficiency with which real power is delivered to customers. Increasing reactive power production at certain locations (usually near a load center) can sometimes alleviate transmission constraints and allow cheaper real power to be delivered into a load pocket [1]

In the other hand Estimation of reactive power more than requirement value will cause additional costs and will occupy capacity of reactive power sources that have other capabilities, like generators. Evaluation of requirements of reactive power for system is very important.

In this paper by using probabilistic load flow with Monte Carlo method we gain the exact value of requirement reactive power of system in a traditional and a restructured power system.

The remaining section of the paper is organized as follows: In section 2 we will review restructured power system, In section 3 we examine the pool market, probabilistic and optimal load flow, probabilistic load flow using Monte Carlo method in section 4 and the results of the case study will be described in section 5 and 6.

2. Restructured Power System

Power industry are changing and moving in a direction that by allowing producers to competition and market conditions, try to reduce the costs of producing and distributing electricity, eliminate certain inefficiencies, increase customer choice. This evolution towards a competitive electricity market often called deregulation or restructuring and includes below benefits:

1 Providing choice for consumers
2 Creation suitable condition for providing better service
3 Competitive supply of electrical products at different levels and consequently set reasonable prices for consumers
4 Attract investments in the private sector and guidance for public benefit no need for massive public investment
5 Offered products quality enhancement with respect to the competition and…

3. Pool Market Model, Probabilistic and Optimal Load Flow

In the pool market model, there are two main sides of entities participating in the market, i.e. customer and supplier. The pool operator takes electricity transaction bids and offers from these two entities and dispatches them in an economic manner depending on the price and MW biddings. In general, the customers and suppliers do not directly interact to each other, but only indirectly through the pool operator. [2] In the pool market, participants post selling offers for a certain amount of electrical energy at a specified price and in a period of time. These offers are sorted by price ascending. And produce market supply curve. And also, the market demand curve is obtained from amount and price of customer requests which are arranged descending. Intersection of supply and demand curves is represented market equilibrium.

In order to implement the restructured power system in this paper we use pool market model.

Probabilistic load flow is a macroscopic random method under steady-state operating conditions of the power system, which considers different random factors on the performance of power system.[3] To consider the uncertainty in the power system probabilistic load flow is a proper way that in it uncertainty in the system parameters are considered as random variables. In this paper probabilistic load flow using Monte Carlo simulation is done.

Optimal load flow is a nonlinear optimization to determine the optimum parameters in a power system that has an objective function and number of constraints. Here the objective function is total cost of production of active or reactive power or both. These costs may be patchy or linear functions of the output of the generator polynomial functions are defined. The formulation is as follows:

$$\min_{Pg,Qg} \sum f_{1i}(P_{gi}) + f_{21}(Q_{gi}) \tag{1}$$

$$P_{gi} - P_{Li} - P(V,\theta) = 0 \tag{2}$$

$$Q_{gi} - Q_{Li} - Q(V,\theta) = 0 \tag{3}$$

$$\widetilde{S_{ij}^{f}} \leq S_{ij}^{max} \tag{4}$$

$$\widetilde{S_{ij}^{t}} \leq S_{ij}^{max} \tag{5}$$

$$V_t^{min} \leq V_t \leq V_t^{max} \tag{6}$$

$$P_{gi}^{min} \leq P_{gi} \leq P_{gi}^{max} \tag{7}$$

$$Q_{gi}^{min} \leq Q_{gi} \leq Q_{gi}^{max} \tag{8}$$

In this paper, by using probabilistic optimal load flowproduction changes, load changes and active and reactive cost changes in a power system are considered. And also Monte Carlo simulation method is used to determine amount of reactive power demand in a pool market .

4. Probabilistic Load Flow Using Monte Carlo Method

Monte Carlo simulation is a process that occurrence of modes in all its aspects is purely coincidental. In Monte Carlo simulation, a random number generator is used to model the occurrence of events. This behavior is desirable, because by repeating the simulation, a probability distribution of random numbers will be obtained. And the mean, median, variance and statistical parameters will be calculated.[8] In this paper, by using probabilistic load flow and probabilistic optimal load flow production changes , load changes and active and reactive cost changes in a power system are considered. And also Monte Carlo simulation method is used to determine amount of reactive power demand in a power system.

To consider the uncertainty in the power system probabilistic load flow is a proper way that in it uncertainty in the system parameters are considered as random variables. In this paper, using Monte Carlo method, uncertainty of the system production and loads are modeled as random variables And with running the program and probabilistic load flow and probabilistic optimal load flow for about 8000 to 10000 times, probability distribution of parameters of system be obtained and the amount of requirements reactive power of the system in a traditional and restructured power system are obtained. MATLAB software was used to generate normal random numbers and the required random numbers are generated.

Probabilistic load flow using Monte Carlo method is the most comprehensive tool to evaluate different scenarios for uncertainty in a system. In Monte Carlo simulation at first the base network that in it the network structure is characterized by the production of generator and load bus is selected, then to the following, load flow is done:

Using a random normal distribution function a random amount of load for a load bus is selected.

The selected bus is modified with the new value selected

Using Newton-Raphson load flow network status can be evaluated and the amount of output parameters is characterized.

If the stopping criterion is provided by the simulation, the simulation ends, otherwise, items 1 to 3 are repeated. Stopping criterion can determine the number of iterations for Monte Carlo simulation or the specific coefficient of variation is.[9]

With the finishing Monte Carlo simulation, average value of the output parameters obtained from the following equation can be used as the best estimation:

$$\overline{A} = \frac{1}{N}\sum_{i=1}^{N} A_i \tag{9}$$

And the variance of output parameters can be obtained from the following equation:

$$V = \frac{1}{N}\sum_{i=1}^{N}(V_k - \overline{V})^2 \tag{10}$$

Where $\overline{M}$ average output parameters, M_k simulation output parameter at the k th iteration and N is the number of samples tested.

In this paper that the output parameter is the system reactive power, Formulas 1 and 2 are as follows:

$$\overline{Q} = \frac{1}{N}\sum_{i=1}^{N} Q_i \tag{11}$$

$$V = \frac{1}{N}\sum_{i=1}^{N}(Q_i - \overline{Q})^2 \tag{12}$$

Where $\overline{Q}$ average output reactive power, Q_i amount of reactive power at the i th iteration and N is the number of samples tested.

5. Case Study

The study is done on 30 buses IEEE system that has 6 generator buses and 24 load buses. It is shown in figure 1:

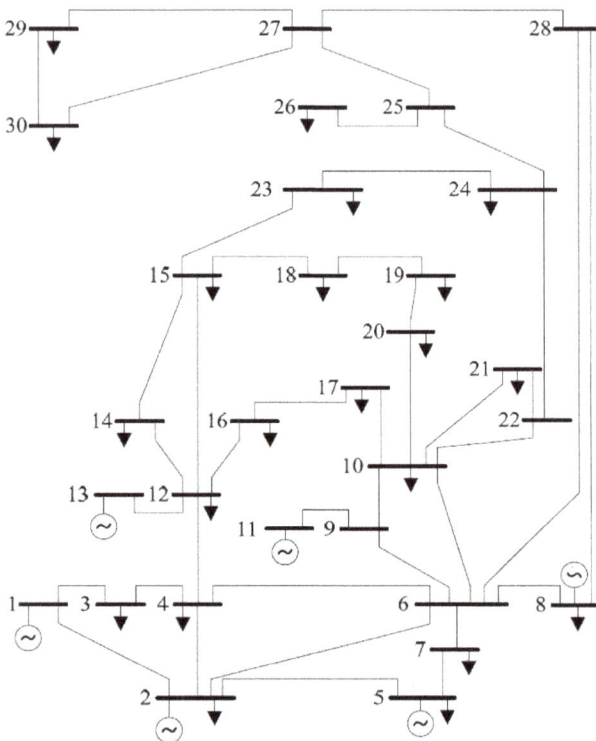

Figure 1. 30 buses network diagram

Algorithm based on Monte Carlo simulation is as follows:
Count=0

generate uncertainty random numbers of production, load, or cost

Insert random numbers   in the original system and modify system now

Do normal load flow or optimal load flow
Count+1.
If converged, go to step 8.
If did not converge, count-1, now and Go to Step 2
Do Monte Carlo simulation to saturate system parameters
If converged, go to the last step
If not converged, go to step 2
The flowchart of proposed method is illustrated in figure 2.
Compute the average reactive power requirements of the network

To evaluate the system in a traditional system we used Newton Raphson load flow, with considering different probabilities in load and production and doing 10000 iterations for each mode, the amounts of the requirement reactive power at each bus and all system obtain.

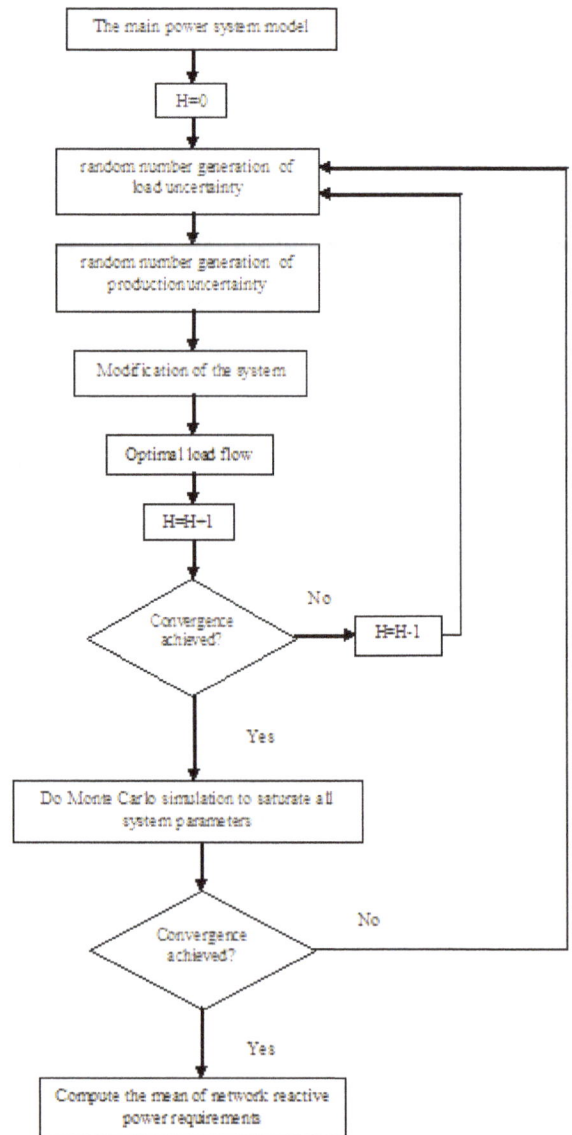

Figure 2. Flowchart of proposed method

Loads of power system is variable and uncertain. Different probability distribution functions may be used to show uncertainty. In this paper we used normal distribution

Figures 3, shows Random variation of active load consuming at bus 2. for generating normal random number," randn "rule in MATLAB software is used. For example in order to generate random numbers of active load consuming at bus 2 following expression is used:

$$x = 21.7 + 7.23 randn\ (1) \tag{13}$$

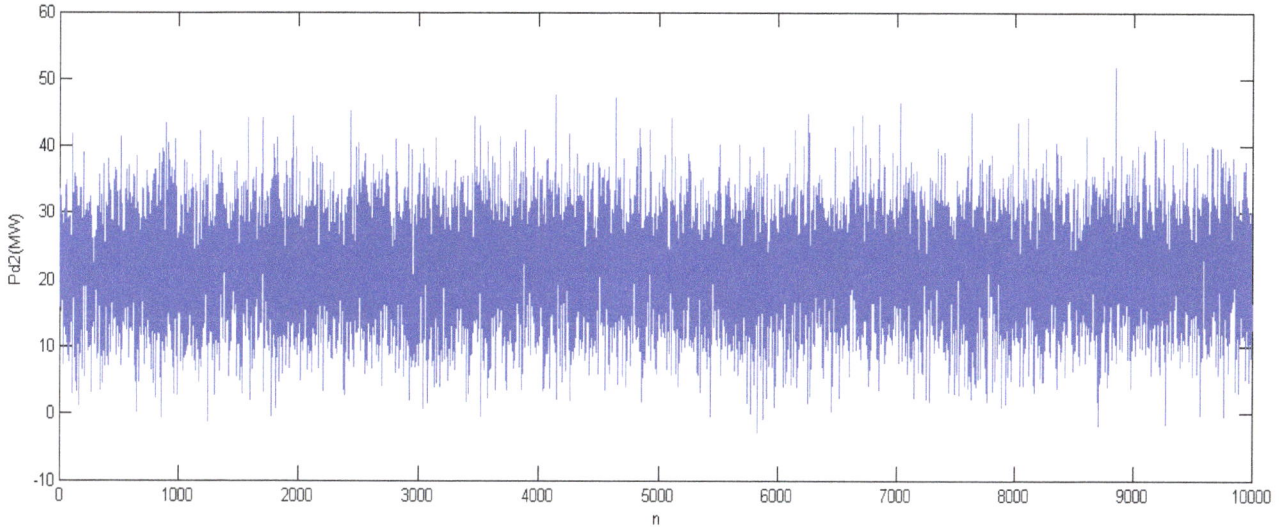

Figure 3. *Diagram of random variation of active load consuming at bus 2*

To evaluate the system in a restructured system and in a pool market, in the above algorithm, optimal load flow is used instead of the ordinary load flow. And moods of production, load and price changes are considered as a normal random distribution. This time the number of iterations of simulation is 8000 times.

Figures 4 and 5 are shown examples of the diagrams of mean changes of reactive power generation at bus 1, 2, 5, 8, 11, 13, and the whole system, result of random changes in the parameters. Results of a number of moods are shown in tables 1 and 2.

Table 1. *The results of the required reactive power in a traditional power system*

Random States	Average of Reactive power random changes	Bus5	Bus8	Bus13	Total of System
Change of all loads		39.3955	36.5619	10.5669	136.6483
Heavy loading		39.9907	39.8906	11.1937	147.0116
Result of Monte Carlo simulation		35.9175	36.3037	10.4899	134.7411
Optimal load flow with a medium value		35.66	36.11	10.45	133.93

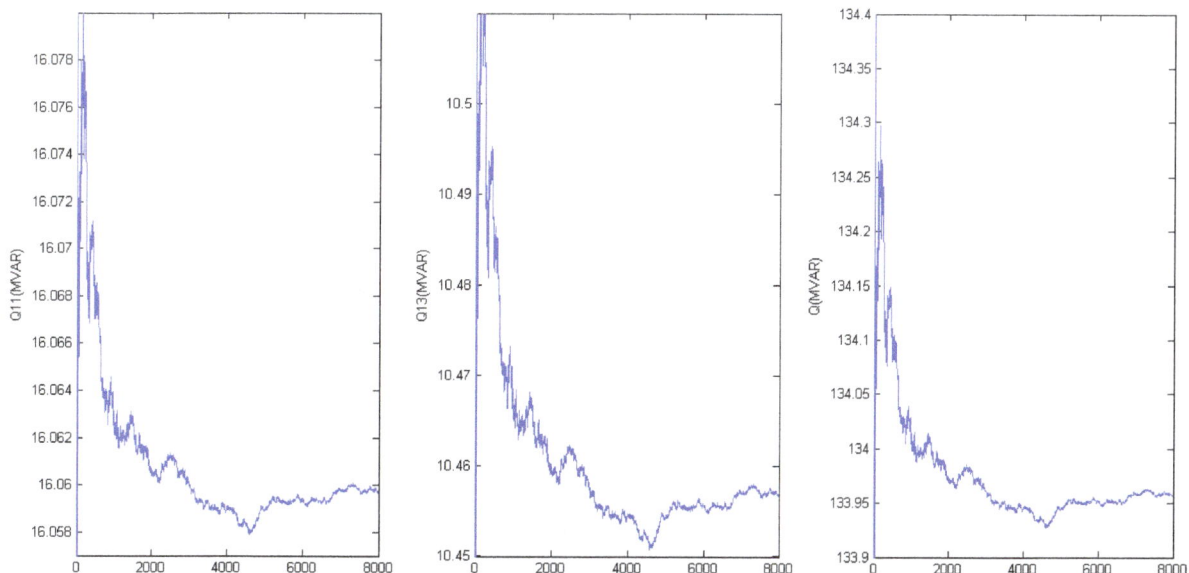

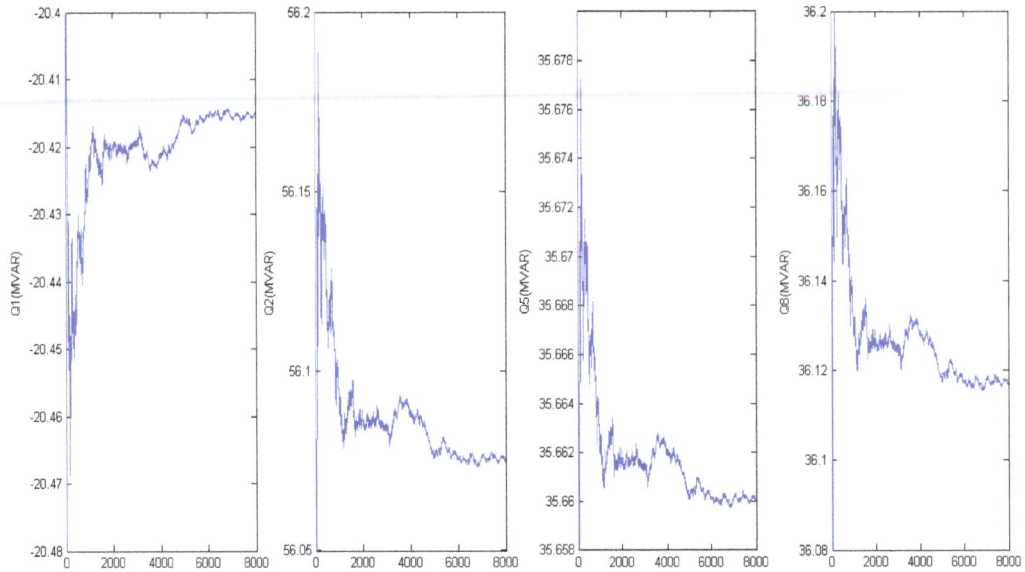

Figure 4. *Diagram of the system reactive power changes in a traditional power system due to random variations in load at bus 15*

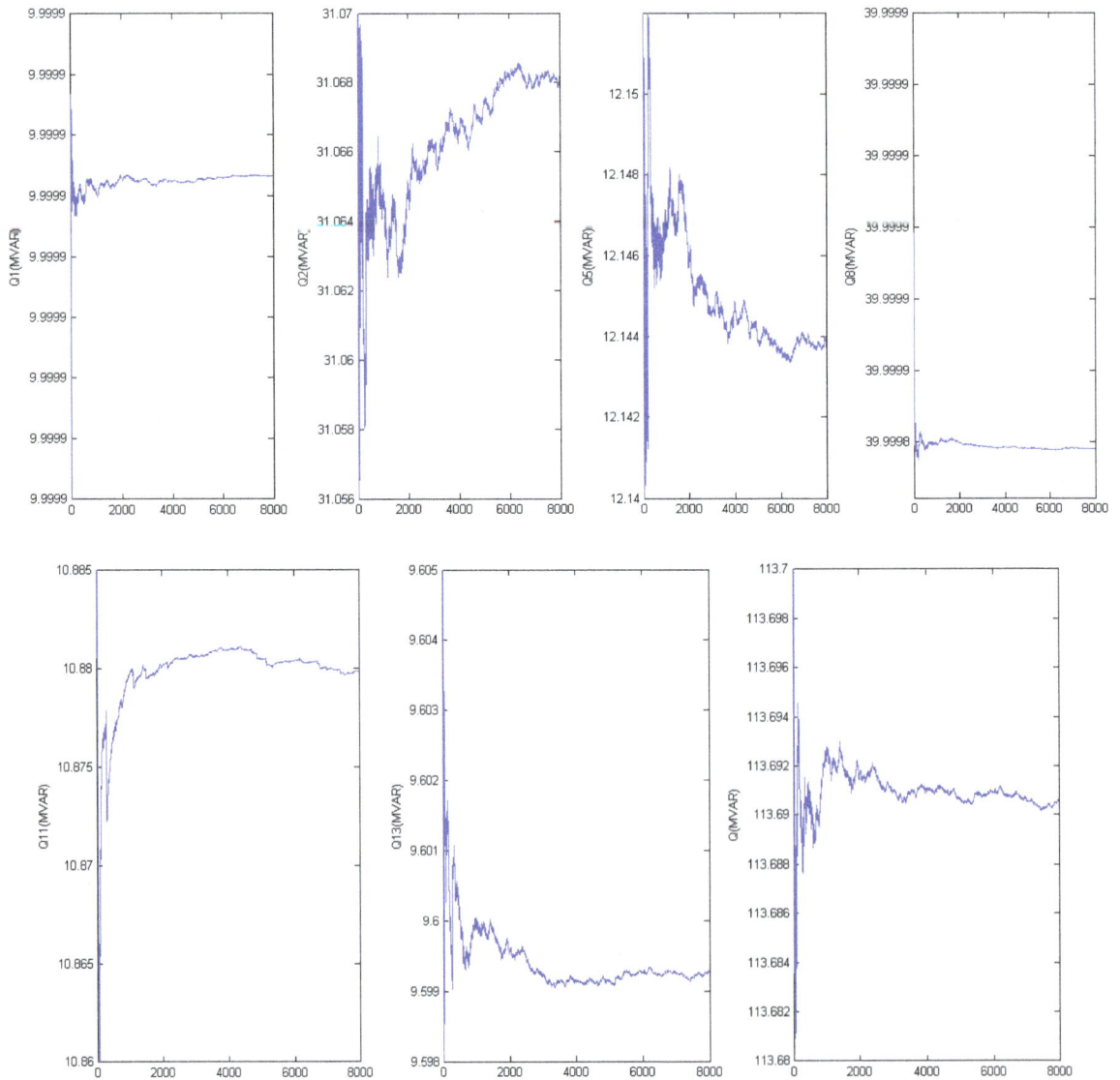

Figure 5. *Diagram of the system reactive power changes in a pool market power system due to random variations in active power cost at bus 11*

Table 2. The results of the required reactive power in a pool market

Average of Reactive power random changes / Random States	Bus5	Bus8	Bus13	Total of System
Change of all loads	12.137	39.9999	9.5826	113.8114
Heavy loading	31.4943	40.2744	11.8362	146.2001
Result of Monte Carlo simulation	12.7307	39.9945	9.6846	114.7774
Optimal load flow with a medium value	12.14	40	9.6	113.69

Cost function is a function of polynomial system (14) considered. Where the coefficients are shown in Appendix Tables 3 and 4.

$$f(p) = C_3 P^3 + C_2 P^2 + C_1 p + C_0 \qquad (14)$$

Table 3. Active power generation cost coefficients

Bus	Startup	Shutdown	C_2	C_1	C_0
1	0	0	0.038432	20	0
2	0	0	0.25	20	0
5	0	0	0.01	40	0
8	0	0	0.01	40	0
11	0	0	0.01	40	0
13	0	0	0.01	40	0

Table 4. Reactive power generation cost coefficients

Bus	Startup	Shutdown	C_2	C_1	C_0
1	0	0	0.02	0	0
2	0	0	0.0175	0	0
5	0	0	0.0625	0	0
8	0	0	0.00834	0	0
11	0	0	0.025	0	0
13	0	0	0.025	0	0

6. Discussion and Conclusion

In this paper by using Monte Carlo simulation, that is a flexible method, we found the exact amount of reactive power is needed for a restructured and a traditional power system. To do this we used the method of probabilistic optimal load flow and considered the uncertainties of production, load and cost and obtained the results of reactive power changes of different buses and also all buses that is the amount of total requirement of reactive power for system. One of states of uncertainty considered is a variable heavy loading of system. That is shown the maximum need of reactive power in a critical condition and without collapse.

Simulation is down on 30 buses IEEE system. The results are shown in section 5. Result of comparison between restructured and traditional is shown in tables 1 and 2.

References

[1] "Principles for Efficient and Reliable Reactive Power Supply and Consumption" STAFF REPORT Docket No. AD05-1-000 February 4, 2005

[2] Surachai Chaitusaney, and Bundhit Eua-Arporn, "Actual Social Welfare Maximization in Pool Market" IEEE 2002.

[3] X.Wang , J.R.McDonald   "Modern power system planning" IEEE 2002

[4] Majid Oloomi Buygi, Gerd Balzer, Hasan Modir Shanechi, Mohammad Shahidehpour,"Market-Based Transmission Expansion Planning" IEEE Transaction On Power Systems, VOL. 19, NO. 4 Nov.2004

[5] Walid El-Khattam, Y.G.Hegazy, M.M.A.Salama"Investigating Distributed Generation Systems Performance Using Monte Carlo Simulation" IEEE 2006.

[6] Roy Billinton- Ronald Allen"Reliability Evaluation of Engineering Systems" Plenum Press 1994.

[7] Richard E. Brown "Electric Power Distribution Reliability" Marcel Dekker, Inc. 2002.

[8] Mehrdad Hojjat, Habib Rajabi Mashhadi Optimal Dispatch of Power System Restructuring in considering the impact of the independent random variablesranian conference on electrical engineering, Tabiat Modares University 2008.

[9] Hasan Abniki, Hasan Monsef, Mina Jafari Inanlo "Flash estimate voltage using a Monte Carlo method" International Power System Conference, Tehran 2011.

[10] Maryam Ramezani,Mohammad Reza Khalghani, Hamid Falaghi "Probabilistic load flow in the power systems, wind power plants based on classified data" International Power System Conference, Tehran 2011.

[11] Laleh Haddadi, Abolfazl Pirayesh Neghab, Seid Babak Mozafari "Analysis of requirement reactive power for a restructured power system using Monte Carlo method" Journal of Basic and Applied Scientific Research, ISSN: 2090-4304, sep.2013

24

Adaptive noise cancellation for eliminating artifacts of life signals using fuzzy and neural networks

Mohammad Seifi*, **Aliakbar Kargaran Erdechi, Ahmad Hajipour**

Department of Electrical and Computer Engineering, Hakim Sabzevari University, Sabzevar, Iran

Email address:

mohamad_saifi@yahoo.com (M. Saifi), aliakbar.kargaran@gmail.com (A. kargaran), a.hajipour@hsu.ac.ir (A. Hajipour)

Abstract: Electroencephalogram (EEG) is one of the commonly used non-invasive techniques for understanding the brain functions. This paper presents a method for removing electroocular (EOG) artifacts in the electroencephalogram (EEG). A new adaptive radial-basis function- networks- (RBFN-) for the adaptive noise cancellation (ANC) problem is proposed. Also, the algorithm of structure identification and parameters adjustment is developed. The proposed RBFN approach implements Takagi-Sugeno-Kang (TSK) fuzzy systems, functionally. Simulation results demonstrate that the proposed adaptive RBFN can remove the noise successfully and efficiently with a parsimonious structure.

Keywords: RBFN, ANC, Adaptive Filtering, Neural Network

1. Introduction

Electroencephalogram (EEG) is one of the commonly used non-invasive techniques for understanding the brain functions. As EEG reflects the mental activity, it can also be used as a communication tool for people who cannot control their environment through normal neuromuscular channel. This new mode of communication involves direct connection with brain and computer and development of algorithms to translate features extracted from EEG to commands which are employed to control some output devices in real time [1].

But Eye movements and blinks produce electrical potentials that propagate over the scalp creating significant electrooculographic artifacts in the recorded electroencephalogram. These artifacts often complicate the interpretation of the EEG. Thus potentials created from noncerebral origins may degrade the performance of Brain Computer Interface (BCI). It is very important to ensure that electric signals caused by eye are removed from EEG as they are the most important sources of contamination in BCI systems [2]. The eye is equivalent to an electric dipole with cornea and retina as positive and negative polarity, respectively. When the eye moves, electric field around it changes and it produces an electric signal called electrooculogram (EOG). It propagates over the scalp and appears in the EEG as noise or artifact. Normally EOG

signals are of high amplitude, low frequency and affect the lower band power of EEG. The simplest linear filtering method cannot be applied here as frequencies of artifacts and neurological phenomena in EEG overlap. So artifacts are serious problems in the interpretation and analysis of EEG [2]. There are many methods to avoid ocular artifacts from EEG such as Independent Component analysis (ICA) [4], principal component analysis [7], time and frequency domain regression methods [8, 9] and adaptive filtering [3]. Among these methods, adaptive filtering gives better results and can be applied in real time [3, 4]. From [5, 6], it can be concluded that the adaptive filtering algorithm using REOG as the third reference input can remove the artifacts more efficiently. In this paper modification of adaptive filtering algorithm is done by using two reference inputs.

2. Adaptive Noise Cancellation

The basic method used in this paper for elimination of artifacts is Adaptive Noise Cancellation (ANC). It is a process by which the interference signal can be filtered out by identifying a model between a measurable noise source and the corresponding immeasurable interference. The method uses a noisy signal as primary input and a reference input that consists of noise correlated in some unknown way with the primary noise. It is also assumed that the desired clean signal is uncorrelated with noise source and interference signals. By adaptively filtering and subtracting the reference input from the primary input, the output of the adaptive filter will be the error signal, which acts as a feedback to the adaptive filter. With this setup, the adaptive filter will be able to cancel the noise and obtain an estimate of the less noisy signal [6].

Figure 1 shows the basic concept of adaptive noise cancellation using the proposed adaptive RBFN filter The primary input of the canceler is the measurable EEG signal $q(k)$, which is naturally contaminated by artifacts. In other words, the artifact signal $n(k)$ (noise source signal) goes through unknown nonlinear dynamics f (the route in human body from the artifact generating source to each EEG electrode on the scalp) and generates a distorted noise $d(k)$, which is then added to clean EEG signal $s(k)$, in the place of electrodes to form $q(k) = s(k) + d(k)$ which is the very signal measured by the EEG electrodes. The reference input to the canceler, which is the input to the adaptive filter in fact, is the noise source signal n(k), measured directly from the artifact generating origin i.e. EOG for ocular artifacts. The noise n(k) is filtered to produce $\hat{d}(k)$ which is as close a replica as possible of interference signal, $d(k)$ the distorted and delayed version of $n(k)$. The aim is to retrieve clean EEG, $s(k)$, from the measured EEG signal, q(k), that is procurable by estimating the $\hat{d}(k)$ using the adaptive filter, which is then subtracted from measured q(k) to produce the output of the system, $\hat{s}(k)$ that would be close to the required signal $s(k)$. In the noise canceling application, the objective is to produce an error signal (output) that is a best fit in the least squares sense to the signal $s(k)$ This is accomplished by feeding back the system output to the adaptive filter to minimize the error signal until it reaches the value $\hat{s}(k) = s(k)$. As a result, the output of the noise canceler $\hat{s}(k)$ is the estimation of corrected or clean EEG. $\hat{s}(k) = s(k) + d(k) - \hat{d}(k)$ that is $\hat{s}(k) \cong s(k)$.[6]

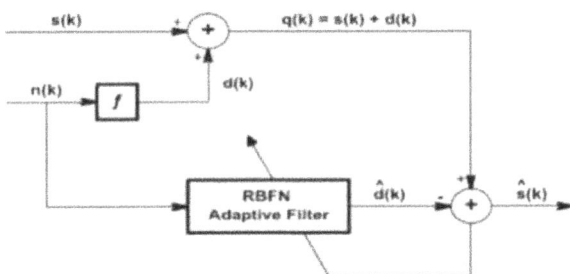

Figure 1. *Adaptive Noise Cancellation*

3. Adaptive RBFN-Based Filter

The adaptive RBFN-based filter to perform TSK inference is depicted in Figure 2. Basically, it is a multi-input single-output (MISO) system [10]. It is straightforward to extend the result to multi-input multi-output (MIMO) systems. The functions of the nodes in each of the five layers are described as follows:

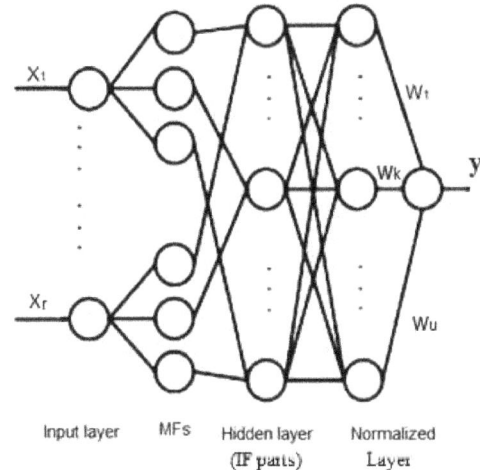

Figure 2. *Structure of radial-basis function- networks*

Layer 1: Each node in layer 1 represents an input node and it corresponds to one input linguistic variable. These nodes simply transmit input signals to the next layer directly.

Layer 2: Nodes in this layer stand for input terms of an input linguistic variable. Each one acts as a one-dimensional membership function (MF) which is a Gaussian function of the following form:

$$\mu_{ij}(x_i) = \exp\left(-\frac{(x_i - c_{ij})^2}{\sigma_j^2}\right) \quad (1)$$

Where i=1,2,...r, r is the number of input variables ,j=1,2,...u, u is the number of membership (MF) functions ,μij, jth membership function of the ith input variable xi ,cij center of the jth MF of xi ,σj width of the jth MF of xi

Layer 3: Each node in layer 3 represents a possible IF part of a fuzzy rule. The number of rules in this system is exactly the number of RBF neurons. For the jth rule Rj, its output is

$$\emptyset_j = \exp\left(-\frac{\sum_{i=1}^{r}(x_i - c_{ij})^2}{\sigma_j^2}\right) \quad (2)$$

$$= \exp\left(-\frac{\|X - c_j\|^2}{\sigma_j^2}\right)$$

where X = (x1, ..., xr) and Cj is the center of the jth RBF neuron.

Layer 4: Nodes in this layer are called normalized nodes. The number of normalized nodes is equal to that of RBF neurons. The output of the normalized nodes is given by

$$\psi_j = \frac{\emptyset_j}{\sum_{k=1}^{u}\emptyset_k} \quad (3)$$

$$= \frac{\exp\left(-\frac{\|X - c_j\|^2}{\sigma_j^2}\right)}{\sum_{k=1}^{u}\exp\left(-\frac{\|X - c_k\|^2}{\sigma_k^2}\right)}$$

The nodes in layer 4 are fully connected with the nodes in layer 3 for normalization.

Layer 5: Each node in this layer represents an output variable which is the summation of incoming signals from layer 4. Its output is given by

$$y(X) = \sum_{k=1}^{u} w_k \cdot \psi_k \tag{4}$$

$$= \frac{\sum_{k=1}^{u} w_k \exp\left(-\frac{\|X - c_k\|^2}{\sigma_k^2}\right)}{\sum_{k=1}^{u} \exp\left(-\frac{\|X - c_k\|^2}{\sigma_k^2}\right)}$$

where y is the value of the output variable and w_k is essentially the consequent part of the kth fuzzy rule. In the TSK fuzzy inference system, we have

$$w_k = t_{k0} + t_{k1}x_1 + \cdots + t_{kr}x_r \tag{5}$$

where k = 1, 2, ...u. The parameters $t_{k0}, t_{k1}, ...t_{kr}$ are the linear parameters in the consequent part of the kth fuzzy rule. From Fig.2 and the detailed functions of the nodes, it can be seen that the following conditions hold to make the adaptive RBFN-based filter equivalent to a TSK fuzzy inference system.

1) The number of links from layer 2 to each node in layer 3 is equal to the number of input variables. That is, a set of r input term nodes (one for each input linguistic node) in layer 2 is connected to a rule node in layer 3.

2) Each RBF neuron is a fuzzy rule essentially.

3) For the RBF neurons, no bias is considered.

4) The weights between layer 4 and layer 5 are the linear functions instead of a real constant.

4. Results and Analysis

EEG data with ocular artifacts are taken from [11] for testing the proposed method. Figure 3 shows the EEG that eliminated with EOG signal and then we applied this signal to the proposed filter.

Figure 3 indicates the performance of the proposed filter in eliminating the ocular artifact. Figure3.a and Figure3.b show the measured EEG signal and directly measured EOG signal, respectively, the estimated ocular interference signal using the proposed adaptive filter is shown in Figure 3.c, and ultimately Figure 3.d illustrates the clean EEG resulting from subtracting the estimated interference signal from the measured EEG as shown in the ANC system. Considering the part marked by arrows that indicate the severe presence of ocular artifacts, it is obvious that the proposed filter has removed the artifacts successfully and has desirable performance in this area. The result is also achieved in frequency domain using power spectrum magnitude (PSM) criterion. Figure 4.a shows the power spectrum of the EEG signal contaminated with EOG, the power spectrum of EOG signal is shown in Figure 4.b and the power spectrum of cleaned EEG is illustrated in Figure 4.c. Note that the absence of low frequencies of EOG in output signal indicates the removal of EOG artifact.

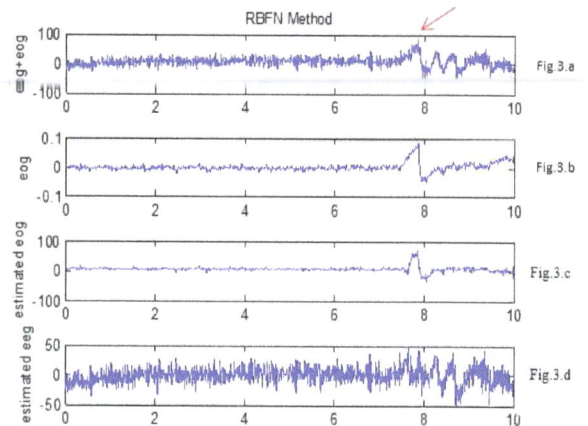

Figure 3. EOG artifact cancellation

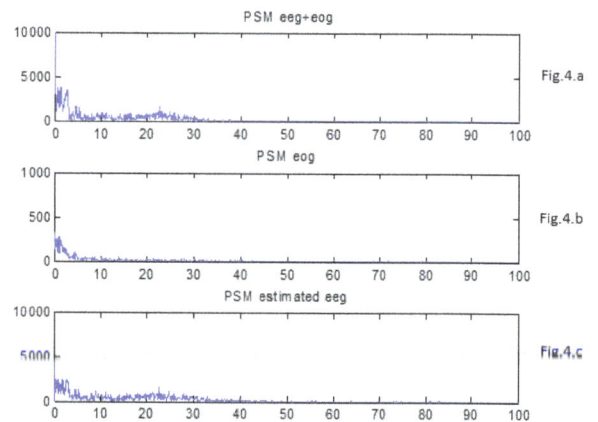

Figure 4. power spectrum of signals

5. Discussion and Conclusion

An adaptive noise cancellation technique was studied and used to de-noise life signals. Simulation results show that the proposed adaptive RBFN-based filter can cancel the noise signal from the distorted information signal successfully.

In this research, the results of method were discussed and analyzed.

References

[1]   R. Jonathan, J. M. Dennis, B. Niels, P. Gert and M. Theresa, *"Brain computer interfaces for communication and control,"* pp.767-791, Clinical Neurophysiology, 2002.

[2]   M. Fatourechi, A. Bashashati, K. W. Rahab and E. B. Gary, *"EMG and EOG artifacts from BCI systems: a Survey,"* pp. 480-494, Clinical Neurophysiology, 2007.

[3]   P.He, M.Kahle, G.Wilson and C.Russel, *"Removal of ocular artifacts from EEG: A comparison of Adaptive filtering method and regression method using simulated data"*, IEEE engineering in medicine and Biology 27th annual conference, 2005.

[4]   Ehsan Zezhadaraya, Mohammed B Shamosollahi. " *EOG artifact removal from EEG using ICA and ARMAX modeling"*, Private communication, 2004.

[5] P. He, G.Wilson and C.Russel, *" Removal of Ocular artifacts from Electro encephalogram by adaptive Filtering,"* pp.407-412, Medical &Biological Engineering & Computing, 2004.

[6] B.W.Jervis, M.Thomlinson, C.Mair, J.M. Lopez and M.I.B Garcia, *"Residual ocular artifacts subsequent to ocular artifact removal from the electroencephalogram"*, pp.293-298, IEE Proceedings-Volume 146, Issue 6, Nov. 1999.

[7] T. D. Lagerlund, F. W. Sharbrough, and N. E. Busacker, *"Spatial filtering of multichannel electroencephalographic recordings through principal component analysis by singular value decomposition," J. Clin. Neurophysiol.*, vol. 14, pp. 73–82, 1997.

[8] R. Verleger, T. Gasser, & J. Möcks, *"Correction of EOG artifacts in eventrelated potentials of the EEG: Aspects of reliability and validity"*, Psychophysiology, vol. 19, pp 472-480, 1982.

[9] J. L. Whitton, F. Lue, & H. Moldofsky, "A spectral method for removing eye movement artifacts from the EEG", *Electroenceph. Clin. Neurophysiol.* vol. 44, pp. 735-741, 1978.

[10] J.S.R. Jaug, *"Anfis: Adaptivenetwork-based fuzzy inference system,"* IEEE Trans. on Systems,Man,and Cybernetics, vol. 23, pp. 665-685, 1993.

[11] www.physionet.org

25

Pitch aircraft control with type-2 FLC and PD controller

Syed Mohamad Reza Haji Mirzaie[1, *], Hodeiseh Gordan[2], Jalil Shirazi[3], Amin Nikbakht[1]

[1]Electrical Engineering Department, Khavaran Institute of Higher Education, Mashhad, Iran
[2]Department of Electrical Engineering, Sobhan Institute of Higher Education, Neyshabur, Iran
[3]Department of Electrical Engineering, Islamic Azad University, Gonabad Branch, Gonabad, Iran

Email address:

Mohamad.hajimirzaie@gmail.com (S. M. R. H. Mirzaie), Hodeisehgordan@gmail.com (H. Gordan), Shirazi_stu@yahoo.com (J. Shirazi),
aminikbakht@yahoo.com (A. Nikbakht)

Abstract: Aircraft systems are inherently unstable systems. The equations governing the motion of an aircraft are a very complicated set of six non-linear coupled differential equations. The linearized equation around the operating point is simulated in simulink MATLAB software. Also, the linear part of the system of nonlinear equations is simulated in simulink MATLAB software. In this study, combinations of PD controllers with fuzzy controller in a unity feedback system has been employed. This paper gives a comparison between the two types of FLC type-1 and type-2 in order to show the great effect of the new type of FLCs in reducing overshoot of the step response and improving the robustness of the system.

Keywords: PD Controller, Pitch Aircraft Angle, Type-1 Fuzzy Logic Controller, Type-2 Fuzzy Logic Controller

1. Introduction

For linear systems and some of non-linear systems, classic controllers such as PD and PID have been widely used in industrial control processes because of their simple structure and robust performance in a wide range of operating conditions. Several numerical approaches such as Fuzzy Logic Controller (FLC) algorithm and evolutionary algorithms have been used for the optimum design of PID controllers [1]. FLS constructed based on type-1 fuzzy systems (T1FS), referred to as T1FLS, have demonstrated their ability in many applications, especially for the control of complex nonlinear systems that are difficult to model analytically [2, 3]. However, researchers have shown that T1FLS have difficulty in modeling and minimizing the effect of uncertainties [4]. One reason for this is that T1FS are certain in the sense that for each input there is a crisp membership grade. T2FS, characterized by membership grades that are themselves fuzzy, were first introduced by Zadeh in 1975 to account for this problem [5]. However, there were some obstacles about the implementation of the T2FLSs on real world problems such as characterization of type-2 fuzzy sets, performing operations with type-2 fuzzy sets, inferencing with type-2 fuzzy sets and obtaining the defuzzified value from the output of a type-2 inference engine.

Since the publication of Karnik and Mendel [2], in which they proposed new concepts to overcome the difficulties mentioned above the number of literatures regarding T2FLSs have increased rapidly [6, 7]. In Satish et. al, [8], a new hybrid Fuzzy PD+ I controller (FPD+I) has been proposed and implemented.

In this paper, a pitch displacement of aircraft has been controlled and shows that the T2FL and PD controller with respect to T1FLC and PD controller improves the performance of a system.

This paper is organized as follows. Section 2 deals with the approximation of pitch aircraft control system model. In section 3, we present the analytical design of type-2 fuzzy control. Later, some simulations are executed to verify the validity of the proposed approach in section 4. Finally, section 5 concludes the paper.

2. Modeling a Pitch Aircraft Controller

The active and influential forces on an aircraft are shown in the figure 1.

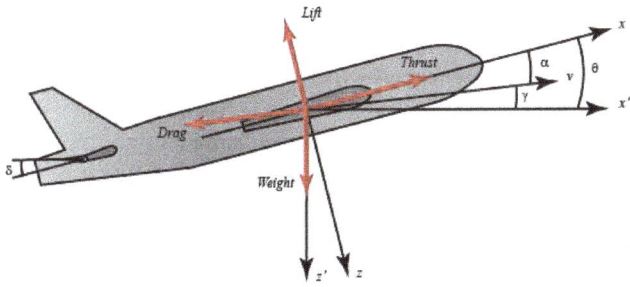

Figure 1. Forces acting on an aircraft.

The equations governing the motion of an aircraft are a very complicated set of six non-linear coupled differential equations. This article assumes that airplanes in the sky along the horizon is moving at constant speed and altitude. In this case, the resultant forces weight and lift and the lift and the resultant forces thrust where drag is equal to zero. Also, it is assumed that change in pitch angle does not change the speed of an aircraft. This assumption might seem unrealistic but simplifies the problem. Under these assumptions, the longitudinal equations of motion of an aircraft can be written as:

$$\dot{\alpha} = \mu\Omega\sigma[-(C_L + C_D)\alpha + (1/(\mu - C_L))q - (C_w\sin\gamma_e)\theta + C_L] \quad (1)$$

$$\dot{q} = \mu\Omega/2i_{yy}\{[C_M - \eta(C_L + C_D)]\alpha + [C_M + \sigma C_M(1 - \mu C_L)] + (\eta C_w \sin\gamma_e)\delta_e]\} \quad (2)$$

$$\dot{\alpha} = \Omega q \quad (3)$$

With replaced some numerical values of a Boeing 777 in equation (1), (2) and (3) we reach the following equations [10].

$$\dot{\alpha} = -0.313\alpha + 56.7q + 0.232\delta_e \quad (4)$$

$$\dot{q} = -0.0139\alpha - 0.0426q + 0.0232\delta_e \quad (5)$$

$$\dot{\theta} = 56.7q \quad (6)$$

Where α, θ, q and δ, are attack of angle, pitch angle, rate of change of pitch angle and the elevator deflection angle, respectively. To model systems, three state variable for the system are considered and the matrices A and B was ob-tained, as following:

$$\begin{bmatrix} \dot{\alpha} \\ \dot{q} \\ \dot{\theta} \end{bmatrix} = \begin{bmatrix} -0.313 & 56.7 & 0 \\ -0.0139 & -0.042 & 0 \\ 0 & 56.7 & 0 \end{bmatrix} \begin{bmatrix} \alpha \\ q \\ \theta \end{bmatrix} + \begin{bmatrix} 0.232 \\ 0.0232 \\ 0 \end{bmatrix} [\delta_e] \quad (7)$$

According to the output, matrices C and D derived as follows:

$$y = \begin{bmatrix} 0 & 0 & 1 \end{bmatrix} \begin{bmatrix} \alpha \\ q \\ \theta \end{bmatrix} + [0][\delta_e] \quad (8)$$

After few steps of algebra, you should obtain the following transfer function:

$$\frac{\theta(s)}{\delta_e(s)} = \frac{1.151s + 0.1774}{s^3 + 0.739s^2 + 0.921s} \quad (9)$$

Figure 2 presents the open-loop system step of response [11].

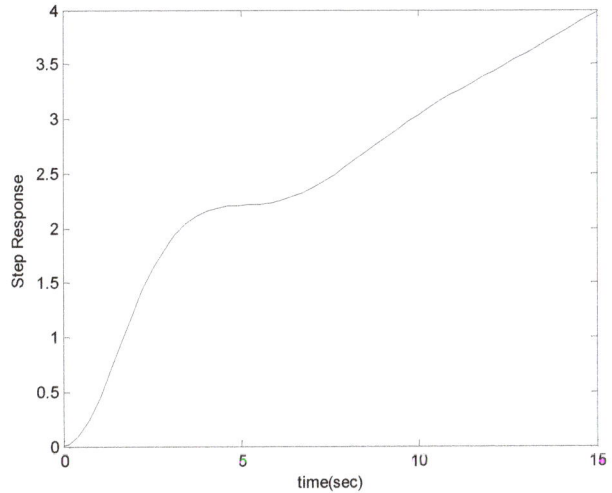

Figure 2. Open-loop step of response.

3. Design Type-2 Fuzzy Logic Controller

Fuzzy logic systems (FLCs) are based on the knowledge in the form of "IF-THEN" rules. Figure 3 shows a type-1 fuzzy logic system consisting of the fuzzifier and the defuzzifier. As illustrated in figure 3, a type-1 fuzzy logic system consists of 4 components: the rule base, the fuzzy inference engine, the fuzzifier and the defuzzifier [4]. In the type-1 fuzzy logic systems, for every input in the type-1 collection of fuzzy there is degrees of membership with real recently, a type of fuzzy sets characterized by membership grades that are themselves fuzzy has been introduced. This fuzzy logic systems is type-2 fuzzy logic systems [4].

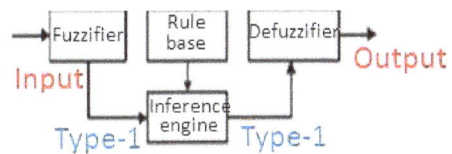

Figure 3. The main structure of type-1 fuzzy logic system.

T2 FLSs are characterized by the shape of their MFs. The membership function of general type-2 fuzzy is shown in figure 4 [6]. The extra mathematical dimension is provided by the blurred area. The FOU is bounded by upper and lower MFs.

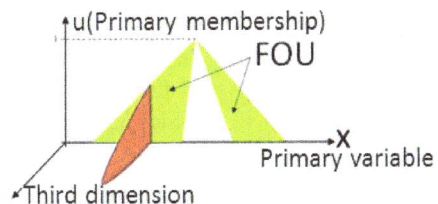

Figure 4. Type-2 fuzzy set.

The structure of a type-2 FLS is shown in figure 5 [4].

This structure is very similar to the type-1 fuzzy systems. Since the main difference between type 1 and type 2 fuzzy systems is the fuzzy sets. The only difference between type-1 and type-2 fuzzy logic system structure is in the block "type-reducer " .

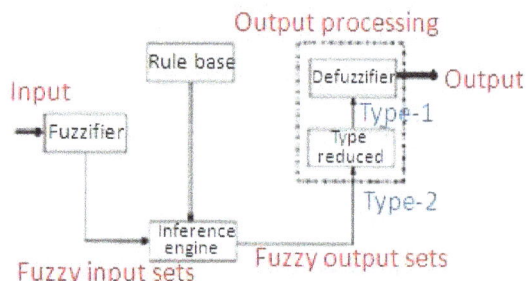

Figure 5. *The main Structure of a T2 FLS*

Each block is introduced in the following section:

3.1. Fuzzifier

The fuzzifier maps the input variables with the actual values to a type-1 or type-2 fuzzy set or a singleton fuzzy set in input.

3.2. Rule base

Fuzzy rule base is a set of fuzzy "if - then" rules. I-th law type 2 fuzzy systems is shown as follow:
R^i: IF x_1 Is $\tilde{F}_1^i$ and x_2 Is $\tilde{F}_2^i$ and ... and x_n Is $\tilde{F}_n^i$,

$$\text{Then } y \text{ is } \tilde{G}^i \tag{5}$$

Where $i = 1, ..., M$ and N is the number of rules. where " $\sim$ " implies that the fuzzy set is a type-2 fuzzy set. Where x_js are inputs $j = 1, ..., n$, $\tilde{F}_j^i$ are input sets, $\tilde{G}^i$s are output sets and y is the output.

3.3. Fuzzy Inference Engine

The inference engine of a type-1 FLS provides a mapping from input type-1 fuzzy sets to output type-1 fuzzy sets by using all rules. The inference process in a type-2 FLS is very similar to that in a type-1 FLS. Hence a type-2 fuzzy member function mapping from input space to output space based on type-2 fuzzy logic is created.

3.4. Type-reduced

In a type-1 FLS, the output of the inference engine is normally a type-1 fuzzy set but, in a type-2 FLS, the output of the inference engine is normally a type-2 fuzzy set. Hence, using a type-reduced block a type-1 fuzzy set is obtained from the type-2 output sets of the FLS. This operation is called type reduction.

3.5. Defuzzification

In the defuzzification, a type-1 fuzzy set is produced to a crisp output.

4. Simulation and Results

The main goal of this study is to investigate and see whether the T2FLC performs better than similar T1FLC in controling the system. Therefore, in both cases we consider the same situation. The controller is designed mamdani type that in the number of rules and fuzzy sets for input and output are the same. Block diagram of the control structure using T2FLC and control of PD in the figure 6 is shown. In this paper, the parameters of PD controlling is set with trial that k_P and k_d are considered 25 and 4, respectively.

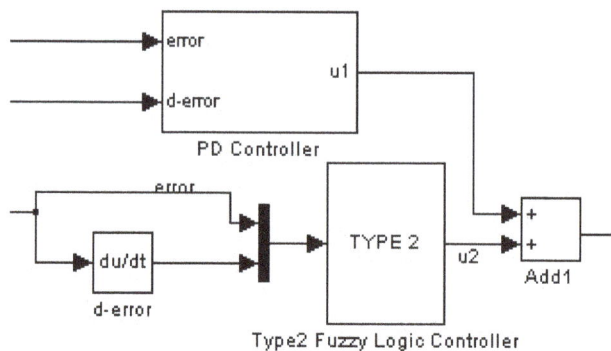

Figure 6. *Block diagram of controller structure*

In this study, the fuzzy controller has two inputs and one output. The error e(t) and the change of error ((de(t))/dt) signals are used To design the fuzzy controller. In type 1 and type 2 fuzzy controller, gaussian functions are used as the input and output. Input variable are marked with three belong with the label of N, Z and P and output variables are marked with the label GN , N , Z , P , GP GN. All the membership functions of the two FLCs inputs and outputs are shown in figures 7, 8, 9, and 10.

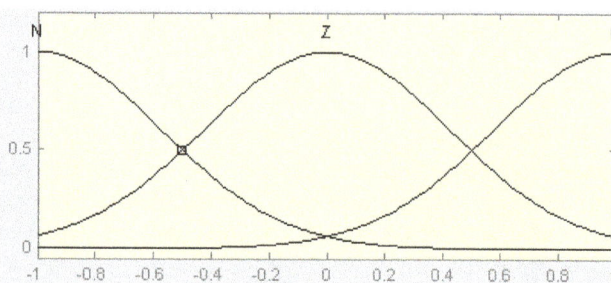

Figure 7. *Type-1 fuzzy sets for the input variables.*

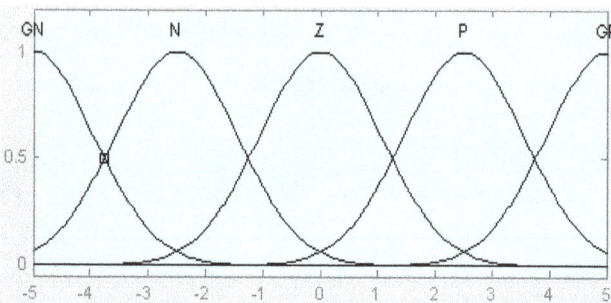

Figure 8. *Type-1 fuzzy sets for the output variable.*

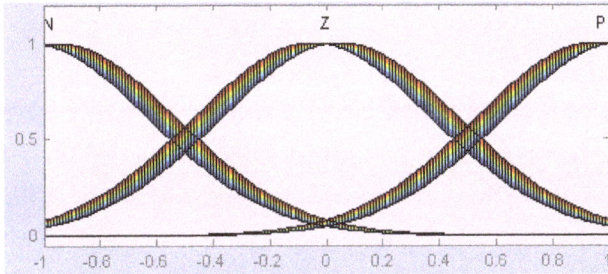

Figure 9. Type-2 fuzzy sets for the input variables.

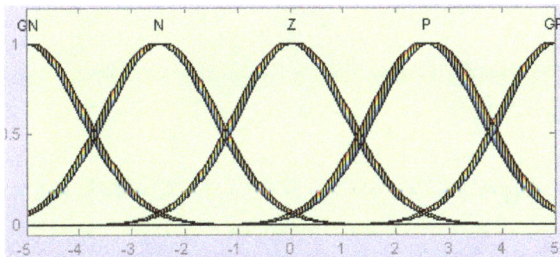

Figure 10. Type-2 fuzzy sets for the output variable.

Since we have considered for each input three fuzzy sets, we can evaluate the whole nine. All of the fuzzy rules collection are available in table 1.

Table 1. Rule base of the two FLCs

d-error / error	N	Z	P
N	GN	N	Z
Z	N	Z	P
P	Z	P	GP

In the fuzzy controller type -1 centroid method for defuzzification mechanism and in type-2 fuzzy controller to reduce the type of center method and center of gravity method for defuzzification is used for building. These methods using toolbox fuzzy type 2 is selected in the MATLAB environment [9]. Results are depicted in figure 11 and a comparison between the two types of FLCs and PD controller is done. The properties of each controller is shown in table 2.

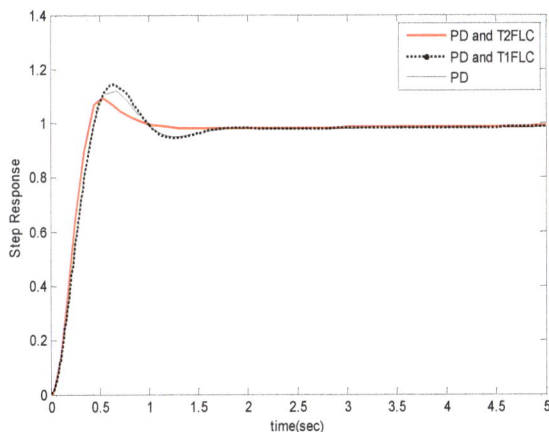

Figure 11. System response with T1FLC and T2FLC

Table 2. The properties of each controller

M_P%	t_p	t_r	t_s	Properties Controller
12%	0.6675	0.4514	1.76	Controller PD
14.6%	0.6562	0.445	1.7	Controller PD and T1FLC
9.4%	0.5338	0.401	1.5	Controller PD and T2FLC

5. Discussion and Conclusion

In this study, a PD controller with fuzzy controller is used. Fuzzy controllers are compared in two states: type 1 and type 2. As mentioned in table 2, we can conclude type-1 fuzzy controller does not always improve response. PD controller with the type- 2 fuzzy controller percent of rate overshoot step response other than the controller reduces. Also, we can conclude that, with given the uncertainty of fuzzy functions, the output response of the system is robust. These results are implemented using the IT2FLS toolbox in MATLAB software environment.

References

[1] Y. John and R. Langari, Fuzzy Logic Intelligence, Control and Information. NJ: Prentice Hall, 1998.

[2] P. King and E. Mamdani, The application of fuzzy control to industrial process, Automatica, vol. 13, pp. 235–242, 1997.

[3] L. A. Zadeh, Outline of a new approach to analysis of complex systems and decision processes. IEEE Trans. Syst., Man, Cybern., vol. 3, no. 1, pp. 28–44, 1973.

[4] J. M. Mendel, Rule-Based Fuzzy Logic Systems: Introduction and New Directions. NJ: Prentice-Hall, 2001.

[5] Zadeh, L. A., The concept of a linguistic variable and its application to approximate reasoning, Part I. Information Sciences, 1975, 8(3), pp. 199–249.

[6] Saba T. Salim, Design of Type-2 Fuzzy Logic Controller for a Simple Furnace System. Eng. & Tech. Journal, Vol.30 , No. 8, 2012

[7] Ismail ATACAK, Design of a hybrid type-2 fuzzy logic/proportional integral controller for single-phase three-wire inverter system. Scientific Research and Essays Vol. 6(23), pp. 5049-5064, 16 October, 2011

[8] Satish. R. Vaishnav, Zafar. J. Khan, Performance of tuned PID controller and a new hybrid fuzzy PD + I controller, World Journal of Modelling and Simulation, Vol. 6, No. 2, pp. 141-149, 2010.

[9] M. B OZEK, Z. H. Akpolat, A Software Tool: Type-2 Fuzzy Logic Toolbox. Wiley Periodicals Inc. 2008.

[10] A. Torabi, A. Karsaz and S. K. M. Mashhadi, "Design of Modern and Predictive Controllers for Aircraft Control System", Latest Trends in Circuits, Systems, Signal Processing and Automatic Control, pp. 344–348, 2014.

[11] http://ctms.engin.umich.edu/CTMS/index.php?example=AircraftPitch&section=SystemModeling

26

Open-loop voltage control voltage source inverter for a linear load & non-linear load

Akhilesh Sharma[1], Neeraj Kumar[2], Gunjan Gupta[3]

[1]NERIST, Nirjuli, Arunachal Pradesh, India
[2]DIET, Rishikesh, Uttarakhand, India
[3]Invertis University, Bareilly, U.P., India

Email address:

akhil_ful@rediffmail.com (A. Sharma), neer_81@rediffmail.com (N. Kumar), gunjan.g@invertis.org (G. Gupta)

Abstract: The conventional sources of energy are limited which are unable to meet the demand. There is always need of generation of electrical energy worldwide, may be in terms of few kilowatts to hundreds of kilowatts. The power requirement at the domestic level is terms of few kilowatts so there should be a way to meet the electrical demand at domestic level during load shading or fault. A subtitle to this could be a Voltage controlled voltage source inverters (VCVSIs). Now a day, such inverters are widely used in many applications like power supplies, power quality controllers, renewable energy, marine and military to meet the demand. Wherever an ac supply is need from a DC source, such inverters may be used. They are the heart and soul of converting DC into an ac source. Hence, such inverters should be designed so that they are robust and efficient. Inverter models can be design and simulated with software like Matlab Simulink to check suitability of inverter before implementing the hardware. Loads at the domestic level may be linear or non- linear hence a simple approach to design a voltage source inverter using open loop is presented in this paper.

Keywords: Voltage Controlled Voltage Source Inverters, MOSFET, Thyristors, Open-Loop Control System

1. Introduction

Exponential growth of population and industries has lead to more demand of electrical power. The generating stations are unable to meet the power demand. Even the frequent power failure, hampers the living of people in cities, towns and villages. There is misbalance in utilization and generation of electrical power.

So it has become mandatory to switch engineers to find other alternative to meet so portion of the demand. Non-conventional sources have their own limitations. For example, solar power plant depends on factors like weather condition, altitude and solar radiation etc. If sky is clear, the places located at higher altitude will receive high intensity of solar radiation than the places located at low altitude during the day. The luminous intensity also varies from forenoon to afternoon. Hence, it is difficult to harness electrical energy from such sources at our will. One way to do so is to develop fast switching devices MOSFET, IGBT, BJT and Thyristors etc which help in converting DC into ac source called inverter [1].

A fixed DC source can be converted to ac through an inverter. The function of an inverter is to change a fixed dc input to an ac output voltage of desired magnitude and frequency hence it can be said that the output voltage obtained from the device could be fixed or variable one without change in frequency. A variable output voltage can be obtained by varying the input dc voltage and maintaining the gain of the inverter constant. On the other hand, if the dc input voltage is fixed and it is not controllable, a variable output voltage can be obtained by varying the gain of the inverter, which is normally accomplished by pulse width modulation control within the inverter. [4 and 6].

An ideal inverter should have output voltage exactly similar to a sine wave. But this is not so when a practical inverter is considered. The output voltage wave may be non sinusoidal containing harmonics of different order. This may due to the switching on and off of an inverter. For low- and medium- power applications, square – wave or quasi – square wave voltages may be acceptable; this is not so for high power applications, where low distorted sinusoidal wave forms may be accepted. In order to achieve sinusoidal wave of low distortion, the operating speed of power semiconductor devices should be high. Further reduction in harmonic is possible if proper switching technique is applied. [1 and 5].

2. Principle of Operation of a Single Phase

The principle of operation depends upon the switching of power electronic devices. Accordingly it can either be half wave or full wave inverter. A half wave inverter is one which requires two choppers whereas a full wave inverter requires four choppers. The former is discussed as under [1 and 6].

2.1. Half-Wave Inverter

The principle of single phase half wave inverters is explained on the basis of figure 1 as under:

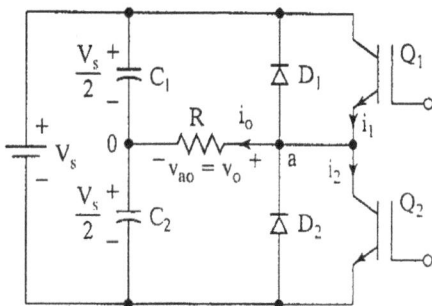

Figure 1. A single phase half wave inverter

A single phase half wave inverter consists of two choppers, say IGBTs in this case. When only one IGBT Q_1 is turned on for a time $T_0/2$, the instantaneous voltage across the load v_0 is $V_s/2$. If only the IGBT Q_2 is turned on for a time $T_0/2$, - $V_s/2$ appears across the load. The logic circuit should be designed in such a way that both the IGBTs do not turn at a time i.e. if Q_1 turn on, Q_2 should be off and vice –versa. Figure 2 shows the wave forms for the output voltage and currents with resistive load.

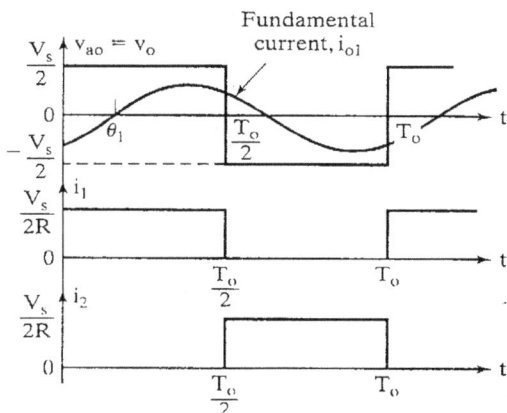

Figure 2. Voltage and current waveform with resistive load

The root- mean- square (rms) output voltage can be found from

$$V_0 = (\frac{2}{T_0}\int_0^{T_0/2} \frac{V_s^2}{4} dt)^{1/2} = \frac{V_s}{2} \qquad (2,1)$$

The instantaneous output voltage can be expressed in Fourier series as

$$v_0 = \frac{a_0}{2} + \sum_{n=1}^{\infty}(a_n \cos(n\omega t) + b_n \sin(n\omega t))$$

$$b_n = \frac{1}{\pi}\left[\int_{-\frac{\pi}{2}}^{0} \frac{-V_s}{2}d(\omega t) + \int_{0}^{\frac{\pi}{2}} \frac{V_s}{2}d(\omega t)\right] = \frac{4V_s}{n\pi}$$

which gives the instantaneous output voltage v_0 as

$$v_0 = \sum_{n=1,3,5...}^{\infty} \frac{2V_s}{n\pi}\sin n\omega t = 0 \qquad n\ 2,4,6 ... \quad (2.2)$$

where $\omega = 2\pi f_0$ is the frequency of output voltage in radians per second. Due to the quarter wave symmetry of the output voltage along the x- axis, the even harmonics voltages are absent. The fundamental component of the rms output voltage is obtained for n=1, so equation (2.2) reduces to

$$V_{01} = \frac{2V_s}{\sqrt{2}\pi} = 0.45V_s \qquad (2.3)$$

For a highly inductive load, the load current cannot change immediately with the output voltage i.e. if Q_1 is turned off at $t = \frac{T_0}{2}$, the load current would continue to flow through D_2, load, and lower half of the dc source until the current falls to zero. Similarly, when Q_2 is turned off at $t = T_0$, the load current flows through D_1, load, and upper half of the dc source. When either of the diode D_1 or D_2 conducts, energy is fed back to the dc source and these diodes are known as feedback diodes. Figure 3 shows the load current and conduction intervals of the device for a purely inductive load.

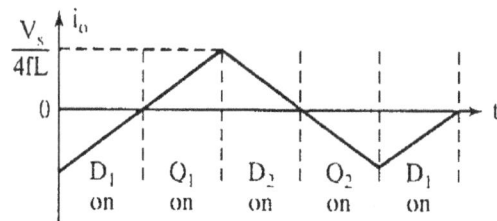

Figure 3. Current waveform with inductive load

IGBTs can be replaced by any other switching devices like MOSFET, BJT, etc as per the requirement. If t_{off} is the turn off time of a device, there should be a minimum delay time of $t_d = (t_{off})$ between the outgoing device and triggering of the next incoming device. Otherwise, short circuit condition would result through the two devices. In order to avoid this condition, the conduction time is reduced. The maximum conduction time of a device would be $t_{on} = \frac{T_0}{2} - t_d$.

All practical devices require a certain turn - on and turn – off time.

2.2. Principle of Operation of a Full-Wave Single Phase Inverter

A single phase bridge voltage source inverter (VSI) is shown in figure 4. It consists of four choppers. When IGBTs Q_1 and Q_2 are turned on simultaneously, the input voltage Vs appears across the load. If IGBTs Q_3 and Q_4 are turned on at the same time, the voltage across the load is reversed and is – Vs. the wave for the output voltage is shown in figure 5.

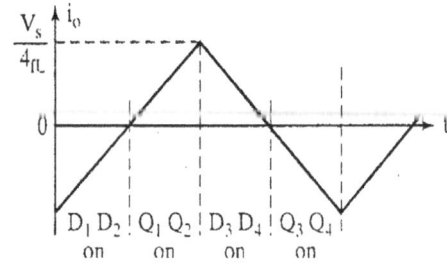

Figure 5. *Load current with highly inductive load*

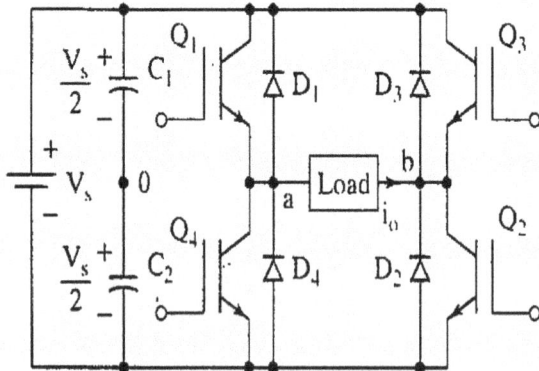

Figure 4. *A single phase full wave inverter*

Table 1 shows the five switch states. IGBTs Q_1, Q_4 in figure 4 act as the switching devices S_1, S_4, respectively. If two switches: one upper and one lower conduct at the same time such that the output voltage is$\pm V_s$, the switch state is 1, whereas if these switches are off at the same time, the switch state is 0. Figure 5 shows the load current for highly inductive load.

The rms output voltage can be found from the expression as under

$$V_0 = \left(\frac{2}{T_0}\int_0^{T_0/2} V_s^2 \, dt\right)^{1/2} = V_s \qquad (2.4)$$

Equation 2.2 can be extended to express the instantaneous output voltage in a Fourier series as

$$v_0 = \sum_{n=1,3,5\ldots}^{\infty} \frac{4V_s}{n\pi}\sin n\omega t \qquad (2.5)$$

Table 1. *Switch State for Single Phase Full Voltage Source Inverter (VSI)*

State	State No.	Switch State	v_{a0}	v_{b0}	v_0	Component Conducting
S_1, and S_2 are on and S_4, and S_3 are off	1	10	$\frac{V_s}{2}$	$\frac{-V_s}{2}$	V_s	S_1, and S_2 if $i_0>0$ D_1 and D_2 if $i_0<0$
S_4, and S_3 are on and S_1, and S_2 are off	2	01	$\frac{V_s}{2}$	$\frac{-V_s}{2}$	V_s	D_4 and D_3 if $i_0>0$ S_4 and S_3 if $i_0<0$
S_1, and S_3 are on and S_4, and S_2 are off	3	11	$\frac{V_s}{2}$	$\frac{V_s}{2}$	0	S_1 and D_3 if $i_0>0$ D_1 and S_3 if $i_0<0$
S_4, and S_2 are on and S_1, and S_3 are off	4	00	$\frac{-V_s}{2}$	$\frac{-V_s}{2}$	0	D_4 and S_2 if $i_0>0$ S_4 and D_2 if $i_0<0$
S_1, S_2 S_3 and S_4 all are off	5	Off	$\frac{-V_s}{2}$ $\frac{V_s}{2}$	$\frac{V_s}{2}$ $\frac{-V_s}{2}$	$-V_s$ V_s	D_4 and D_3 if $i_0>0$ S_4 and S_3 if $i_0<0$

The fundamental rms component of the output voltage is obtained by substituting n=1. Therefore the fundamental voltage is

$$V_1 = \frac{4V_s}{\sqrt{2}\pi} = 0.90 V_s \qquad (2.6)$$

3. Simulation

The entire simulation has been carried out in MATLAB 7.0.1 SIMULINK. Both linear and non linear loads have been considered for open loop control of voltage controlled voltage source inverter. The parameters of the inverter are given in table 2 [2].

Table 2. *Parameters of Inverter*

PARAMETER	LABEL	VALUE	UNIT
Rated power	-	3.5	kVA
Rated output frequency	F_{load}	50	Hz
Rated output voltage	V_{load}	230	V
Battery voltage	V_{batt}	24	V
Battery and lead wire resistance	R_{batt}	30	$m\Omega$
DC filter capacitance	C_{dc}	4	mF
Inverter Switching frequency	f_{sw}	20	kHz
Filter inductor	L_f	1	μH
Filter inductor Resistance	R_{Lf}	1	$\mu \Omega$
Transformer turns ratio	N	18	---

The Matlab Simulink model of the single-phase voltage control voltage source inverter is shown in Figure 6. The model developed using the Simulink power system blockset, comprises of components such as power electronic devices (full-bridge and rectifier) and elements such as inductors, capacitors and resistors. The DC model used comprises of the battery (V_{batt}) and its respective resistance and lead wire resistance (R_{batt}) as well as the filter capacitor (C_{dc}) and a DC bus current measurement resistor (I_{dc} measure), which is of the order of micro-ohms as it is only used for DC bus current measurement. The output from the full-bridge block comprises of the filter inductor and its resistance (L_f, R_{Lf}) and filter capacitor with damping resistor (C_f, R_{Cf}). Also included is the step-up transformer (T_x). The resistor '$R_{measure}$' is of the order of micro-ohms and is only used for load current measurement which is similar to 'I_{dc} measure.

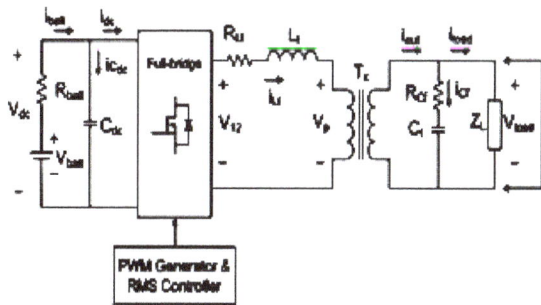

Figure 6. *Physical components of single phase VCVSI*

A system model showing the physical components of the single-phase Voltage controlled voltage source inverter modeled using Matlab Simulink is shown in Figure 6. This

inverter uses a low-voltage DC bus ($24V_{DC}$), which is stepped up to 240VAC using a step-up transformer (T_x). The transformer provides galvanic isolation and is a simple solution for the stepping up of a low-voltage DC bus. The DC bus in the model comprises of the battery (V_{batt}), lead wire and battery resistance (R_{batt}), and DC filter capacitor (C_{dc}). The full-bridge uses IGBT switching devices with the full-bridge output filtered using a low-pass LC filter (L_f and C_f). The inductor filter resistance is represented as R_{Lf} with the LC filter-damping resistor being R_{Cf}. The load connected to the inverter (Z_L) is considered arbitrary (linear and/or non-linear).

The PWM generator provides the switching signals for the full-bridge. The inverter system modeled with Matlab Simulink was achieved using the power system and standard simulink blocksets. The system was analyzed by sending the required signal to scopes and the workspace, which allowed for analysis and design of the inverter model. The PWM signals for each of the power electronic devices in the full-bridge come from the PWM generator block. This block allows for the switching frequency and number of inverter legs to be selected with all PWM signal multiplexed on a single bus into the full-bridge block (pulses). The input to this block (signal(s)) is the sinusoidal reference for the inverter. For the open-loop control, the reference signal (sinusoidal reference) is generated from a generator in Simulink. The simulink models for open- loop control of a linear load and a non-linear load are shown in figures 7 and figure 8 respectively [2 and 3].

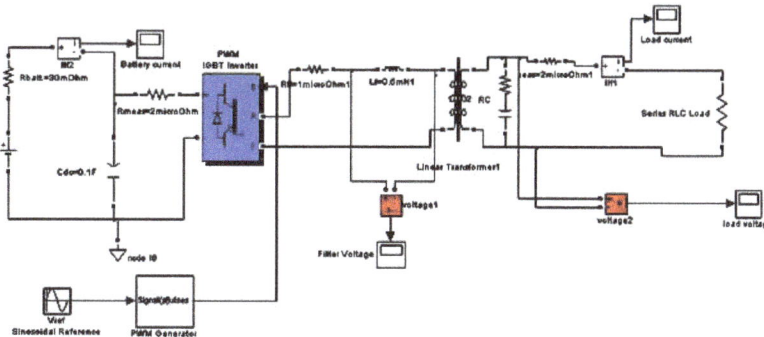

Figure 7. *Simulink Model for linear load*

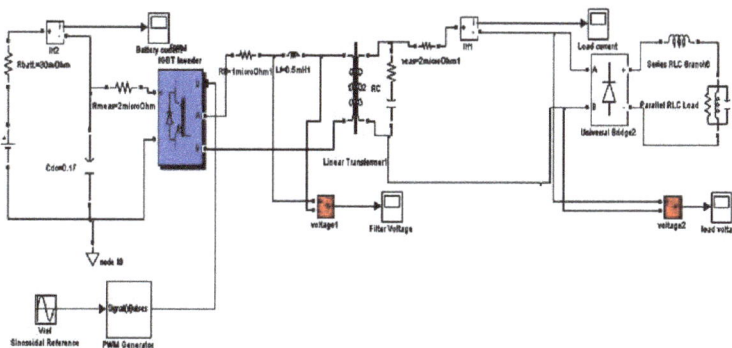

Figure 8. *Simulink Model for non-linear load*

4. Results

The simulation results for liner and non-linear loads have been obtained and are shown in figures.

4.1. Linear Load

Figures 9.1, 9.2, 9.3, 9.4 and 9.5 are the simulation results of the Voltage Controlled Voltage Source Inverter with open loop control having a linear load of 3.5 kVA. From the figure 9.1, initially, the load current is maximum, about 6.2 A and settles around 4 A after 0.33 seconds. The maximum load voltage is 90V and it decreases until it settles down at 60V as seen in figure 9.2. The filter current is quite high, and is about 3700 A initially and settles around 2000 A. The high in filter current is due to too small impedance offered by the filter inductor L_f as seen in figure 9.3. The filter voltage is shown in figure 9.4. The maximum filter voltage is 4V initially. It settles to 2V after 0.33 second. The battery current settles at 500 A, starting from zero as seen from figure 9.5. In the beginning, when the inverter is switched on, the battery current starts building up. Because of the presence of LC in the circuit, battery current settles to 500 A after 0.33 seconds. Both, load current and load voltage are similar, except their amplitudes.

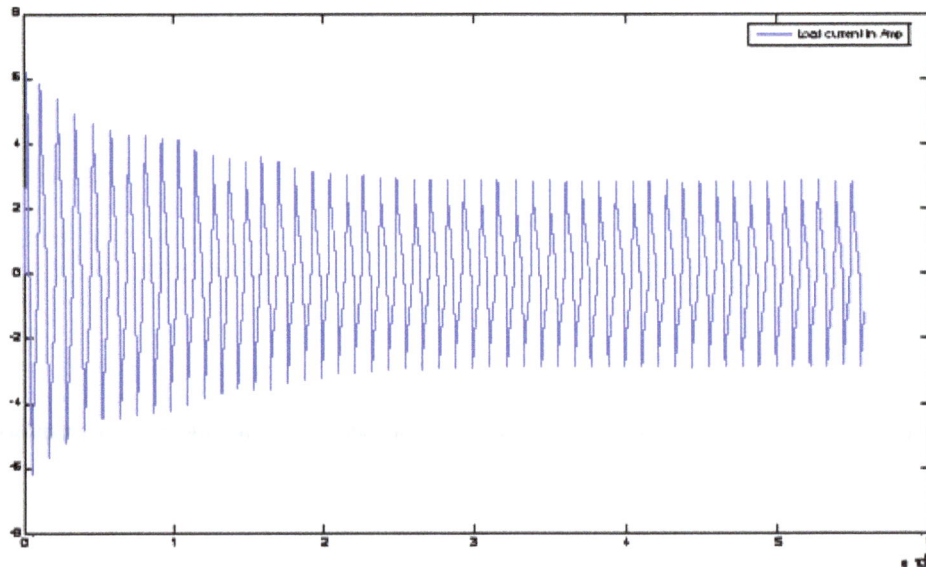

Figure 9.1. Load current

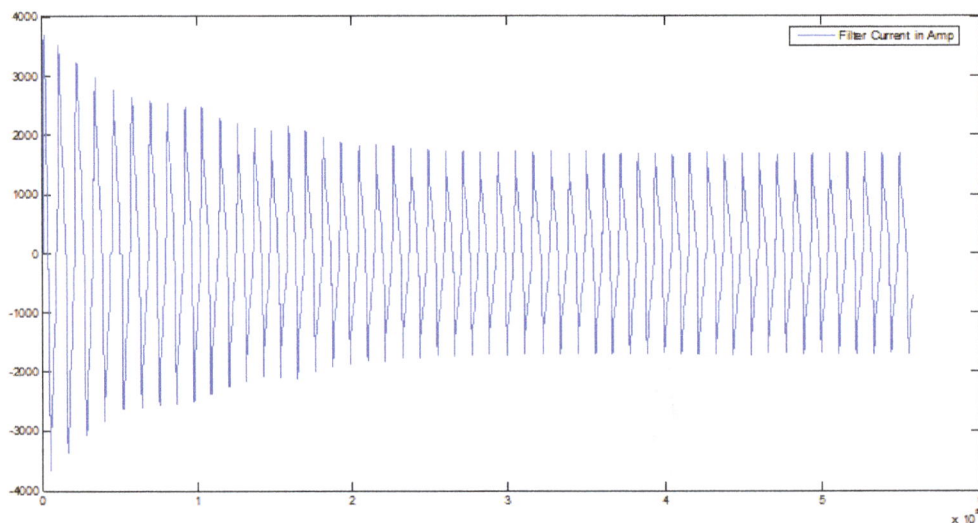

Figure 9.2. Filter current

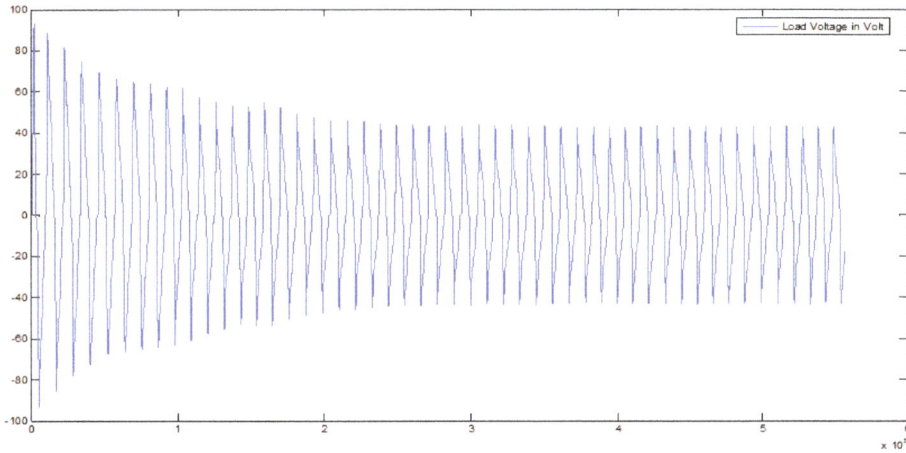

Figure 9.3. *Load voltage*

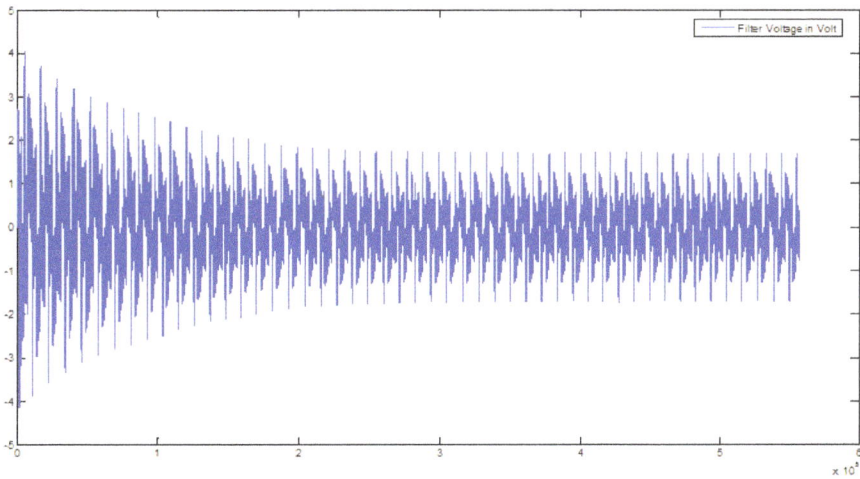

Figure 9.4. *Filter Voltage*

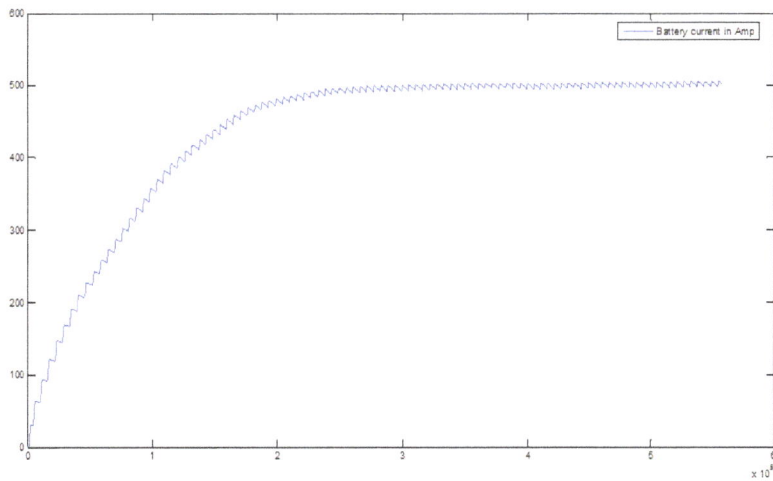

Figure 9.5. *Battery current*

4.2. Non-Linear Load

The various plots obtained through open loop control of a non-linear load are shown in figure 10.1, 10.2, 10.3, 10.4 and 10.5 respectively. As it is seen from the figure 10.1 that the current waveform is not sinusoidal as was in the case of linear load where both load current and load voltage were similar. It is due to non-linearity in the load, hence current wave form is not a complete sinusoidal. The amplitude of the load current becomes constant after 0.03 seconds and is about 1.26 Amps. The amplitude of the voltage wave is 60V as seen in figure 10.2. The steady amplitude of filter current is 1700A and maximum amplitude of the current is 3700 A as seen in figure 10.3. The filter amplitude of voltage is small nearly about 2V and the battery current rises from zero to 450 A in 0.05 seconds as seen in figure 10.4 and figure 10.5 respectively.

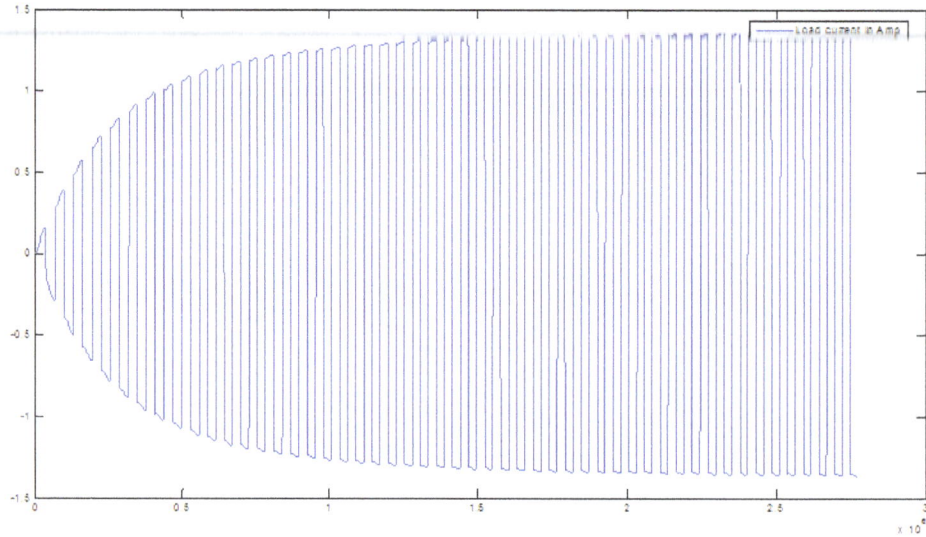

Figure 10.1. Load current

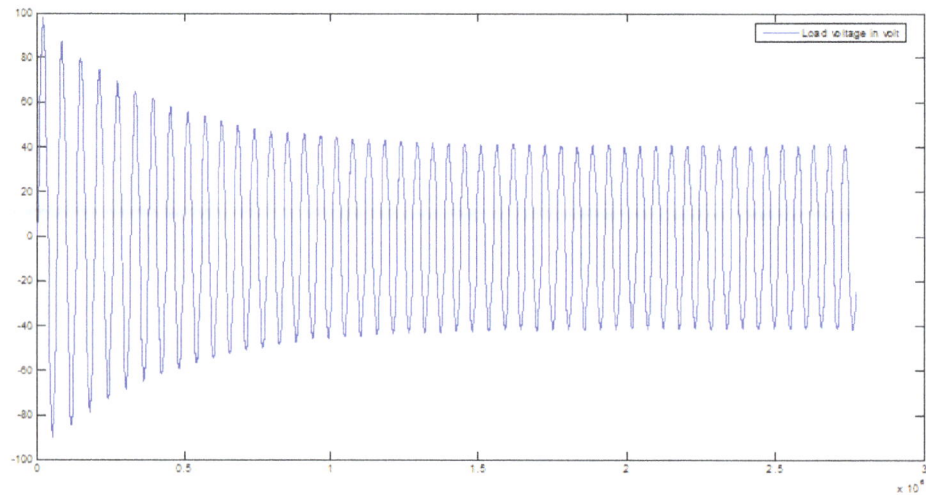

Figure 10.2. Load voltage

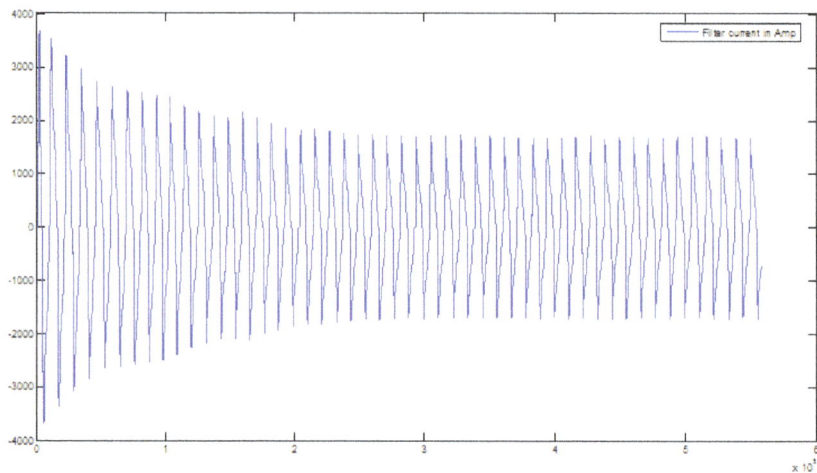

Figure 10.3. Filter current

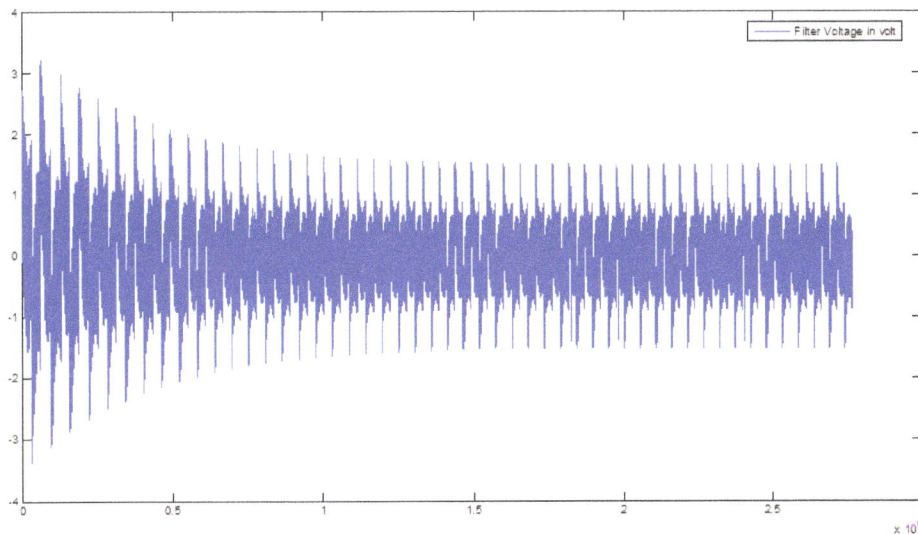

Figure 10.4. Filter voltage

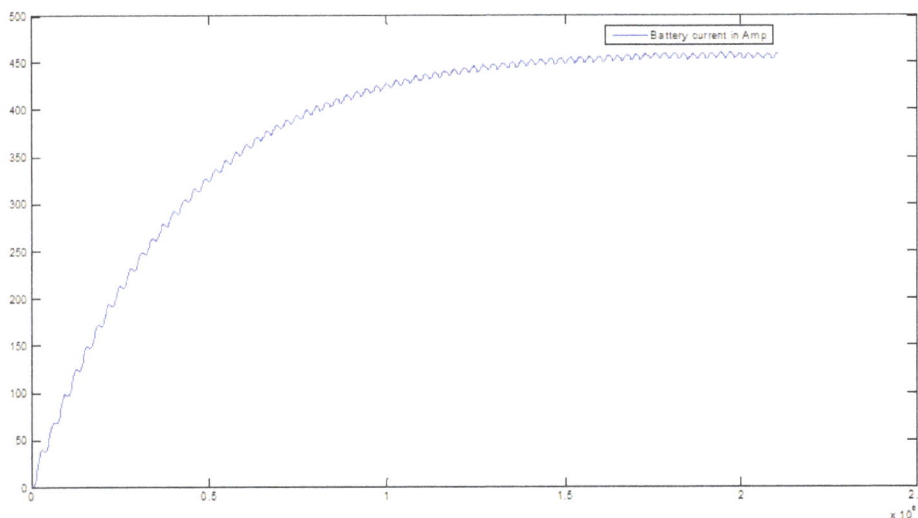

Figure 10.5. Battery current

5. Conclusions

A Matlab Simulink model of a single-phase 3.5 kVA Voltage Controlled Voltage Source Inverter with open-loop control has been developed for linear and non-linear loads.

When the nature of load is purely linear, the load current has constant frequency and amplitude after 0.33 sec but this is not so in case of non-linear load, where the load current is not sinusoidal with constant amplitude although there is decrease in steady time. When the load is non-linear, the filter current decreases. There is also decrease of as source current, thus allow to use non-linear load. Irrespective of the nature of load, the filter voltage in both cases is independent of the load current. So it can be concluded that for a open loop system, the non-linear load is preferred.

About the Authors

Akhilesh Sharma is working as Assist. Prof. in North Eastern Regional Institute of Science and Technology (under Ministry of Human Resource & Development, Govt of India), Nirjuli, Arunachal Pradesh, India in electrical engineering department. He did his Master's degree in Power Electronics & Drives from Madan Mohan Malaviya Engineering College Gorakhpur in 2009 under Uttar Pradesh Technical University, Lucknow, Uttar Pradesh, India.. His area of interest is Neural Networks, Inverter and Electric Drives.

Neeraj Kumar has done post graduation in electrical engineering from Madan Mohan Malaviya Engineering College Gorakhpur in 2009 under Uttar Pradesh Technical University, Lucknow, Uttar Pradesh, India. He is working as Associate professor in DIET, Rishikesh, under Uttarakhand Technical University, Dehradun, Uttarakhand, India. His field of research includes electric drives and control.

Gunjan Gupta is working as assistant professor in Invertis University, Bareilly, India. He has done post graduation from NITTTR, Chandigarh, India. His area of research includes Fuzzy Logics, Instrumentation and Power Electronics.

References

[1] N. Mohan, T. M. Undeland, and W. P. Robbins, Power Electronics - Converters, Applications, and Design, 2nd ed: John Wiley & Sons, Inc., 1995.

[2] "Matlab 6, Release 12.1." Natick, Massachusetts: The MathWorks (www.mathworks.com), 2006. "PSIM Version 7.0." Woburn, MA: Powersim Inc. (www.powersimtech.com), 2006.

[3] M. Trigg, "Digital Sinusoidal PWM Generation using a Low-cost Micro-controller Based Single-Phase Inverter," presented at ETFA 2005, Catania, Italy, 2005.

[4] M. C. Trigg, H. Dehbonei, and C. V. Nayar, "Digital Sinusoidal PWMs for a Microcontroller based Single-Phase Inverter. Part 1: Principles of digital sinusoidal PWM generation," IJE Power electronics and instrumentation hardware, 2005.

[5] H. Dehbonci, M. C. Trigg, and C. Nayar, "A Novel Sinewave Inverter for Harsh Environment," presented at AUPEC 2005, Hobart, Tasmania, Australia, 2005.

[6] M. H. Rashid," Power Electronics circuits, devices, and applications ", Pearson Education India, 3rd Edition 2004.

27

Performance comparison for face recognition using PCA and DCT

Mozhde Elahi[1, *], Mahsa Gharaee[2]

[1]Department of Control Engineering, Gonabad University, Gonabad, Iran
[2]Department of Electronic Engineering, Gonabad University, Gonabad, Iran

Email address:
Mozhde.elahi@gmail.com (M. Elahi), mahsagharaee@gmail.com (M. Gharaee)

Abstract: In this paper Performance of Principle Component Analysis and Discrete Cosine Transform methods for feature reduction in face recognition system is compared. In face recognition system, feature extraction is based on wavelet transform and Support Vector Machine classifier for training and recognition is employed. According to experimental results on ORL face dataset the PCA method gives better performance compared to using DCT method.

Keywords: Face Recognition, Wavelet Transform, PCA, DCT

1. Introduction

Over the past few years, the user authentication is important because the security control is required everywhere. ID cards and passwords are traditional way for authentication although the security is not so reliable and convenient. Recently, biological authentication technologies through fingerprints, iris print, retina, palm print, face, etc is playing an important role in modern personal identification systems [1].The face is chosen for the suggested system.

The key to face recognition is discriminant feature extraction and classification designing [2]. The extraction of image features is one of the fundamental tasks in image recognition [3].

Wavelet Transform is a popular tool in image processing and computer vision. Many applications, such as compression, detection, recognition, image retrieval et al have been investigated [4].

A problem in face recognition system is the curse of dimensionality in the pattern recognition literature. A common way of dealing with it is to employ a dimensionality reduction technique such as Principal Component Analysis 'PCA' to pose the problem into a low-dimensional feature space such that the major modes of variation of the data are still preserved [5].

Another way to reduce dimensionality of feature vector is Discrete Cosine Transform 'DCT'. The DCT converts high-dimensional face images into low-dimensional spaces in which more significant features are maintained [3].

Support vector machines (SVMs) have been recently proposed as new kinds of feedforward networks for pattern recognition [6]. It belongs to a family of generalized linear classifiers. The main idea of a SVM is to construct a hyperplane as the decision surface in such a way that the margin of separation between positive and negative examples is maximized.The separating hyperplane is defined as a linear function drawn in the feature space [7].

This paper is organized as follows. Section 2 reviews the background of Wavelet decomposition of an image. In section 3, two ways to reduce dimensionality of feature vector (PAC-DCT) are defined. The multi-class SVMs are presented in section 4. The proposed method and experimental results are presented in section 5 and finally, section 6 gives the conclusions.

2. Feature Extraction

2.1. Discrete Wavelet Transform

Wavelet Transform (WT) has been a very popular tool for image analysis in the past ten years [6].The wavelet decomposition of an image can then be interpreted as a set of independent, spatially oriented frequency channels [8].

The DWT is closely related to multi-resolution analysis and sub-band decomposition The 1D-DWT recursively decomposes the input signal,So(n) , in approximation and detail at the next lowest resolution stages. Let $S_i(n)$ and $W_i(n)$ be the approximation and detail, respectively, of the signal at level i. The approximation of the signal at level i+ 1 is computed using[9]:

$$S_{i+1}(n) = \sum_{k=0}^{L-1} g(k)S_i(2n-k) \quad (1)$$

and the detail of the signal at level i+ 1 is computed using:

$$W_{i+1}(n) = \sum_{k=0}^{L-1} h(k)S_i(2n-k) \quad (2)$$

Where g(k) and h(k) are, respectively,the low-pass and high-pass filter coefficients, and L is the size of filters. This technique for computing the DWT is often referred to as the pyramid algorithm or Mallat's algorithm[10]. Output of the low-pass filter is approximation coefficients and high-pass filter is detail coefficients.

The two-dimensional wavelet transform is got by applying one-dimensional wavelet transform to the rows and columns of two-dimensional data. An approximation image is derived from 1-level wavelet decomposition of an image and three detail images in horizontal, vertical and diagonal directions respectively. The approximation image is used for the next level of decomposition[11].Fig.1show the process of decomposing an image.

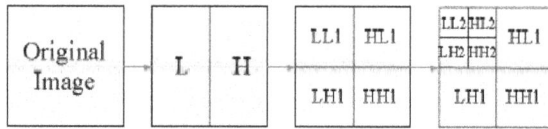

Figure 1. The process of decomposing an image

The LL sub band is the low frequency sub band of the original image. This sub band contributes to the global description of a face. LL sub band will be most stable sub band, so we use it as the feature representation of a face, in this paper.

3. Feature Reduction

3.1. Principal Component Analysis

The Principle Component Analysis method, also called Eigenface method, is the most classical,and commonly used algorithm in modern face recognition field [12]. PCA is used to find a low dimensional representation of data [13]. Let us have a brief view of the principle of PCA [11]:

Step1: A set of M images with size N*N can be represented by vectors of size N^2

$$\Gamma_1, \Gamma_2, \Gamma_3, \dots, \Gamma_M$$

Step2: The average training set is defined by

$$\varphi = \frac{1}{m}\sum_{i=1}^{M} \Gamma_i \quad (3)$$

Step3: Each face differs from the average by vector

$$\varphi_i = \Gamma_i - \varphi \quad (4)$$

Step4:A covariance matrix is constructed as follows:

$$c = A^T A \quad (5)$$

Step5: Finding eigenvectors of N^2 x N^2 matrix is very difficult. Therefore, we use the matrix $A^T A$ of size M x M and find eigenvectors of this small matrix.

Step6: If v is a nonzero vector and λ is a number such as Av=λv, then v is an eigenvector of A with eigenvalue λ.

Step7: Consider the eigenvectors v_i of $A^T A$

$$A^T A v_i = u_i v_i \quad (6)$$

Step8: Multiply both sides by A, we can obtain the result:

$$AA^T(Av_i) = u_i(Av_i) \quad (7)$$

Step9: A face image can be projected into this face space by

$$\Omega_k = U^T(\Gamma^k - \varphi)k = 1, \dots, M \quad (8)$$

3.2. Discrete Cosine Transform

The DCT is a popular technique in imaging and video compression, which transforms signals in the spatial representation into a frequency representation [14]. The DCT of an image basically consists of three frequency components namely low, middle, high each containing some detail and information in an image [15]. DCT is conceptually similar to Discrete Fourier Transform (DFT)[16].

The forward 2D-DCT [20] of a M x N block image is defined as

$$C(u,v) = \alpha(u)\alpha(v)\sum_{m=0}^{M-1}\sum_{n=0}^{N-1} f(x,y) * \cos\frac{\pi(2x+1)u}{2M}\cos\frac{\pi(2y+1)v}{2N} \quad (9)$$

The inverse transform is defined as

$$f(x,y) = \sum_{u=0}^{M-1}\sum_{v=0}^{N-1} \alpha(u)\alpha(v)C(u,v)\cos\frac{\pi(2x+1)u}{2M}\cos\frac{\pi(2y+1)v}{2N} \quad (10)$$

where

$$a_p = \begin{cases} \frac{1}{\sqrt{M}} u = 0 \\ \sqrt{\frac{2}{M}}u = 1,2,\dots,M-1 \end{cases}$$

$$a_q = \begin{cases} \frac{1}{\sqrt{N}} v = 0 \\ \sqrt{\frac{2}{N}}v = 1,2,\dots N-1 \end{cases}$$

and, x and y are spatial coordinates in the image block, and u and v are coordinates in the DCT coefficients block.

4. Support Vector Machine

Support Vector Machine(SVM) is a known supervised learning methods used for classification and regression. The

SVM is widely used in face detection and recognition. The support vector machine views the training data as two sets of vectors in an n-dimensional space, and then it will find a hyperplane which maximizes the margin between the two closest points in the training set which are termed as support vectors, and the calculation of SVM is only related with the support vectors. The optimal hyperplane can be computed as a decision of the form [2].

$$f(x) = sgn\left(\sum_{i=1}^{l} y_i \alpha_i K(x_i, x) + b\right) \qquad (11)$$

Where x_i is the training vectors, x is the testing vector and K(.,.) is the kernel function that must satisfy Mercer's condition. The coefficient α_i and b can be determined by solving the following quadratic programming (QP) problem:

$$\min_\alpha \frac{1}{2}\sum_{i,j} y_i y_j \alpha_i \alpha_j K(x_i, x_j) - \sum_j \alpha_j \qquad (12)$$

$$\text{s.t} \qquad \sum_i y_i \alpha_i = 0$$

$$0 \le \alpha_i \le C, \quad \forall i$$

The unbound C is the penalty parameter that represents the tradeoff between minimizing the training set error and maximizing the margin [2].

The basic SVM is a binary classifier, so there are tree developed methods for c-class recognition (C >2). The first method is one against all approach, which needs to construct c SVM classifiers and each one separates one signal class from all other classes. The second method is one against one approach and the third method is based on tree algorithm. In this paper, we use first method.

5. Experimental Resultsand Analysis

To evaluate the effectiveness of the proposed method, we used Cambridge ORL [1] face database.This face database contains 40 individuals, and each individual has 10 images with variations in pose, illumination, facial expression and accessories [11].

The procedure of the algorithm design in this paper is as follows:

Step1: Perform wavelet transform on ORL database,

Step2: LL sub-image of wavelet decompositionlevels are selected,

Step3: For each person three images are selected as training images, and seven images as testing images,

Step4: Apply reduce methods on extracted features:
- Apply PCA on features vector,
- Apply DCT on features matrix and the next convert it to vector,

Step5: Train SVM with sample feature,

Step6: Exert test vectors on SVM to calculateerror.

In this paper, application of different mother wavelet such as, Haar, Db1 and Sym1were tested. Also we assay

differentlevel of wavelet transform to find best result. Result shows the third level of Haar wavelet is the best.

In order to find efficient feature by DCT, this algorithm applied on different level. The resolution of images is changed in from of 48x48 to 6x6 using the third level of wavelet decomposition so, the size of DCT coefficients matrix is 6*6. Fig 2 and Table 1 show result of choice different level of DCT.

Table 1. Error from different level of DCT

Size of DCT block	Error percent
1*1	93/3333
2*2	73/3333
3*3	25
4*4	14/1667
5*5	5/8333
6*6	5/7776

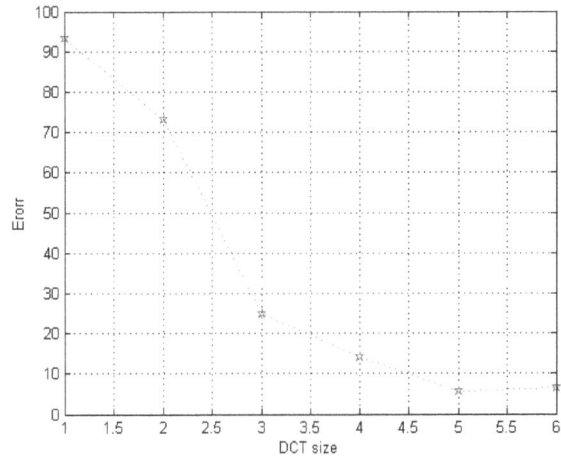

Figure 2. Changes of error by increasing the size of the DCT block

In order to find efficient feature by PCA, we employ PCA algorithm for distinct number of feature. Fig 3 shows changes of error by differencenumber of feature that selected by PCA. Reduce the feature in from of 36 to 22 bringing best result that is 95% recognition rate.

To show effect of proposed combination algorithm, each method tested alone. DWT, PCA and DCT algorithms are tested. The result is shown in table 2.The exported resultis the best effect after do algorithms with different number of feature.

Table 2. Recognition rate

Algorithm	recognition rate
Original image	%83/33
DWT	%87/5
PCA	%90
DCT	%87/5
DWT/PCA	%95
DWT/DCT	%94/17

Table 2 shows that our method performs better than other algorithms.

[1]Oliverli Research Laboratory

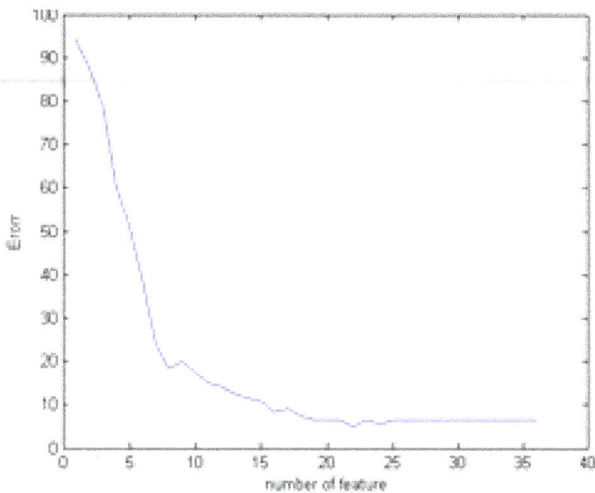

Figure 3. Changes of error by increasing the number of feature that reduction by PCA

6. Discussion and Conclusion

Our experimental results can be summarized as follows: The use of DWT as a feature extractor and DCT-PCA as featurereduction methodsto train and test support vector machine. The combination of DWT and DCT-PCA methods shows a very satisfactory result.

In the future, we plan to continue our research by investigating other possible methods (both feature extractors and classifiers) to achieve a better system performance.

Reference

[1]   M.Sharkas, 'Application of DCT Blocks with Principal Component Analysis for Face Recognition' Proceedings of the 5th WSEAS Int. Conf. on SIGNAL, SPEECH and IMAGE PROCESSING, Corfu, Greece, August 17-19, 2005 (pp107-111)

[2]   X.M. Wand, CH. Huang, G.Y. Ni, J.G. Liu, 'Face Recognition Based on Face Gabor Image and SVM' 978-1-4244-4131-0/09/$25.00 ©2009 IEEE

[3]   M. Wang, H. Jiang, and Y. Li, ' Face Recognition Based on DWT/DCT and SVM' 2010 International Conference on Computer Application and System Modeling (ICCASM 20ID)

[4]   B. Luo, Y. Zhang, and Y.H. Pan, ' Face Recognition Based on Wavelet Transform and SVM' Proceedings of the 2005 IEEE, International Conference on Information Acquisition ,June 27 - July 3, 2005, Hong Kong and Macau, China

[5]   M. S. Sarfraz, O. Hellwich and Z. Riaz, ' feature Extraction and Representation for Face Recognition' ISBN

978-953-307-060-5, Published: April 1, 2010 under CC BY-NC-SA 3.0 license

[6]   M. Mazloom, S. Kasaei, and H. Alemi, 'Construction and Application of SVM Model and Wavelet-PCA for Face recognition' 2009 Second International Conference on Computer and Electrical Engineering

[7]   I.K. Timotius, I. Setyawan, and A.A. Febrianto, 'Face Recognition between Two Person using Kernel Principal Component Analysis and Support Vector Machines'International Journal on Electrical Engineering and Informatics - Volume 2, Number 1, 2010

[8]   G. Yu, S.V. Kamarthi, 'A cluster-based wavelet feature extraction method and its application' , Engineering Applications of Artificial Intelligence 23 (2010) 196–202, Accepted 27 November 2009

[9]   R.C. Palero, R.G. Girones and F. Ballester-Merelo, 'Flexible architecture for the implementation of the two-dimensional discrete wavelet transform (20-0WT) oriented to FPGA devices', Microprocessors and Microsystems, vol. 28, pp. 509-518,2004

[10] S. G. Mallalt, 'Multifrequencychannal decompositions of images and wavelet models', IEEE Transactions on Acoustics, Speech, and Signal Processing, vol. 37, pp. 209 1-2 1 10, 1989.

[11] L. Xian, Y. Sheng, W. Qi, and L. Ming, 'Face Recognition Based on Wavelet Transform and PCA' 2009 Pacific-Asia Conference on Knowledge Engineering and Software Engineering

[12] H. Wang, S. Yang, and W. Liao, 'An Improved PCA Face Recognition Algorithm Based on the Discrete Wavelet Transform and the Support Vector Machines' 2007 International Conference on Computational Intelligence and Security Workshops

[13] M. Mazloom, S. Kasaei, 'Combination of Wavelet and PCA for Face Recognition' GCC Conference (GCC), 2006 IEEE , E-ISBN 978-0-7803-9591-6, E-ISBN 978-0-7803-9591-6, INSPEC Accession Number 11747056

[14] V.p.Vishwakarma, S. Pandey, and M.N. Gupta, 'A Novel Approach for Face Recognition Using DCT Coefficients Re-scaling for Illumination Normalization' 15th International Conference on Advanced Computing and Communications, 0-7695-3059-1/07 $25.00 © 2007 IEEE, DOI 10.1109/ADCOM.2007.12

[15] K. Manikantan, V. Govindarajan, 'Face Recognition using Block Based DCT Feature Extraction', Journal of Advanced Computer Science and Technology, 1 (4) (2012) 266-283

[16] A.R. Chadha, P.P. Vaidya, and M.M. Roja, 'Face Recognition Using Discrete Cosine Transform for Global and Local Features', Proceedings of the 2011 International Conference on Recent Advancements in Electrical, Electronics and Control Engineering (IConRAEeCE) IEEE Xplore: CFP1153R-ART; ISBN: 978-1-4577-2149-6

28

Design and realization of a miniaturized low loss iris bandpass filter on substrate integrated waveguide configuration in 2.4GHZ band

Mona Sameri[1, *], Farokh Hojat Kashani[2]

[1]Electrical and Communication Engineering department, Islamic Azad University - South Tehran Branch, Tehran, Iran
[2]Department of Electrical Engineering, Iran University of Science & Technology, Tehran, Iran

Email address:
Sameri.mona@yahoo.com (M. Sameri), kashani@iust.ac.ir (F. H. Kashani)

Abstract: In this paper, a fourth order bandpass Chebyshev filter based on Iris discontinuities and using SIW technology is designed. The structure of the filter consists of SIW cavity resonators that are connected to each other with Iris discontinuities, which operate as impedance inverters. This filter is implemented on a substrate of RO4003C with a central frequency of 2.45GHz and a fractional bandwidth of %4.08 and is designed to be used in Wireless Local Area Networks (WLAN) or Bluetooth. The main idea is to reduce the size of the filter with a factor of two without any distortion in the frequency response. Simulations are done using full-wave HFSS simulator.

Keywords: Substrate Integrated Waveguide, Wireless Local Area Networks (WLAN), Iris Discontinuity, Bandpass Filter, Impedance Invertor

1. Introduction

Designing filters in millimeter-wave and microwave systems with high quality factor and appropriate performance has always been a challenging subject. WLAN compatible with the IEEE 802.11 b/g standard work on a carrier frequency of 2.4 GHz. These frequency bands cover radio frequency heating processes, Bluetooth, wireless digital phones, and medical diathermic machines [1]. Many bandpass filter structures in the unlicensed band of 2.4-2.5 GHz have been suggested so far. In [2], an appropriate method is proposed to minimize the bandpass filter dimensions based on the artificial quasi-TEM transmission lines using multiple platform technology. These transmission lines are separated by meshed planes so as to reduce the coupling effect of the layers, and thus, to improve the performance of the system. A microstrip bandpass filter with parallel coupling has been discussed in [3] in the same frequency band. A bandpass filter with serial structure using.

multilayer technology "Low Temperature Co-fired Ceramic" has been suggested in [4]. This filter has a grounded capacitor connected to two LC resonators, which are parallel to the filter, and results in two Transmission zeroes in the

frequency response. Bandpass filters based on acoustic wave surface technology with high selectivity and low loss are used widely in communication systems. Bandwidth constraint and the necessity of using redundant matching elements are the main disadvantages of acoustic wave filters[5].On the other hand, LTCC technology can be exploited to integrate some more bandpass filters in a substrate [2,4,6],but due to the low quality factor in LTCC technology, these filters are not suitable from sensitivity and selectivity viewpoints. Combining these two technologies, one can gain advantage from both of them in the sense of selectivity, amount of loss, and filter's dimension. The design procedure of an acoustic wave surface filter based on LTCC technology which is to be used in WLAN and Bluetooth is proposed in [7]. Microwave filters using waveguide technology are used in aerospace systems mostly due to low loss and high power transmission capability [8, 9]. But since these structures are too bulky, they cannot be used for mass production. Moreover, assembly is a way time-consuming and costly process which sets barriers for using such structures [10, 11]. With the appearance of integrated waveguide substrate, many circuits have been designed and implemented based on this structure [11, 12]. Most components of the flat waveguide have been realized using SIW technology, since the integrated waveguide substrate structure have the same propagation characteristics as the rectangular waveguide. This solution can significantly decrease the size and weight of the components compared to the rectangular waveguide. Furthermore, the loss of SIW components is less than those of microstrip components and the radiation and packaging problems are also resolved. The SIW components make a good compromise with rectangular waveguide and microstrip line structure [11]. In this paper, a fourth-order bandpass filter with Chebyshev frequency response and symmetric metallic walls, known as Iris filter is designed and simulated. This filter is implemented on the integrated waveguide substrate structure with unlicensed 2.4-2.5 GHz frequency band for WLAN, which is also compatible with IEEE 802.11 b/g standard. It contains TE_{101} half-wave waveguide resonators that are separated from each other by the Iris discontinuity. Using impedance invertors (k) and a circuit structure for the distance between the wall, the physical dimensions of the structure are extracted [13]. In the second section, an appropriate approach for designing an Iris bandpass filter along with its simulation results are discussed. The third section is dealt with reducing the filter's dimensions by a factor of two without any distortion in the frequency response. The paper is concluded in section four.

2. Filter Design Process

The SIW structure is realized by two lines of metallic vias that are periodically embedded inside the insulator substrate [11]. Because of these metallic layers that surround SIW, only TE_{n0} modes are propagated in this structure [14]. The most important parameters are the distance between adjacent vias and their diameter. For specific values of $^{p}/_{\lambda c_{c}}$ and $^{d}/_{\lambda c_{c}}$, the integrated waveguide substrate structure has the same behavior as the rectangular waveguide and its radiation loss can be approximately ignored[15].Applying usual discontinuities in the waveguide structure, one can realize the waveguide filters. But due to the constraints in the fabrication process, the only ways to make a reactive element in the SIW are to create totally empty and metal lined cavities or to pattern its upper and lower metallic planes. Hence, it is not possible to create capacitor discontinuities in the one-layer SIW structure [16].

Figure (1-a) shows one of such realizable discontinuities in the SIW structure, which the filter is known as an Iris filter. Iris discontinuities create a symmetric induction window with width $W_i (i = 1,2,...,n)$ and the cavity resonators that have a length of $L_i (i = 1,2,...,n+1)$ are embedded between these discontinuities. The coupling between cavity resonators are controlled by the width distance of the Iris discontinuities [17].

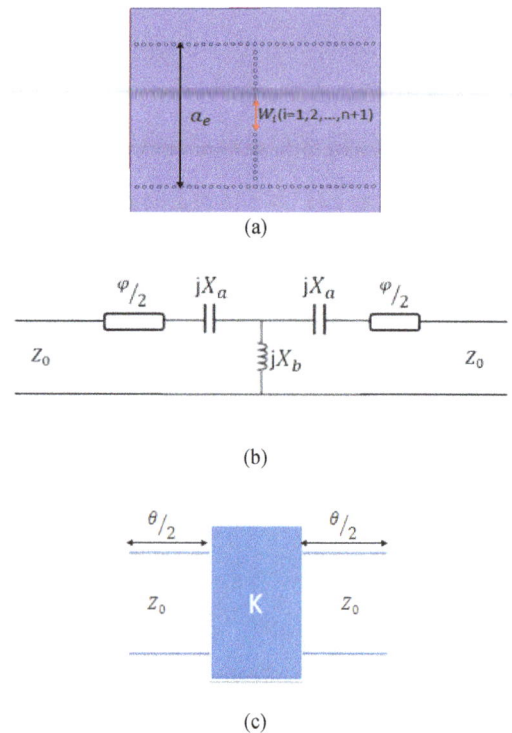

(a)

(b)

(c)

Figure (1). *(a) Iris discontinuity in SIW structure, (b) T network equivalent circuit of Iris discontinuity in the SIW structure, (c) Iris discontinuity equivalent circuit using impedance inverter in the SIW structure*

According to figure (1-c), these induction posts that connect SIW transmission lines to each other, act as impedance inverters [9]. Hence, the design process of SIW filter with Iris discontinuity reduces to finding the inverter characteristic impedance and the width distance between the symmetric metallic walls[13].The impedance inverter normalized characteristic impedance for a bandpass filter with Chebyshev frequency response can be calculated as the following [17].

$$\Delta = \frac{\lambda_{g1} - \lambda_{g2}}{\lambda_{g0}}$$

$$\frac{K_{0,1}}{Z0} = \sqrt{\frac{\pi}{2} \frac{\Delta}{g_0 g_1 \Omega}}$$

$$\frac{K_{j,j+1}}{Z0} \bigg|_{j=1 \; to \; n-1} = \frac{\pi\Delta}{2\Omega} \frac{1}{\sqrt{g_i g_{i+1}}} \qquad (1)$$

$$\frac{K_{n,n+1}}{Z0} = \sqrt{\frac{\pi}{2} \frac{\Delta}{g_n g_{n+1} \Omega}}$$

$$\lambda_{g0} = \frac{\lambda_{g1} + \lambda_{g2}}{2}$$

In (1), g_i's are the values for the lowpass filter components. with Chebyshev response , Δ is the relative bandwidth of the guided wavelength , Ω is the normalized lowpass frequency, Z_0 is the characteristic impedance of the transmission line, λ_{g1} and λ_{g2} are the guided wavelengths in the upper and lower edges of the bandpass filter, respectively [17]. Determining the characteristic impedance values of the impedance inverters, one can calculate the physical length of the cavity resonators as below [17]:

$$\frac{X_{j,j+1}}{Z_0} = \frac{\frac{K_{j,j+1}}{Z_0}}{1 - (\frac{K_{j,j+1}}{Z_0})^2}$$

$$\theta_i = \pi - \frac{1}{2}\left[tan^{-1}\left(\frac{2X_{j-1,j}}{Z_0}\right) + tan^{-1}\left(\frac{2X_{j,j+1}}{Z_0}\right)\right] \quad Li = \frac{\theta_i \lambda_{g0}}{2\pi} \quad (2)$$

Figure (2-b) shows the T equivalent circuit of this discontinuity. The values of its elements depend on frequency, dimensions, and Iris discontinuity position. The T equivalent circuit parameters are derived from the following equations [18]:

$$\varphi = -tan^{-1}(2X_p + X_s) - tan^{-1}(X_s)$$

$$jX_s = \frac{1 - S_{12} - S_{11}}{1 - S_{11} + S_{12}} \quad (3)$$

$$jX_p = \frac{2S_{12}}{(1 - S_{11})^2 - S_{12}S_{12}}$$

In equation (3), $X_s = \frac{X_b}{Z_0}$, $X_s = \frac{X_b}{Z_0}$, and $S_{ij}\big|_{i,j=1,2}$ are the corresponding scattering parameters in each Iris discontinuity.

After determining the physical parameters of the Iris waveguide filter by equations (1) to (3), the equivalent SIW parameters are calculated as [19]:

$$L_{SIW} = L + \frac{d^2}{0.95p}$$

$$a_{SIW} = a_e + \frac{d^2}{0.95p} \quad (4)$$

Note that to obtain a filter with suitable frequency response, optimum design of the transmission converter from microstrip to SIW becomes particularly important. One of the most common converters used between SIW structure and microstrip is tapered converter which is shown in figure (2). This converter converts the dominant propagation mode of microstrip line to dominant propagation mode of SIW [20].

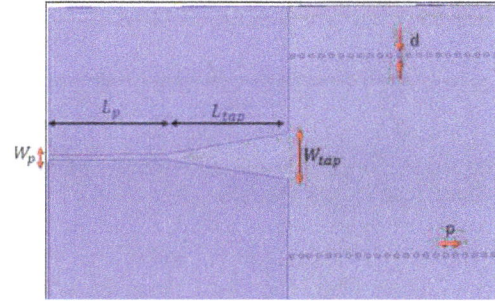

Figure (2). microstrip to SIW converter

We design a fourth-order SIW bandpass filter with Chebyshev frequency response and a ripple of 0.03 dB. The substrate is RO4003C with 0.508 mm thickness, the loss tangent is 0.0027, and the dielectric constant is chosen to be 3.38. The cutoff frequency of the dominant mode is 1.8 GHz in this design. Moreover, in order to minimize the radiation loss of the SIW structure, we set d=0.5mm and p =1mm. Figure (3) shows the structure of a fourth order SIW Iris bandpass filter. The optimum parameters for this filter calculated in 2.4 GHz frequency and %4.08 fractional bandwidth are given in Table (1).

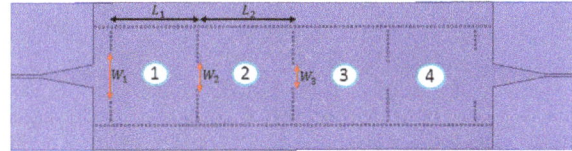

Figure (3). The structure of an SIW fourth-order Iris bandpass filter realized by direct coupling

Table (1). physical dimensions of the designed SIW bandpass filter (unit: mm)

p	1	L_{tap}	25.45
d	0.5	L_1	40.8009
a_{SIW}	44.4919	L_2	45.1924
W_p	.2531	W_1	24.1518
L_p	13.3476	W_2	13.1308
W_{tap}	10.64	W_3	11.6328

The frequency response of the designed filter is obtained by HFSS as illustrated in Figure (4).

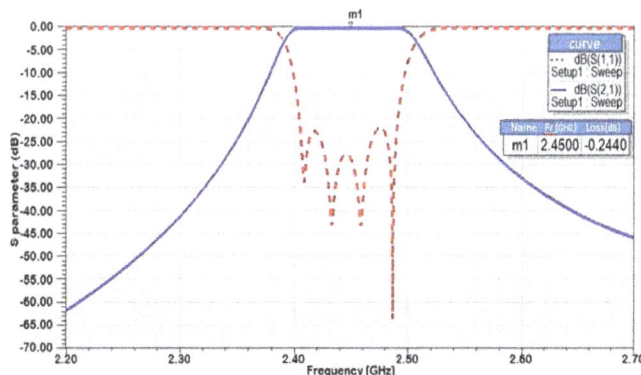

Figure (4). The frequency response of an SIW fourth-order Iris bandpass filter realized by direct coupling

3. Filter Dimension Reduction

Since the designed filter in Figure (3) has a symmetric structure, using a simple procedure at the discontinuity point W_3 that is placed in the center of the structure, it is possible to fold the filter. But in order to prevent the degradation in the frequency response of the filter, it is required to regulate the discontinuity point W_3 such that the guided wave between the two discontinuities W_2 and W_3 travel the same length L_2 . Figures (5-a) and (5-b) show the procedure of designing the folded filter. Finally, the frequency response of the folded SIW filter is shown in Figure (6).

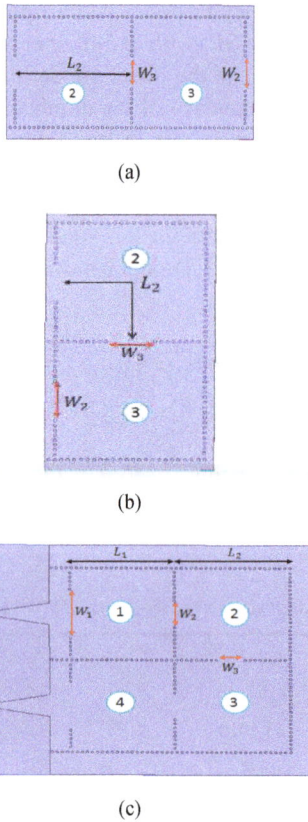

(a)

(b)

(c)

Figure (5). (a) Direct coupling between cavity resonators 2, 3, (b) Cross coupling between cavity resonator 2, 3, (c) Folded fourth-order SIW Iris bandpass filter

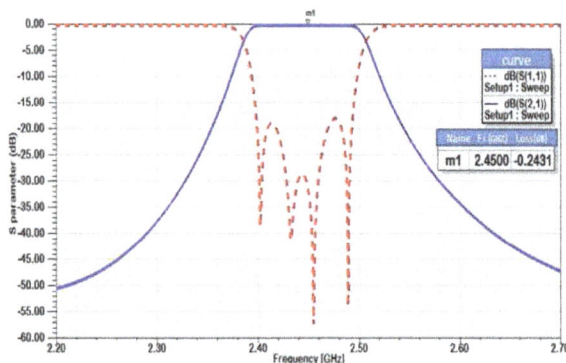

Figure (6). The frequency response of the folded fourth-order SIW Iris bandpass filter

Similarly, the cavity resonator 1, 4 can be folded to form a T structure, which is illustrated in Figure (7). The frequency response of the T structure filter is shown in Figure (8).

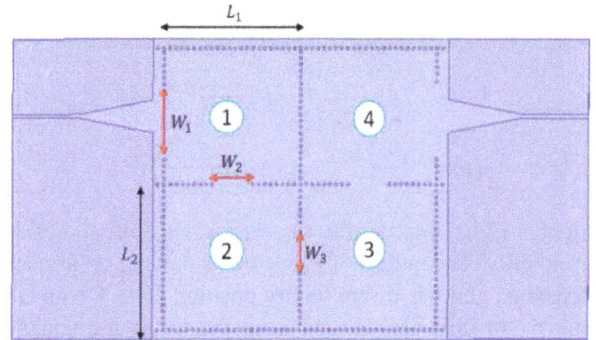

Figure (7). Fourth-order SIW Iris bandpass T-structured filter

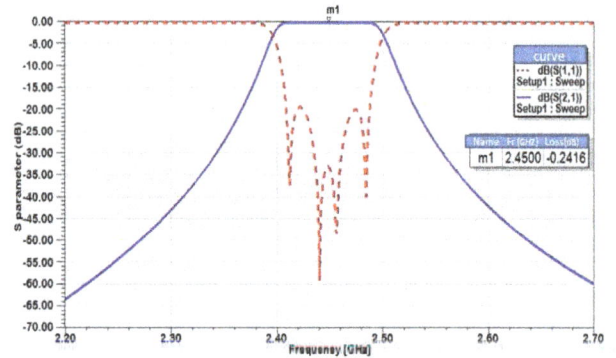

Figure (8). The frequency response of the fourth order SIW bandpass T-structured filter

4. Discussion and Conclusion

In this paper, an appropriate approach for designing an Iris bandpass filter using the concepts of impedance inverters and TE_{101} half-wave resonators has been proposed. This filter is implemented on the integrated waveguide substrate structure with unlicensed 2.4-2.5 GHz frequency band for WLAN, which is also compatible with IEEE 802.11 b/g standard.

The dimensions of the filter are reduced in our design. High selectivity, low loss, proper performance, lumped dimensions, low cost fabrication, and the ability to be integrated with other planar circuits are the main advantages of the proposed filter.

References

[1]   J.Horne and S.Vasudevan. "Modeling and Mitigation of Interference in the 2.4 GHz ISM Band", Applied Microwave and Wireless, vol. 9pp. 59-71, April 1997.

[2]   H.S. Wu, H.J. Yang, C.J. Pang and C-K.C. Tzuang, "Miniaturized Microwave Passive Filter Incorporating Multilayer Synthetic Quasi-TEM Transmission Line, " IEEE Transactions on Microwave Theory and Techniques, vol. 53, no. 9, pp. 2713-2720, September 2005.

[3]	Wang Wensong, Xiao Hang, Yang Shuhui and Chen Yinchao, "Experimental Study of 2.4 GHz Bandpass Filter for WLAN, "2011 Fourth International Conference on Intelligent Computation Technology and Automation , vol.2, pp.544-547, March 2011.

[4]	C.F. Chang and S.J. Chung, "Bandpass Filter of Serial Configuration With Two Finite Transmission Zeros Using LTCC Technology," IEEE Transactions on Microwave Theory and Techniques, vol. 53, no.7, pp. 2383-2388, July2005

[5]	G.L.Matthaei, "Acoustic surface-wave transversal filters, " IEEE Transactions on Circuit Theory ,vol. 20, no.5 , pp. 459-470,1973.

[6]	Y. H. Jeng , S-F. R. Chang, and H. K. Lin, "A high Stopband-Rejection LTCC Filter With Multiple Transmission Zeros, " IEEE Transactions on Microwave Theory and Techniques, vol. 54, no. 2, pp.633-638 , February 2006.

[7]	Novgorodov, V., B. Freisleben, J. Hornsteiner, M.Schmachtl, B.Vorotnikov, P.Heide,andM.Vossiek, "Compact low-loss 2.4 GHz ISM-band SAW bandpass filter on the LTCC substrate, " In IEEE Microwave Conference, APMC 2009. Asia Pacific, pp. 2072-2075, 2009.

[8]	Cameron, Richard J., Chandra M. Kudsia, and Raafat R. Mansour. "Microwave filters for communication systems," fundamentals, design and applications. Vol. 1. Wiley-Interscience, 2007.

[9]	Ian Hunter "Theory and Design of Microwave Filters," The Institution of Electrical Engineers, Vol.48.Iet, 2001.

[10]	Deslandes, Dominic, and Ke Wu. "Integrated microstrip and rectangular waveguide in planar form." Microwave and Wireless Components Letters,vol.11, no.2 , pp. 68-70, February 2001.

[11]	K.Wu,D.Deslandes,and Y.Cassivi, "The Substrate Integrated Circuits-A New Concept for High Frequency Electronics and Optoelectronics, "Telsiks 2003,Serbia and Montenegro,Nis, vol.1,no.4, pp. P-III-P-X , October 2003.

[12]	M.Bozzi,A.Georgiadis, K. Wu, "Review of substrate-integrated waveguide circuits and antennas," IET Microw. Antennas Propag.,Vol. 5, no. 8, pp 909–920, December 2011.

[13]	Hao, ZhangCheng, Wei Hong, Hao Li, Hua Zhang, and Ke Wu. "A broadband substrate integrated waveguide (SIW) filter, " In Antennas and Propagation Society International Symposium, 2005 IEEE, vol. 1, pp. 598-601., 2005.

[14]	F. Xu, and K. Wu,"Guided-wave and leakage characteristics of substrate integrated waveguide", IEEE Transactions on Microwave Theory and Techniques, vol .53, no. 1, pp 66-73, January 2005.

[15]	D. Deslandes, and K. Wu, "Accurate modeling, wave machanisms, and design considerations of a substrate integrated waveguide," IEEE Transactions Microwave Theory and Techniques, vol. 54, no. 6, pp. 2516–2526, January 2006.

[16]	N. Marcuvitz, "Waveguide handbook", Piscataway, NY, IEEE electromagnetic waves series, 1986.

[17]	G. L. Matthaei, L. Young, E. M. T. Jones, "Microwave filter, impedance matching network, and coupling structures, " Artech house, Norwood,MA, 1980.

[18]	Shi Yin, Protap Pramanick, "Use of 3D Field Simulators in the Synthesis of Waveguide Capacitive Iris Coupled Lowpass Filters, " International Journal of RF and Microwave Computer-Aided Engineering vol. 10, no. 3, Apr. 2000, pp. 190 -198.

[19]	Y. Cassivi, L. Perregrini, P. A. M. Bressan, K. Wu, and G. Conciauro,"Dispersion Characteristics of Substrate Integrated Rectangular Waveguides," IEEE Microwave Wireless Componets Letters, vol. 12, no.9,pp. 333-335, September 2002.

[20]	D. Deslandes, and K. Wu, "Integrated microstrip and rectangular waveguide in planar form," IEEE Microwave Wireless Componets Letters., vol. 11, no. 2, pp. 68-70, February 2001.

29

VoIP codec selection for digital PLC systems

Maryam Shabro, Behnam Gholamrezazadeh Family, Vahid Hammiaty Vaghef

Communication Department, Niroo Research Institute, Tehran, Iran

Email address:

mshabro@nri.ac.ir (M. Shabro), behnam_gh_f@yahoo.com (B. G. Family), vvaghef@nri.ac.ir (V. H. Vaghef)

Abstract: IP-based networks developments highly integrate the communication systems and significantly reduce the installation cost. Digital PLC systems with the capability of Ethernet and VoIP services are designed because of the demand of new communication services in the electric power industry. Also due to the limited bandwidth in digital PLC systems, there are several techniques for optimal use of transmission capacity. In this paper, the possibility of using VoIP services in digital PLC systems has been investigated. Many conventional standards are studied for several parameters such as total algorithm delay, voice payload size/duration, packets per second, and bandwidth to select the best codec. Finally, a proper codec is recommended to employ in the digital PLC's VoIP service according to the two important factors in quality of voice services, i.e. available bandwidth and latency.

Keywords: Voice codec, Bandwidth, Delay, Digital PLC, PLC Link Transition, VoIP

1. Introduction

Power Line Carrier (PLC) technology has been widely used by electrical power utilities for telecommunication purposes. It utilizes high voltage transmission lines as a communication media. Reputable companies have presented Digital PLC (DPLC) systems by using digital modulation technology, audio compression and Time Division Multiplexing (TDM) in order to increase the transmission capacity of traditional analog PLCs. Therefore, the DPLC's can provide more voice and data services. New DPLC systems have provided data transmission rates between 20-200kbps at 4 kHz to 32 kHz bandwidth by using compression and signal processing technologies. With higher communication rates by using DPLC systems, it is possible to provide Ethernet data transmission between two substations; in this case PLC acts as an Ethernet bridge [1]. Some applications of Ethernet services in the power utilities are:

- Connecting RTU to SCADA centre or TCP-IP-based network.
- LAN interconnection for extracting data from the HV substation.
- Providing VoIP services.
- Providing E-mail and Internet services.

Ethernet bridging over DPLC is a reliable and economical solution for providing a low throughput Ethernet access for specific operational applications in a HV substation without fiber or microwave coverage. As for other low capacity bridging technologies, it is important to optimize the payload and therefore the size of the Ethernet frames when the Ethernet system includes a DPLC bridge.

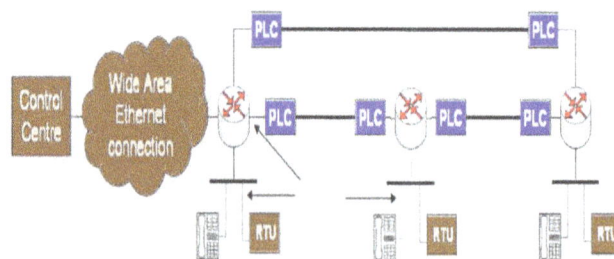

Figure 1. *Typical Ethernet bridging over DPLC.*

In this article, the reliable capacity of DPLCs under real condition, the framing in VoIP services, bandwidth reduction techniques for VoIP services, the possibility of using VoIP services in DPLCs and comparison of various codec regard to their bandwidth requirements are discussed. Also effective factors in quality of VoIP communication in telecommunication networks using DPLCs to choose suitable codec for the VoIP services through DPLC has been examined.

2. Available Bit Rates in Reliable Communication with DPLC System

PLC performance and bandwidth capabilities depend upon the transmission line conditions. The PLC link's maximum transmission rate is determined by the available frequency bandwidth, which depends upon the density of the frequency plan and hence the number of PLC links in the immediate vicinity. The available transmission rate depends also upon the channel's Signal to Noise Ratio (SNR) and consequently upon the line noise and attenuation. The noise across HV PLC links is mainly due to the corona effect (ionization of the air) depending upon atmospheric conditions (air humidity, rain, pollution, and ice over the transmission line). According to Fig. 2, typically, with an SNR of about 35dB, the system can operate at 8bps/Hz (32 kbps in 4 kHz) [2]. In order to assure high availability of bandwidth, the performance of the PLC link must be estimated under adverse atmospheric conditions. So, with an SNR of about 25dB, the system can operate at 6bps/Hz (24 kbps in 4 kHz).

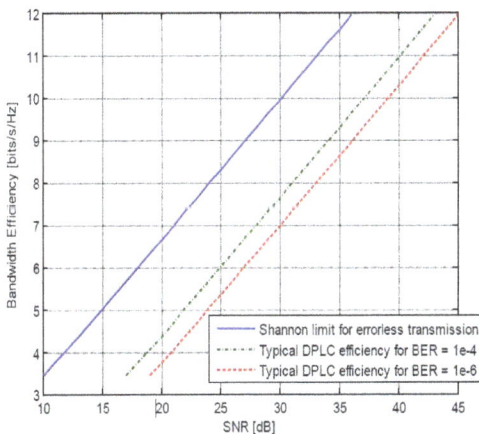

Figure 2. *Typical DPLC bandwidth efficiency.*

Clearly, these are the gross data rate. The specific value of this Gross Bit Rate depends on the implementation. In DPLC this overhead is due to the synchronization and TDM framing. In addition, there are some services which have higher priority to access the channel (such as dispatching voice channels and data from SCADA systems). Therefore, a part of the bit rate is used for these services (typically around 10kbps).

3. VoIP Communication and Framing

In a VoIP service as shown in Fig. 3, the following steps is necessary to generate and use of voice data packets:
- Sampling analog signals and converting them to digital signals by a codec;
- Data compression for optimal use of bandwidth;
- Using a real-time protocol (RTP) to control the sequence of received voice packets;
- A signaling protocol used to invoke user (H.323, SIP);
- Extract audio from received data packet;
- Decompressing received audio information and

converting digital data to analog signal by a soundcard or phone.

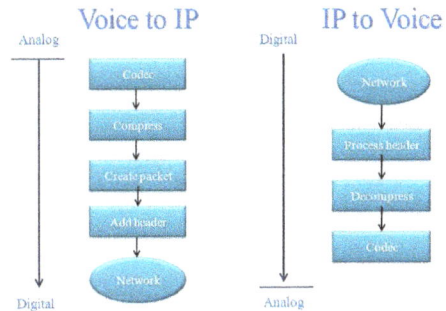

Figure 3. *Blocks in VoIP Services*

Also the RSVP (Resource Reservation Protocol) is used in a VoIP connection. This protocol is responsible for service quality management function. RSVP is a QoS protocol for managing bandwidth allocation for Internet media connections (e.g. a VoIP voice connection). This protocol uses network resources and reserves bandwidth for voice transmission.

In general, VoIP packets consist of two parts, the voice payload and header (Fig. 4). As mentioned before, after converting the analog voice signal to digital data, to optimize the use of available bandwidth, the digital data should be compressed based on a suitable algorithm. The ITU Institute has developed numerous standards in this field. The type of codec determines audio sampling rate and actual required bandwidth. VoIP's overhead includes three protocols IP, UDP and RTP.

The RTP identifies the samples sequence (Fig. 5). The RTP also provides a method for diagnosing delay and Jitter [3]. The UDP specifies packages destination port. This protocol does not have any information about the packet sequence and does not provide any guarantee of delivery. The IP delivers the packets to the host. This protocol like UDP protocol does not guarantee delivering of all packets or delivering them in the same order they were sent.

The IP, UDP, and RTP header add a total of 40 bytes per packet. Besides, the MAC layer has about 42 bytes. So the most amount of communication bandwidth consists of these overheads. Besides the bandwidth required for VoIP depends on the type of network and use of Silence Suppression or Voice Activity Detection (VAD) techniques.

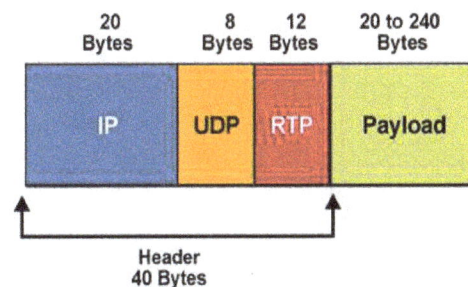

Figure 4. *Structure of VoIP packets.*

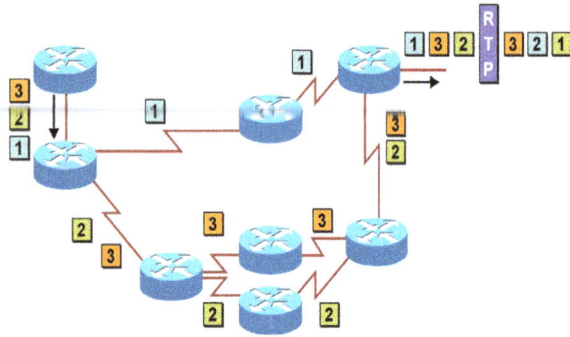

Figure 5. RTP`s role in sorting packets.

4. Reducing Required Bandwidth for a VoIP Communication over PLC Networks

Due to the limited capacity of DPLC systems to provide IP based services, considerations for improving bandwidth efficiency will be essential. In this section some methods for reducing VoIP packet's size and the possibility of using in DPLC systems is investigated.

- In a Point-to-point communications it is possible to decrease 40 bytes of overhead to 2 or 4 bytes with a series of compression methods on overheads of IP, UDP, and RTP. DPLC systems use the point-to-point links, thus this compression method can be applied.
- In DPLC systems according to the TDM method for multiplexing different data services, some overhead will

be added to the data frame. Therefore there is no need to send the MAC layer overheads (42 bytes). For example if HDLC framing is used at least 4 bytes as overhead and CRC is added to the contents of each packet which is much less than the MAC layer overheads.

- Few frames of audio data can be sent in one packet with the same overhead. It should be noted that prolonged packet increases transmission delay of sound. Moreover, regard to the behavior of noise on PLC network, a packet loss, will cause to lose all the frames and the quality of sound service will drop significantly.

Selecting the appropriate voice codec is important. If high compression rate codec is used to reduce bandwidth requirements, the complexity of algorithm and communication delay will be increase.

For example, take the codec G.729 (8 kbps), the minimum bandwidth required for VoIP communication without the compression of header is 31.2 kbps and with the compression of header is 11.6kbps and the delay caused by algorithm is 25ms, also for the codec G.723.1 (5.3 kbps), the minimum bandwidth required for VoIP connection without the compression of header is 20.8 kbps and with the compression is 7.7 kbps and the delay caused by algorithm is 37.5 ms. As described in Part 2 of this article, due to the limited capacity of DPLC systems and required bandwidth by codec for VoIP services, only the standard codec G.723.1 (5.3 kbps), G.723.1 (6.3 kbps)[4], and the G.729 (8 kbps)[5] are suitable for VoIP connection using DPLCs. The Comparison of some voice codecs are listed in Table 1.

Table 1. Comparison of some voice codecs.

Codec Information				Bandwidth Calculations				
Codec & Bit Rate (kbps)	Codec Sample Size	Codec Sample Interval	Total algorithm delay	Voice Payload Size	Voice Payload Duration	Packets Per Second (PPS)	Bandwidth w/cRTP	Ethernet Bandwidth
G.711 (64 kbps)	80 Bytes	10 ms	20 ms	160 Bytes	20 ms	50	67.6 kbps	87.2 kbps
G.723.1(6.3 kbps)	24 Bytes	30 ms	37.5 ms	24 Bytes	30 ms	33.3	8.8 kbps	21.9 kbps
G.723.1(5.3 kbps)	20 Bytes	30 ms	37.5 ms	20 Bytes	30 ms	33.3	7.7 kbps	20.8 kbps
G.726 (32 kbps)	20 Bytes	5 ms	20 ms	80 Bytes	20 ms	50	35.6 kbps	55.2 kbps
G.726 (24 kbps)	15 Bytes	5 ms	20 ms	60 Bytes	20 ms	50	27.6 kbps	47.2 kbps
G.728 (16 kbps)	10 Bytes	5 ms	20 ms	60 Bytes	30 ms	33.3	18.4 kbps	31.5 kbps
G.729 (8 kbps)	10 Bytes	10 ms	25 ms	20 Bytes	20 ms	50	11.6 kbps	31.2 kbps

5. Effective Parameters on a VoIP Quality over PLC Networks

The quality of VoIP connection depends on parameters such as delay, jitter and echo. Telephony services due to the Hybrid sections are faced with echo problem. If the communication delay is increased, echo will be felt more and this will have a considerable decline in the quality of service. ITU-T G.131 standard has described the acceptable echo versus delay [6]. The standard methods for echo cancellation (such as ITU-T G.168) will solve the echo problems in a channel with limited delay. The talker overlap problem will encounter with delay more than 250 ms which diminishes the quality of service.

UDP protocol is used in a VoIP connection so it is not possible to ask for a lost data packet. According to ITU-T standard if the network packet loss rate is 10% or more, the network is unreliable. A Service quality control procedure (RSVP) is used to solve this problem for the VoIP services but it is not possible to use RSVP for channel reservation with the DPLC system, because there are other services (RTU's data, DTS phone, etc) that have higher priority than VoIP. The whole end-to-end delay is an important factor that affects the quality of service. According to ITU-T G.114 standard end-to-end delay less than 150 ms is considered as a clear channel. The end-to-end delay over 400 ms designs is not acceptable for most applications [7].

The most important factors that cause a delay in VoIP

communication are:

Coder delay: Also called processing delay, coder delay is the time taken by the DSP to compress a block of samples.

Packetization delay: Packetization delay is the time it takes to fill a packet payload with encoded or compressed speech.

Queuing delay: Queuing delay is a variable delay and is dependent on the trunk speed and the state of the queue.

Serialization delay: Serialization delay is the fixed delay that is required to clock a voice or data frame onto the network interface.

Network delay: The delay of network equipments that interconnects the endpoint locations is the source of the longest voice-connection delays. These delays are also the most difficult to quantify.

De-jitter buffer delay: Because speech is a constant bit-rate service, the jitter from all the variable delays must be removed before the signal leaves the network. The de-jitter buffer transforms the variable delay into a fixed delay by holding the first sample that is received for a period of time before playing it out. This holding period is known as the initial play out delay [8].

Table 2. Delay of VoIP service with DPLC for two type of codec.

G.729	G.723.1	Codec type
35 ms	97 ms	Coder delay
25 ms	37.5 ms	Packetization delay
30 ms	30 ms	Queuing delay
50 ms	50 ms	Network delay
40-60 ms	60-90 ms	De-jitter buffer delay
180-200 ms	274.5-304.5 ms	Total Communication delay

For instance, Table 2 shows delay for G.723.1 and G.729 codec to use VoIP service with DPLC system.

With increasing number of nodes, the transmission delay of each node should be considered.

Table 3. Delay of VoIP connections based on the number of DPLC links in a network.

G.729 Delay	G.723.1 Delay	Link connection
180-200 ms	274-304 ms	Point-to-Point
260-280 ms	354-384 ms	One transition node
340-360 ms	434-464 ms	Two transition nodes
420-440 ms	514-544 ms	Three transition nodes

Table 3 shows the delay between the two types of VoIP with G.723.1 and G.729 codec based on the number of links contained in DPLC communication. As noted above, based on the ITU-T G.114 standard the end-to-end delay, over 400 ms are not acceptable for most applications. So the audio codec based on G.723.1 standard is suitable for point to point and with one node between DPLCs and the audio codec based on G.729 standard is suitable for a maximum of two node per communication link.

6. Discussion and Conclusions

Available bit rates in DPLC systems due to the limited bandwidth, limited free frequency channel and noise restrictions is far from the theory. Therefore, compression methods for VoIP services in DPLC systems should be employed for optimal use of the channel capacity. Different compression algorithms can increase the delay and degrade the quality of service. Based on the available rates in DPLC and acceptable delay in a telephone communication services, the G.723.1 and G.729 codecs are recommended in this paper as an appropriate selection for VoIP services. Several cases were studied. In case of point to point and one transition node communication links, both G.723.1 and G.729 are recommended while in case of two transition nodes only G729 is the best choice. Thus, despite the fact that G.729 codec requires more bandwidth than G.723.1, the G.729 codec is better for VoIP communication via DPLC due to the shorter latency which leads to more transition nodes.

Acknowledgements

This work was supported by the Niroo Research Institute (NRI).

References

[1] Cigre, Working Group D2.23, "The use of Ethernet technology in the power utility environment", Report 460, 2011.

[2] IEC 62488-1 Ed.1, "Power line communication systems for power utility applications - Part 1: Planning of analogue and digital power line carrier systems operating over EHV/HV/MV electricity grids", IEC, 2012.

[3] Cisco, "Introduction to VoIP", Cisco publications, 2005.

[4] ITU-T Recommendation G.723.1, "Dual rate speech coder for multimedia communication transmitting at 5.3 and 6.3 kbps", ITU-T, 2006. Online at: http://www.itu.int/rec/T-REC-G.723.1-200605-I/en.

[5] ITU-T Recommendation G.729, "Coding of speech at 8kbit/s using CS-ACELP", ITU-T, 2012. Online at: http://www.itu.int/rec/T-REC-G.729-201206-I/en.

[6] ITU-T Recommendation G.131, "Talker echo and its control", ITU-T, 2003. Online at: http://www.itu.int/rec/T-REC-G.131-200311-I/en.

[7] ITU-T Recommendation G.114, "One-way transmission time", ITU-T, 2003. Online at: http://www.itu.int/rec/T-REC-G.114-200305-I/en.

[8] Cisco, "Improving and maintaining voice quality", Cisco publications, 2005.

30

A review of neuro-fuzzy systems based on intelligent control

Fatemeh Zahedi[*], **Zahra Zahedi**

Electrical and Computer Engineering, Shiraz University, Shiraz, Iran

Email address:
Fatemeh.zahedi28@gmail.com (F. Zahedi), Zahra.zahedi28@gmail.com (Z. Zahedi)

Abstract: The system's ability to adapt and self-organize are two key factors when it comes to how well the system can survive the changes to the environment and the plant they work within. Intelligent control improves these two factors in controllers. Considering the increasing complexity of dynamic systems along with their need for feedback controls, using more complicated controls has become necessary and intelligent control can be a suitable response to this necessity. This paper briefly describes the structure of intelligent control and provides a review on fuzzy logic and neural networks which are some of the base methods for intelligent control. The different aspects of these two methods are then compared together and an example of a combined method is presented.

Keywords: Intelligent control, Neural networks, Fuzzy logic, Neuro-fuzzy

1. Introduction

In the last few decades, fuzzy logic has been identified as one of the most active and useful fields of research and study, and has found many uses in many different applications. Of its most important usages are non-linear control systems, time variant systems, and control systems whose dynamics are exactly known, such as servo-motor position control systems and robot arm control systems.

In fuzzy logic systems, decisions are made based on the inputs received in the form of linguistic variables. These variables are grouped up in fuzzy sets and are assigned a degree of membership within that set which are determined by certain formulas that are referred to as membership functions. Fuzzy logic rules are defined as preconditions in the form of if-then rules defined based on the fuzzy sets and the response of each rule is determined through fuzzy implication.

Currently, there are no systematic methods for designing systems based on fuzzy logic. The easiest and the fastest method is to define membership functions and fuzzy rules based on human-operated systems which are then tested until the desired output is obtained [1].

Current research aims at designing systems that work on fuzzy logic and learn from their own experiences. However, there are few articles available that discuss the creating of fuzzy control rules and improvement of those rules from experience. Some of the presented methods can have hopeful results but these methods are mainly subjective and heuristic and finding the proper membership functions are done through trial and error.

On the other hand, applying the neural network's ability to learn, to fuzzy logic systems provides us with more methods that can yield hopeful results. These methods which are referred to as neuro-fuzzy methods create a desirable framework for finding solutions for more complicated problems. If the available knowledge can be expressed in the form of linguistic rules we can have a fuzzy logic system. And if we have the necessary data and are able to learn from that data through simulations, we can use neural networks. Specifying the fuzzy sets, the fuzzy operators and the knowledge base are requirements for building a fuzzy logic system, and specifying the architecture and learning algorithms are requirements for creating a neural network [2].

In the final section of this paper, the cooperative neuro-fuzzy model is reviewed. In this method, the parameters, including the fuzzy rules, rule weights and fuzzy sets are determined using neural networks.

2. Intelligent Control

Generally, modeling most complicated processes can't be achieved by mathematical function or simple physical rules. Intelligent control aims at finding a solution to this problem by combining intelligent and creative characteristics of human controllers. Intelligent control performs under an environment with lots of uncertainties and unexpected situations in a way that keeps its integrity and consistency with that environment and rectifies failures in the system without limits.

Intelligent controls, regardless of what structure or configuration they have, should have the following attributes [3]:

A. Correctness: the ability to perform a specific set of needed operations.

B. Robustness: the ability of the system to maintain its functions under unexpected and unusual situations that can occur internally or externally.

C. Extensibility: the ability of the intelligent control to support planned and unplanned upgrades (hardware or software) without the need for re-design.

D. Reusability & compatibility: the ability to use sub-systems and components in different applications and compatibility with new situations.

There are three basic methods for intelligent control two of which are fuzzy logic and neural networks.

3. Neural network

Control engineers consider neural networks to be large scale, non-linear dynamic systems that are defined within a first order differential equation. Neural networks are in fact new structures for information processing systems that consist of numerous linked processing elements. The links that connect these processing elements are called interconnections. "Fig. 1" is an example of a single processing element. Each processing element has one or more inputs and a single output. Every input has a weight assigned to it and these weights vectors change by learning rules.

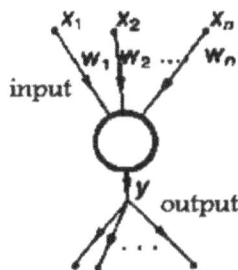

Figure 1. A sample processing element

There can be a question of how such a large dynamic system processes its information? The answer to this question is the use of energy functions for the system. Each non-linear dynamic system has several equilibrium points. These points are the minimum points on the energy landscape. If an arbitrary input pattern is given to the system as its initial state,

the system is capable of approaching one of these equilibrium points subject to the global stability of the system. For example, according to "Fig. 2", E(X) is an energy function with three equilibrium points of P1, P2 and P3. If X(0) is the initial state of the system, X(0) should converge to the closest equilibrium point.

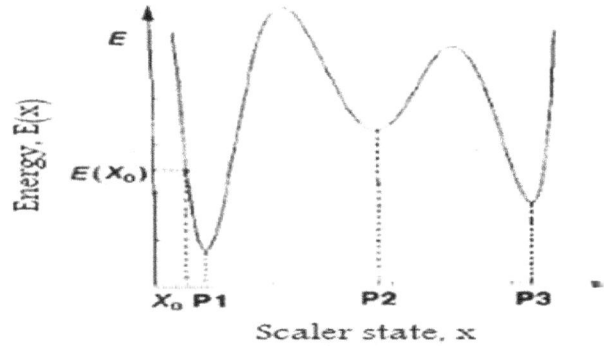

Figure 2. Energy landscape with 3 minimum points

For the defined system in the differential "(1)" and "(2)", the Lyapunov function (energy) would be as the "(3)".

$$\dot{x}_i = a_i(x_i)[b_i(x_i) - \sum_{k=1}^{n} w_{ik} d_k(x_k)] \tag{1}$$

$$\dot{w}_i = w_{ij} + G(w_{ij}, x_i, x_j) \tag{2}$$

$$V(x) = -\sum_{i=1}^{n} \int_0^{x_i} b_i(\varepsilon_i) d_i'(\varepsilon_i) d\varepsilon_i + \frac{1}{2} \sum_{j,k=1}^{n} w_{jk} d_j(x_j) d_k(x_k) \tag{3}$$

4. Fuzzy Logic

4.1. Defining some Fuzzy Operations

We start out by defining some fuzzy operations. If A and B are two fuzzy sets in the universal set of U with membership functions of μ_A and μ_B, then:

Definition 1- Union: The membership function $\mu_{(A \cup B)}$ of the union A∪B for all u∈U is defined as followed:

$$\mu_{(A \cup B)} = \max \{\mu_A(u), \mu_B(u)\} \tag{4}$$

Definition 2- Intersection: The membership function $\mu_{(A \cap B)}$ of the intersection A∩B for all u∈U is defined as followed:

$$\mu_{(A \cap B)} = \min \{\mu_A(u), \mu_B(u)\} \tag{5}$$

Definition 3- Complement: The membership function $\mu_{\bar{A}}$ of the complement A for all u∈U is defined as followed:

$$\mu_{\bar{A}}(u) = 1 - \mu_A(u) \tag{6}$$

4.2. Architecture of Fuzzy Logic Controllers

"Fig. 3" shows the basic configuration of fuzzy logic controllers which is consisted of four parts.

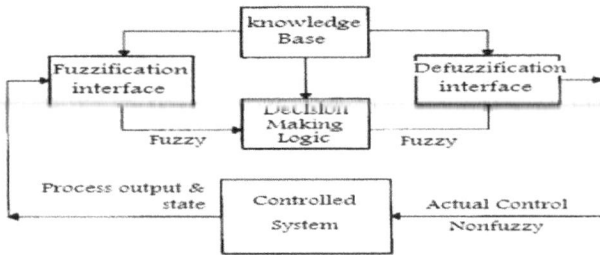

Figure 3. Basic configuration of fuzzy logic controllers

1- Fuzzification interface which is consisted of the following steps:
- Measuring the value of input variables.
- Transfer of a wide range of input variable values to the related universal set.
- Fuzzification, which is in fact the conversation of input data to proper linguistic value.

2- Knowledge base, including "data base" and "linguistic (fuzzy) control rule base".
- The data base provides the necessary definitions that are used to define linguistic control rules and fuzzy data.
- The rule base which is specifies the control goals using a set of linguistic control rules.

3- Decision making logic, which is the kernel of the fuzzy logic controller. In this part human decision making is simulated based on fuzzy concepts and inferring fuzzy control actions by using fuzzy implications and rules of inference.

4- Defuzzification interface which is consisted of the following steps:
- Conversion of a wide range of output variables to universal set.
- Deffuzification of rules and linguistic variables in a way that is retrievable and understandable by the rest of process.

5. Neuro-Fuzzy

5.1. Neuro-Fuzzy Structure

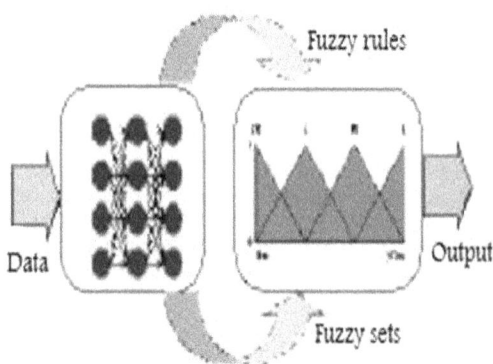

Figure 4. Neuro-fuzzy cooperative model

Neuro-fuzzy is a model in which the neural network uses the training data to determine the membership functions and fuzzy rules of the fuzzy logic system. After specifying the parameters of the fuzzy logic system, the neural network

moves to the margins. Basic rules are usually determined using fuzzy clustering algorithms. The neural network approximates the membership functions from the training data. "Fig. 4" shows a neuro-fuzzy cooperative model.

An example of a neuro-fuzzy structure is as follows [1], [2]:

As it can be seen in "Fig. 5", this structure contains five layers. Two linguistic nodes are present for each output, one for the training data (the desired output), and another for the actual output of the neuro-fuzzy system.

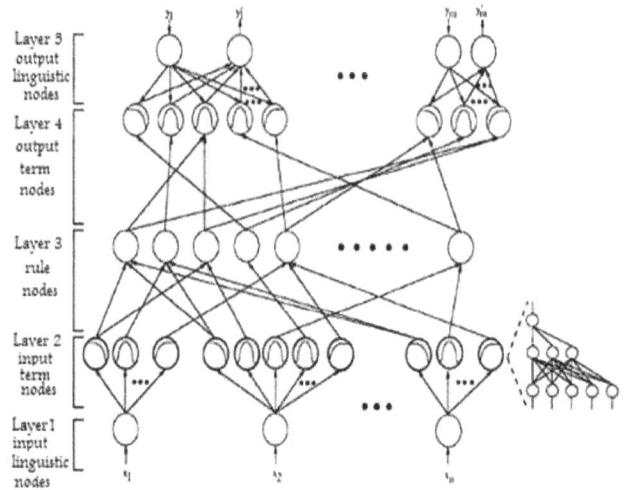

Figure 5. The five layer neuro-fuzzy structure

The first internal or hidden layer is in charge of the fuzzification of the input variables. Each node in this layer can be either singular which identifies a simple membership function, or it can be consisted of multi-layered nodes which calculate the complex membership functions.

The second hidden layer defines the preconditions and rules resulted from the third hidden layer. In this method, the output is generated using the hybrid-learning algorithm consisted of unsupervised learning to retrieve the initial membership functions and rule base.

5.2. Fuzzy Control of an Unmanned Vehicle

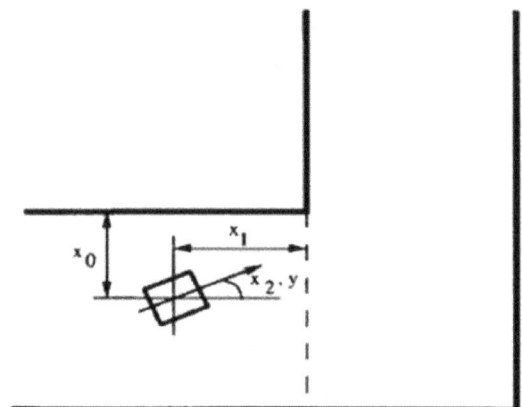

Figure 6. The state variables of the fuzzy vehicle

In this section, an example is presented to show the main application of this model [1]. The vehicle uses examples to learn to automatically move through a track with rectangular turns. X_0, X_1 and X_2 are the linguistic variables that specify the distance from track boundaries from one side and the current angle of the steering wheel. The output linguistic variable Y is the next steering angle. This concept is better explained in "Fig. 6".

"Fig. 7", shows the membership functions of the x_0, x_1, x_2 and y variables, learned after one or two training phases.

"Table, 1" shows the learned fuzzy logic rules. These rules show the hidden layers of the neural network design. In this example, we assume that the vehicle moves at a constant speed and sensors are installed on the vehicle for measuring the x0, x1 and x2 variables. These variables are given to the controller so that it can derive the next steering angle from them. This example has been simulated under different initial conditions for the steering angle and the results have been very close to reality.

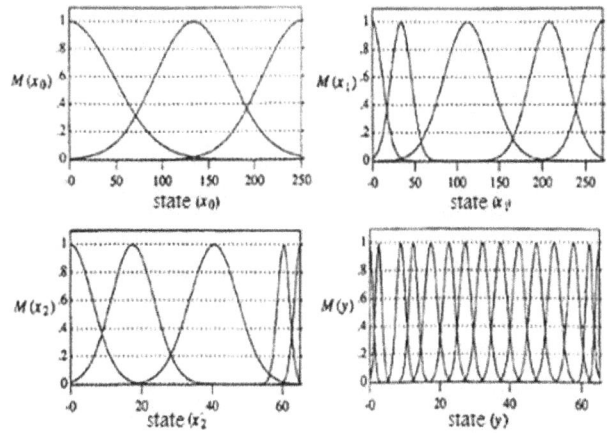

Figure 7. The learned membership functions

Table 1. Learned fuzzy logic rules in this example

Rule	Precondition			Consequence	Rule	Precondition			consequence
	x_0	x_1	x_2	Y		x_0	x_1	x_2	Y
0	0	1	0	6	15	1	2	1	4
1	1	1	0	6	16	2	2	1	2
2	2	1	0	5	17	0	3	1	2
3	0	2	0	3	18	1	3	1	2
4	1	2	0	2	19	2	3	1	2
5	2	2	0	1	20	0	0	2	11
6	0	3	0	2	21	1	0	2	10
7	1	3	0	2	22	2	0	2	12
8	2	3	0	0	23	0	1	2	10
9	0	0	1	9	24	1	1	2	8
10	1	0	1	9	25	2	1	2	13
11	0	1	1	7	26	0	2	2	7
12	1	1	1	7	27	1	2	2	7
13	2	1	1	7	28	2	2	2	13
14	0	2	1	6	29	0	3	2	13

6. Discussion and Conclusion

In this paper, intelligent control and two of its basic methods, neural networks and fuzzy logic were generally represented. Cooperative neuro-fuzzy model was reviewed and the general structure of one of its methods was explained. An example was provided for the two main applications of the model, fuzzy logic controllers and fuzzy decision making systems. This example shows the superiority of the hybrid learning method over the more traditional methods.

References

[1]   C. T. Lin, and C.S.G. Lee. "Neural network based fuzzy logic control and decision system," IEEE. Trans. Comput, vol. 40, pp. 1320-1336, December 1991.

[2]   A. Abraham. "Neuro fuzzy systems: state-of-the-art modeling techniques, connectionist models of neurons," LNCS, vol. 2084, pp. 269-276, Jun 2001

[3]   C. J. Harris, C. G. Moore, and M. Brown. "Intelligent control: aspects of fuzzy logic and neural network," World Scientific Series in Robotics and Automated Systems. UK, vol. 6, 1993.

[4]   B. Bavarian. "Introduction to neural networks for intelligent control," IEEE Control System Magazine, California, vol. 8, pp. 3-7, April 1988.

[5]   C. C. Lee. "Fuzzy logic in control systems: fuzzy logic controller- part I," IEEE. Trans. Syst. Man. Cyber. Syst. California, vol. 20, pp. 404-418, Mar/Apr 1990.

31

Risk Based Management Methods for Maintaining 750 kV Electricity Grid in Northwest China

Li Juan, Liang Naifeng, Zhao Puzhi, Zhou Erbiao, Li Xiaoguang

State Grid Xinjiang Electric Power Compangy, Xinjiang, China

Email address:

773963700@qq.com (Liang Naifeng), 13738674523@163.com (Li Juan)

Abstract: This paper will discuss a different case - a 750 kV transmission grid in China, in which the regulation rules have not been decided and the assets are operating in their early stage. The difficulty in technology and insufficiency of regulation decide the focus of this study: to select the method feasible in this business environment, rather than to develop complicated but unrealistic models.

Keywords: Transmission, Maintenance, Management

1. Introduction

Electricity transmission operators (TSO) in China remain their monopoly status. However, in a national policy made in 2015, the focus of its regulation changes from the end-user price to the transmission tariff. This means that a Chinese TSO is required to clarify its transmission costs, which brings challenges to its maintenance department.

Traditionally, maintenance departments of the TSO's focus on developing repair techniques and performing repair activities in order to guarantee a high reliability of the network. Nowadays, economic and environmental responsibilities have also been assigned to maintenance managers. They are expected to organize the maintenance activities optimally in such a way that the profits and performances of the TSO are maximized. This elevates maintenance management to a new level, namely "asset management".

Meanwhile, the 750 kV transmission network, as the backbone of electricity grid in northwest China, started just within a decade and keeps expanding in a rapid speed in the 2010s. For example, according to the planning, the total length of its 750 kV overhead line will increase for over 150%. In such a rapid growth, maintenance activities should be optimized not only for saving costs, but also for allocating limited human power and maintenance equipment. Moreover, the desert, mountainous and glacial environment in this area has introduced a high technical complexity to its maintenance activities. These are important considerations of establishing risk-based maintenance.

This paper will firstly review existing methods for risk-based maintenance from the perspective of asset managers in Section 2. Then, in Section 3, proper methods will be identified for the preliminary stage of risk-based maintenance on the studied 750 kV grid.

2. General Models of Risk-Based Maintenance

2.1. The Trigger of Introducing Risk-Based Maintenance in the Electricity Transmission Sector

In tens of countries worldwide, the liberalized electricity market launched in the 1980's has brought a complete change to the power delivery sector. During the liberalization processes, governments divested themselves from the utility services and established the trade systems of utility products. Consequently, the corporatized electricity sector is operated by multiple companies whose business areas are defined by the regulator. This means, in the case of an electricity TSO, the involvement in generation and trade is gradually reduced or forbidden in many countries [1].

Before the implementation of the above mentioned regulations, the utility sector was protected by its natural monopoly status. This enabled the system operators to adopt risk-free development strategies such as "expanding the network up to its technical limits", or "enhance the reliability and redundancy with all available budgets". After the market liberalization, an electricity transmission or distribution company can no longer afford such expensive development strategy, since it exposes the asset owner to commercial risks. Coincidently, the growing public concerns regarding the environment and sustainability have led to stricter restrictions on the existence and expansion of the power grid [2] and its subsystems including overhead lines [3], substations [4]and cables [5].

Physical asset management [6], or asset management (AM) in short, is a field of emerging importance in the utility sector. The managed objects are physical assets and asset systems. Distinguished from human, information, financial and intangible assets, physical assets are a subset of fixed assets. They serve as the backbone of the daily operation within a power grid. Maintenance is a collection of the most discussed activities within in the domain of AM. Historically, three major advancements have been achieved and applied in maintenance management [7].

The first advancement is the abandoning of "Corrective Maintenance" (CM). Maintenance was, originally and yet in many occasions, corrective. In CM, components are completely replaced after they fail. In contrast, "Preventive Maintenance" (PM) intends to recover a component from a faulty state through proper repair techniques. The purpose of PM is primarily to avoid major failures and consequent interruptions of routine operation. However, in many cases, the use of PM can lead to the extension of the service life of components.

The second advancement is the introduction of maintenance strategies. In current practice, a maintenance strategy is essentially a method to schedule individual PM tasks within an asset fleet. Proper scheduling controls repair costs significantly, because required resources such as spare parts, transportation services, work force or cash flow are usually lower in costs if arranged in advance rather than requested on site. A basic scheduling approach is ranking health condition indices estimated for individual components. If the index is derived from the usage history of each component, the maintenance strategy is called "Time-Based Maintenance" (TBM). If diagnostic information from inspections, tests and monitors are utilized to assess the index, the maintenance strategy is "Condition-Based Maintenance" (CBM).

The third advancement is the introduction of risk analysis. Generally speaking, risk is an approach to describe the potentiality of an incident such as failure of a certain component. In management, a risk is featured with its probability and consequence and rated with the expected value – the multiplying product of the probability and the consequence.

Maintenance scheduling can be achieved through risk analysis. Risk analysis on an individual component consists of estimating probability and consequences quantitatively. Firstly, the rough level of failure probability can be derived from the health condition index. Secondly, the failure consequence is measured in multiple items[8]. Traditionally in power grids, the network reliability, mainly measured by the customer minute loss of the blackout, is the only item. This is the Reliability-Centered Maintenance (RCM) strategy. Recently, additional aspects such as finance, safety and environment were added in response to regulations and stakeholder requirements. This extends RCM to a complete Risk-Based Maintenance strategy (RBM).

Risk analysis can be performed not only on components to implement a maintenance strategy, but also on asset systems to guide and select a maintenance strategy on its composing components. This technique is frequently referred as the failure mode, effect and criticality analysis (FMECA). In FMECA, the risk analysis is performed on a failure mode. A failure mode describes a function loss on asset systems of different scales, e.g. switchgear, bus, or complete substation. Failure modes are distinguished from each other according to their original faulty component and the physical degradation mechanism. After determining probability, consequence and risk analysis procedures, FMECA is capable to rank failure modes with their expected risk indices. Using this ranking, asset managers can adapt maintenance strategies to the failure mode it tackled. e.g. for a component with highly ranked failure modes, CM is abandoned, frequency of TBM is increased, proper diagnostic tools for CBM is introduced, etc.

RBM and FMECA represent the up-to-date advancement of AM in electricity transmission and distribution. This enables the utility company to optimize the allocation of the maintenance budget.

2.2. Introducing Risk-Based Maintenance at the Tactical Level of Asset Management

Asset managers have widely accepted the triple level model as introduced in Table I. Specifically, for maintenance management, the condition indicator system is mainly implemented at the operational level to schedule maintenance plans, as the second advancement in II.A has stated. Meanwhile, the FMECA is implemented at the tactical level to decide maintenance methods and diagnostic methods, i.e. the maintenance strategies. This is mainly included in the third advancement in 2.1.

Table 1. Features of the three levels of asset management (adapted from [8]).

Level	Time frame	Concerned asset group	Performance	Expenditure	Decision maker
Strategic	Long-term: > a decade	Asset portfolio	Societal, Economic	Capital	Asset owner
Tactical	Mid-term: 3-7 fiscal years	Asset system	Economic, Technical	Budget	Asset manager
Operational	Short-term: < 1 fiscal year	Asset (item)	Technical	Cost	Service provider

Currently, most advancements in maintenance management concentrate on developing diagnostic tools and standard condition indicator systems. These are progresses at

the operation level. Moreover, many TSO's have started to construct their risk-based AM regime, including RBM at the operation level and FMECA at the tactical level. Extensive activities, including life cycle optimization at the operational level and investment decisions at the tactical level have been gradually introduced in specific cases to manage asset systems such as medium voltage cables, overhead lines and power transformers.

3. Selecting Appropriate Methods for Risk-Based Management on 750 kV Assets

3.1. Features of the Studied 750 kV Grid

The studied 750 kV have three features. They will decide our selection of risk assessment method in this section.

- Infant mortality. The studied assets are relatively new and rapidly increasing in population.
- Missing performance benchmarking. The ongoing changes on TSO regulation have not yet indicated any clear consequences on all incidents, especially regarding the environment and safety ones.
- Rich external knowledge on the failure modes, condition indicators of assets. The mother company of the TSO has nationally standardized testing and condition indicator system on major high-voltage assets. However, the effectivity of the standard indicators is undermined by the special operating environment in the wildness in deserts, mountains or glaciers.

Based on the above limitations, the studied risks yield to the failures of high-voltage components and consequent performance losses. At the tactical level of AM, this study should perform FMECA to adapt external knowledge on condition indicators to the specific operating environment of northwest China. The main benefit of such adaption will be the capability to: minimize the failure frequency of high-voltage components of the same type, through prioritizing the maintenance tasks on them.

3.2. Estimation of the Probabilities of Failure Modes

At the tactical level, the probably of a failure mode should be estimated, so that proper diagnostic method can be introduced as countermeasures. The capability to estimate of such probabilities is not automatically ready in a TSO.

A failure is defined as the situation where a certain object cannot function under stated conditions for a specified period of time. Within asset portfolios, a failure can occur at many different levels, such material, subcomponent, component/asset, asset system (including secondary equipment) and asset fleet/portfolio. When failures at a lower level lead to a failure at a higher level, the process is called a failure mechanism or failure mode.

The raw format of failure and life data is a list of failure events. A complete record on a failure event should include (1) the moment, (2) the (initially) failed object, (3) the cause or failure mechanism, (4) the reaction and countermeasures and (5) the consequences. [7]

From the raw format of failure records, two types of parameters can be derived as the conclusion after statistical analysis. They are the failure frequency, and the lifetime/time-to-failure (TTF). In order to convert time series of failure events to TTF, asset manager should record the population, namely the total number of investigated objects from which the failure events have been collected from. In addition, when the investigated objects are significantly different in their physical sizes, size information should also be included in population. (e.g. cables of different length). Such information comes from the installation/deployment data, which is typically the calendar date/time at which each object within the population started to operate/be stressed.

For the studied asset portfolio, the installation and population data is probably ready. However, since the assets are in their infant mortality stage, the TTF cannot be creditably estimated as in those aged asset populations. Therefore, the age-independent failure frequency is the main input to failure mode analysis.

3.3. Risk Management: FMEA and ABC Classification Method

In FMEA [9], as most standards have specified, the Risk Priority Number (RPN) is calculated for each failure mode. RPN is the product of the severity score, the probability score and the detectability score. As mentioned above, the probability score is the frequency of failure mode. The severity score represent the consequences of failure modes, which can be acquired from the external knowledge. The detectability score is not used in this preliminary stage.

ABC analysis method [10] is a method originally applied for quality control. It follows simple principles as follows. 20% products can have high values, which in total contain 70% value of the whole population. They form Class A which should be the focus of management. In our cases, the products are the failure modes, and the value is the RPN. The focus means that innovated condition diagnosis method, e.g. condition monitoring should be implemented. The next 30% products are of average values, i.e. in total contain 25% value of the whole population. They form Class B which should a certain degree of attention should be paid. In our case, such attention means standard diagnosis proposed by the national standards and external knowledge. The remaining 50% products are of low values, i.e. in total contain 5% value of the whole population. This means a minimally requested maintenance scheme.

The FMEA and ABC analysis is applicable when the performance benchmarking of managing electricity transmission assets (see examples in [11]) is still missing in China. The severity in FMEA and values in ABC analysis can be firstly narrowly defined as reliability and a few aspects of safety or environment, but later extended to other business values imposed by future laws.

3.4. Continuous Improvements

As time passes, the infant mortality of the studied asset portfolio will gradually stop. In this situation, the risk assessment proposed above can be improved in several aspects below:

- Optimize the condition indexing, namely the composing rules of condition indicators, for the special operating environment of northwest China.
- Specify different failure mechanisms on the same type of component in failure data, and treat them as different risks accordingly.
- Include detectability of diagnostic methods and preventability of repair methods in the RPN.
- Include not only the consequences on reliability, but also the consequences on safety and environment in the severity score of RPN.

4. Conclusions

Risk-based management on 750 kV electricity transmission assets has been stimulated by the Chinese deregulation policy as well as the fast expansion of the asset population. When the assets are in their infant mortality and the definition on asset values are still missing, it is appropriate to balance costs and risks of failures, firstly through time-constant failure rate model, FMEA and ABC analysis method. These methods are compatible with future developments in aspects such as knowledge on assets, asset performance benchmarking, etc.

References

[1] Mohamed M Saied, Canceling the Power Frequency Magnetic and Electric Fields of Power Lines, IETE Journal of Education, 2013, pp.90-99.

[2] Strategic Environmental Assessment for Power Developments, *Cigré Technical Brochure 487,* 2012.

[3] Lin Che-Yun, Wang Alan X, Lee Beom Suk, Zhang Xingyu, Chen Ray T, High dynamic range electric field sensor for electromagnetic pulse detection, Optics Express, 2011, pp.17372-7.

[4] Improving the impact of existing substations on the environment, *Cigré Technical Brochure 221,* 2013.

[5] LIANG Cai, LIU Wenying, ZHOU Xichao, DAN Yangqing, Analysis on Role of 750 kV Power Network Played in Loss Reduction of Gansu Power Grid, Power System Technology, 2012, pp.30-32.

[6] Specification for the optimized management of physical assets, British Standardisation Institute, 2008.

[7] Qikai Zhuang, Manaing Risks in Electrical Infrastructure Assets from a Strategic Perspective, Next Generation Infrastructure Foundations, Delft, 2015.

[8] E. Rijks, G. Sanchis, and P. Southwell, Asset Management Strategies for the 21st Century, *ELECTRA, Cigré,* vol. 248, 2010.

[9] ZHOU Peipeng, XIANG Zutao, DU Ning, BAN Liangeng, XIE Guoping, Analyhsis on blocking of Qinghai-Tibet DC System Caused by Transformer Energizing in Northwest China 750 kv Grid, Automation of Electric Power Systems, 2013, pp.10-12.

[10] WANG Xiao-liang, DONG Hai-ying, REN Wei, State Evaluation of Secondary Device in 750 kV Power Grid Based on Information Fusion, Proceedings of the Chinese Society of Universities for Electric Power System and its Automation, 2013, pp.24-26.

[11] Asset Management Performance Benchmarking, *Cigré Technical Brochure 367,* 2013.

32

Towards a Precise Direction of Arrival Estimation for Coherent Sources Using EC-MUSIC

Hassan Mohamed EL-Kamchouchi, Ahmed Samir EL-Torgoman

Electrical Engineering, Alexandria University, Alexandria, Egypt

Email address:

ahms20042002@gmail.com (H. M. EL-Kamchouchi), ahms2100@gmail.com (A. S. EL-Torgoman)

Abstract: Smart antennas are used to transmit /receive the signals according smart choices due to DSP, the most important factor in this choices is signal direction for reception or transmission, known as DoA: direction of arrival, there are much many algorithms used to address this issue as MUSIC/ROOT MUSIC/ESPIRIT /MVDR In this paper we will concentrate on MUSIC algorithm are one of the most attractive algorithm regarding its performance, However 2 or more signals are close to each other these algorithm do not perform accurate, as well as under coherent sources the accuracy deteriorates quickly in this paper a reasonable modification for MUSIC had been made by adding coefficients related to number of snapshots and number of element array [1] as well as simple modification for covariance matrix in MUSIC algorithm to capable to handle signals of coherent sources will result in enhancing MUSIC algorithm resolution for coherent sources to 1 degree space which will be called EC-MUSIC (Enhanced Coherent-MUSIC).

Keywords: Smart Antennas, Direction of Arrival, Coherent Sources

1. Introduction

Smart antennas combine multiple antenna elements with signal processing capability to optimize radiation / reception [2] and is divided to two main sections beam forming signal is formed due to signal coming from target which decrease power consumption and nulls interferers jamming signal and frequency reuse will be used within cell space called space division multiple access SDMA [3], the second section of smart antenna is direction of arrival which is so important for finding the direction from which the signal is coming from many algorithms are studied in this area [1, 4, 5, 6], it is important in many applications like sonar/rescue devices wireless communication, algorithms addressing this technique divided into three main categories conventional & beam forming and subspace techniques as [1] MUSIC/ESPIRIT/BARTLETT/CAPON, the implementation is used to eliminate interference & combine the required signals to improve performance, from literature the common problems for this algorithm 2 adjacent signals cannot be detected however MUSIC is more better than Bartlett, MVDR and linear prediction method in this paper we will

address the problem of coherent sources with MUSIC algorithm by inserting transition matrix to correct

covariance matrix de efficient a more simpler way than spatial smoothing method as well as enhancing resulting resolution by inserting enhancement Coefficients as no of snapshots, no of elements, which will result in enhancing MUSIC algorithm resolution for coherent sources to 1 degree space which will be called EC-MUSIC (Enhanced coherent-MUSIC) The paper is organized as follows Road map, Presentation of basic MUSIC algorithm. coherent sources and its problem of detection. how it is solved. presentation of enhancing Coefficients, seeing the results, conclusion and future work

2. Road Map

We can simplify the roadmap in the following points.
A. Coherent MUSIC Overview
B. Basic MUSIC Algorithm
C. Coherent Sources problem in MUSIC
D Solving Methodology
E. Enhancing Resolution
F. Simulation steps and results

3. Coherent MUSIC Overview

Coherent MUSIC based on MUSIC algorithm in which transition matrix is conjugate construction of the data matrix (Covariance matrix) of the MUSIC algorithm.

3.1. Basic MUSIC Algorithm

MUSIC: stands for Multiple SIgnal Classification it has been firstly presented by Schmidt [7] to estimate DOA it depends mainly on covariance Matrix of data from which Eigen vector for signal and noise using the data model in equation.

$$X = S\alpha + n \tag{1}$$

Where

$$S = [s((\phi_1), s(\phi_2), \ldots, s(\phi_M)]$$

$$\alpha = [\alpha_1, \alpha_2, \ldots \alpha_M]^T$$

The Matrix S is N * M where M is steering vector and N is the number of elements assuming signals are to be un correlated the correlation matrix of x is written as

$$R = E \ [XX^H] \tag{2}$$

$$= E[S\alpha\alpha^H S^H] + E[nn^H], \tag{3}$$

$$= SAS^H + \sigma^2 I \tag{4}$$

$$= R_S + \sigma^2 I \tag{5}$$

Where

$$R_S = SAS$$

$$A = \begin{bmatrix} E[|\alpha_1|^2] & \cdots & 0 \\ \vdots & \ddots & \vdots \\ 0 & \cdots & E[|\alpha_M|^2] \end{bmatrix}$$

The signal covariance, R_s is clearly N×N with rank M. It therefore has N-M eigen vectors corresponding to the zero Eigen value let q_m such eigen vector therefore

$$R_S q_m = SAS^H \ qm = 0 \tag{6}$$

$$q_m^H \ SAS^H \ qm = 0$$

$$S^H q_m = 0 \tag{7}$$

Where this final equation is valid since the matrix A is clearly positive definite. Equation implies that all N-M eigen vectors (q_m) of R_S corresponding to the zero eigenvalue are orthogonal to all M signal steering vectors.

This is the basis for MUSIC. Call Qn the N ×(N −M) matrix of these eigenvectors. MUSIC plots the pseudo-spectrum

$$P_{MUSIC}(\phi) = \frac{1}{\sum_{m=1}^{N-M} |S^H(\phi) q_m|^2} =$$

$$\frac{1}{S^H(\phi) Q_n Q_n^H S(\phi)} = \frac{1}{\|Q_n^H S(\phi)\|2} \tag{8}$$

Note that since the eigenvectors making up Qn are orthogonal to the signal steering vectors, the denominator becomes zero when ϕ is a signal direction.

3.2. Coherent Sources Overview

The problem involving coherent sources is a fatal problem for subspace algorithms [10]. When there is a coherent signal in the signal source, the signal covariance matrix is rank defective. In this case, the original super-resolution algorithm will fail. Therefore, it will greatly affect the performance of DOA estimation.

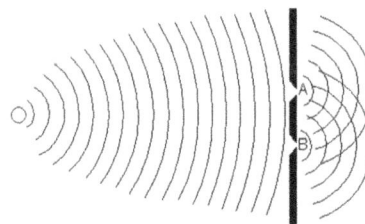

Points A and B are coherent sources

Figure 1. Coherent Sources illustration [11].

3.3. Coherent Sources Problem in MUSIC

Highly correlated or coherent source signals are common in multipath propagation environments due to the reflection and refraction of source signals in practical. Based on such scenario, the coherent sources facilitate the rank loss of the covariance matrix, which could result in the failure of the conventional high-resolution estimation algorithms.

3.3.1. Problem Addressing

Covariance matrix rank loss is addressed by many methods to overcome its bad effect in Direction of arrival estimation. A simple method which was proposed by [9] conjugate reconstruction of covariance matrix of MUSIC algorithm.

3.3.2. Methodology

Make a transformation matrix J, J is an Mth-order anti-matrix, known as the transition matrix, i. e.

$$\begin{bmatrix} 0 & \cdots & 1 \\ \vdots & \ddots & \vdots \\ 1 & \cdots & 0 \end{bmatrix}$$

Let Y=JX*, where X* is the complex conjugate of X, then the covariance of data matrix Y is

$$R_y = E \ [YY^H] = JRX^*J. \tag{9}$$

From the sum of Rx and Ry, the reconstructed conjugate matrix can be obtained.

$$R = Rx + Ry = AR_s A^H + J[AR_s A^H]^* J + 2\sigma^2 I \tag{10}$$

According to matrix theory, the matrices Rx, Ry and R have the same noise subspace. To conduct characteristic decomposition of R and get its eigenvalue and eigenvector, according to the estimated number of signal source, separate the noise subspace, and then use this new noise subspace to construct spatial spectrum and obtain the estimated DOA value by finding the peak.

4. Enhancing Resolution

The last phase in this method to enhance resolution of very close angles of DoA When the separation angles between sources are very small, MUSIC algorithm couldn't estimate the angles correctly. Thus an improvement for that algorithm was proposed by adding new coefficients to achieve this goal and estimate the adjacent angles correctly. This was achieved even with separation of degree only between sources. These coefficients are related to signal and antenna parameters which are affecting the accuracy of MUSIC algorithm. Those coefficients represent the ratio by which each corresponding parameter should be increased in order to get the required super high resolution. Those coefficients will virtually maximize the values of the corresponding parameters during calculation and enhance the estimation accuracy of the algorithms. Those coefficients are defined as follows as

4.1. Snapshots Coefficient (SC)

In order to increase the efficiency of estimation and obtain sharper peaks in MUSIC. A higher number of snapshots is required. Incrementing snapshots by snapshots Coefficient which is greater than one and multiplied by basic number of snapshots. The estimation variance for covariance matrix will decrease and thus sharper peaks for MUSIC are obtained. However, very high values for SC will lead to processing delay since the processing time will also increase. Hence, suitable value for it should be selected according to the requirements the new number of snapshot N′ can then be written as

$$N' = N \times SC \qquad (11)$$

4.2. Elements Coefficient (EC)

For more signal component in estimation result to clarify signal peaks for adjacent coherent signals. Element coefficient if multiplied by original elements number of array M. Thus, it gives more signal component for the algorithm. Here the importance of this modification appears during manufacturing process. By selecting the suitable EC value, the suitable number of elements required for the antenna design can be selected easily. Then we can substitute the new number of element M′ in MUSIC. M′ can be expressed by:

$$M' = M \times EC \qquad (12)$$

5. Simulation Steps & Results

Simulation done with MATLAB ®2015a release to clarify obtained results we have to see it compared with results obtained without any improvement with the raw MUSIC algorithm as a main frame. We will begin with incoherent sources and Raw MUSIC algorithm with 2 direction of arrival

incidence (30°, 70°) degrees as in figure 2, for snapshots mainly =200 and number of element array (Uniform linear array)= 10.

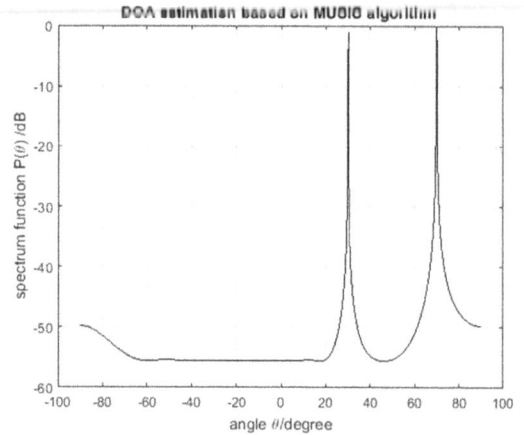

Figure 2. Incoherent sources and raw MUSIC.

Then coherent sources with the same frequency and same incidence angles as last case and raw MUSIC algorithm we find performance degradation as in figure 3

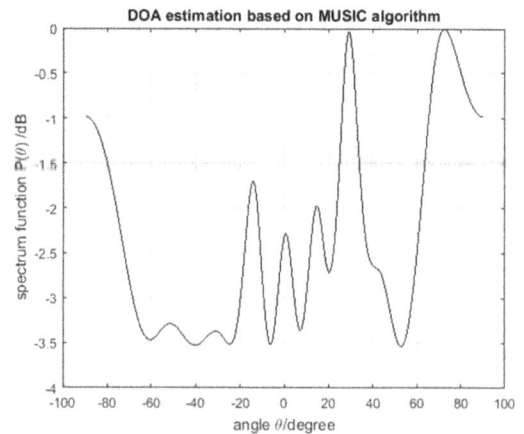

Figure 3. Coherent raw with basic MUSIC.

Then check again with coherent MUSIC algorithm the peaks appear only at the estimated angles with high accuracy as seen in figure 4

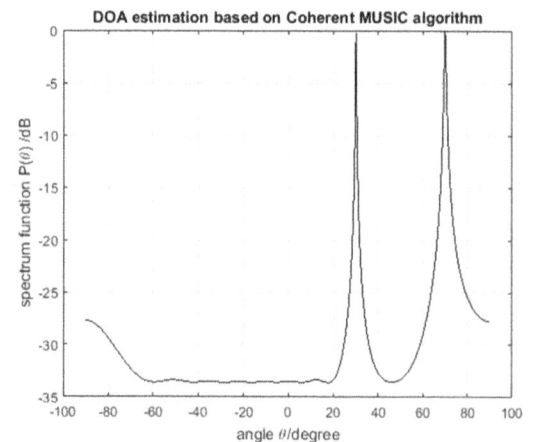

Figure 4. Coherent sources with coherent MUSIC.

But when incident angles is adjacent (30°, 31°) the coherent MUSIC algorithm failed to discriminate them as in figure 5

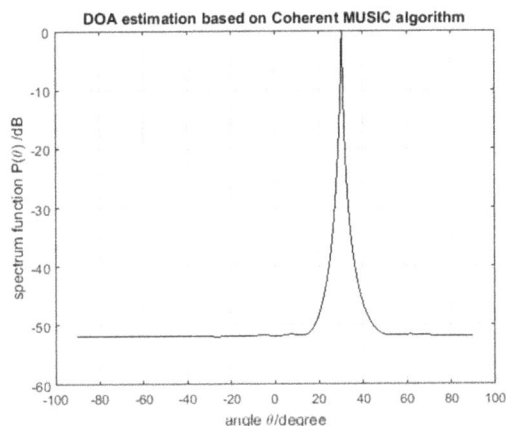

Figure 5. Adjacent coherent sources with coherent MUSIC without enhancing resolution.

But when we use Enhanced coherent MUSIC with enhancing resolution Coefficients we find incredible accuracy as we see in the zoomed figure 6.

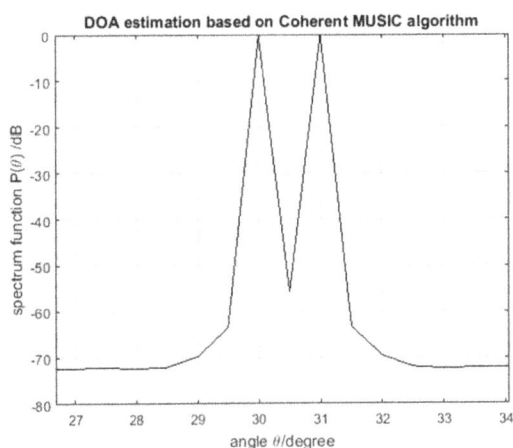

Figure 6. Adjacent coherent signal estimation with EC-MUSIC (Zoom)with EC=3 ;and SC=1.5.

6. Conclusion

In this paper the theory of smart antennas and the basic concept of MUSIC DOA algorithms had been illustrated and discussed. That algorithm when implemented in smart antenna will lead to efficient use of networks. To achieve this goal, the estimation of source direction should be accurate even if sources are too close to each other's in angles and coherent which is achieved through covariance matrix correction and adding enhancing coefficients that led at last to accurate estimation for direction of arrival for adjacent coherent signals.

However, excessive increment for coefficients values will lead to degradation in performance so that; suitable values for coefficients should be selected.

7. Future work

Enhancing resolution of DoA for coherent signals is now an important demand in today's world of smart antennas, this work can be modified to update root MUSIC algorithm as well as Mixed signals identification in Multipath environment will be promising land for further research.

Acknowledgements

I would like to thank my Professor and Co-author Hassan. M EL-Kamchouhi For his un conditional scientific support.

References

[1]   Z. Chen, G. Gokeda, and Y. Q. Yu. Introduction to Direction-of-arrival Estimation. *Artech House*, (2010).

[2]   Y. Liao, A. Abouzaid. Resolution Improvement for MUSIC and ROOT MUSIC Algorithms (2015).

[3]   F. Gross. Smart antennas for wireless communications: with MATLAB, *McGraw-Hill*, (2005).

[4]   M. Jalali and B. Honarvar Shakibaei, Angular accuracy of ML, MUSIC, ROOT-MUSIC and spatially smoothed version of MUSIC algorithms, *International Journal of Computer and Electrical Engineering*, vol. 2, no. 3, pp. 1973-8163, (2010).

[5]   Ahmed Khallaayoun, High resolution direction of arrival estimation analysis and implementation in a smart antenna system. *Montana State University-Bozeman*, (2010).

[6]   K. Karuna Kumari, B. Sudheer, and K. V. Suryakiran, Algorithm for Direction of Arrival Estimation in a Smart Antenna. *International Journal of Communication Engineering Applications*, vol. 2, no. 4, (2011).

[7]   Schmidt, R. O. , Multiple emitter location and signal parameter estimation, *IEEE Trans. on Antennas and Propagation*, vol. 34, no. 3, pp. 276-280, (1986).

[8]   A. Gershman. Class slides. Advanced Topics in DSP. McMaster University. Personal Communication.

[9]   D. Kundu. Modified MUSIC Algorithm for estimating DOA signals, (1993).

[10]   H. Tang DOA estimation based on MUSIC algorithm, 2014, un unpublished.

[11]   Dept. of Physics, Colorado university, physics lab, http://www. colorado.edu/physics/phys2020/phys2020LabMan2000/2020l abhtml/Lab5html/lab5. html, 2000, unpublished.

33

Evaluation of a new hybrid technique based on DTMOS and PFA to improve supply voltage and power consumption of a class-AB amplifier

Hossein Movahedi-Aliabad[1, *]**, Akram Norouzi**[2]**, Sepideh Soltanmoradi**[2]**, Mahshid Nasserian**[3]**, Manijeh Shahi**[2]

[1]Department of Electrical and Electronics, Quchan Technical Institute, Technical and Vocational University, Quchan, Iran
[2]Department of Electrical Engineering, Bojnourd Branch, Islamic Azad University, Bojnourd, Iran
[3]Department of Electrical Engineering, Ferdowsi University of Mashhad, Mashhad, Iran

Email address:

h.movahedi@tvu.ac.ir (H. Movahedi), noroziakram93@gmail.com (A. Norouzi), s.soltanmoradi@yahoo.com (S. Soltanmoradi), naserian_mahshid@yahoo.com (M. Nasserian), shahi.a833@gmail.com (M. Shahi)

Abstract: In this paper, two useful techniques of Dynamic Threshold Voltage MOSFET (DTMOS) and Positive Feedback Amplifier (PFA) are investigated separately and are applied simultaneously on a Class-AB Amplifier in the 180 nm CMOS technology. In the first proposed technique, Simulation results show that operating voltage can be limited to ±0.5 V in which the voltage gain and bandwidth are 52.6 dB and 103.51 MHz, respectively. In the second proposed technique, the power consumption is reduced more than 50%, the open-loop gain is enhanced 47% and Common Mode Rejection Ratio (CMRR) improves to 86.5 dB. By applying combination of these two techniques for designing the amplifier, CMRR increases to 92.1 dB and the power consumption reduces to 97 µW with the bandwidth of 59.12 MHz.

Keywords: Class-AB Amplifier, DTMOS, FVF, PFA.

1. Introduction

One of the most important parts of an electronic measurement system is its input section, (i.e., sensor). Detection of the desired input quantity and converting to an electrical signal is the most important part of such a system. Nowadays, due to decreasing supply voltage and power consumption of these circuits, achieving the minimum size and increasing lifetime of batteries, results in continuing designing electrical circuits with this approach [1].

New techniques of generation of energy from physical quantities such as light, temperature difference between two points, mechanical vibration frequency of components, moving variation, speed of rotation, etc in nanometer or micrometer scale leads to replacing batteries and power supplies with these new kinds of energy generators. Since the power provided by these generators is not high, use of low-power and low-voltage circuits is very important [2].

Hence, for detection of the desired quantities, designing operational amplifier (Op-Amp) and electronic circuits should be done with the goal of achieving high voltage gain in low operating voltage and power consumption. Applications such as portable and low power measurement systems, especially communication and wireless applications, exhibit importance of usage of low voltage and low power consumption circuits these systems [3].

In [4], Sarbishaei et al. use positive feedback for increasing the voltage gain. However, they couldn't achieve low operating voltage and power supply. Also, in [5] Lopez-Martin et al., employ combination of local common mode feedback (LCMFB) and super class-AB OTA (operational transconductance amplifier) for increasing voltage gain and bandwidth in three circuits with different types of supply circuits in the internal stage.

Class-AB operational amplifier is one of the circuits that can be used in low voltage and low power consumption circuits [6]. In order to achieve this, in this paper two techniques of dynamic threshold voltage MOSFET (DTMOS) and positive feedback amplifier (PFA) are used separately and their combination that is the main idea of this paper are applied on a class-AB operational amplifier. In continue, simulation results are compared with those reported in [4] and [5].

2. Theorem

Detecting and measuring required quantities and also minimizing the operating supply voltage leads to usage of energy provided by physical components in nanometers sizes. Fig. 1 shows the block diagram of a wireless sensor [3]. In this structure the required power for driving internal parts is provided by an electrical generator and is managed for usage of internal components of the sensor. Structure of the detection part and converting the input quantities to an electrical signal should be designed very well to operate in the following conditions:

- When achieved voltage from energy generator is not enough.
- When size of wireless sensor due to specific applications and low power should be decreased.

Hence using techniques for decreasing supply voltage is useful that are discussed in this paper later.

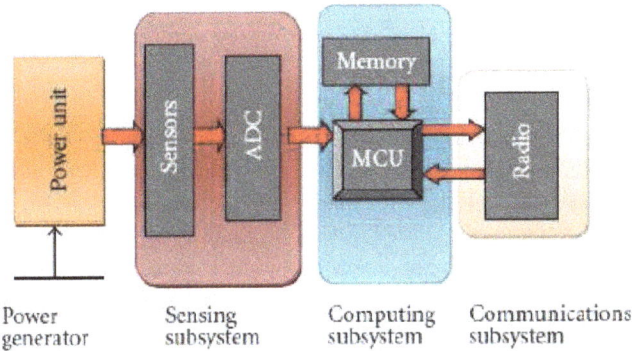

Fig. 1. Internal structure of a wireless sensor[3].

2.1. DTMOS Technique

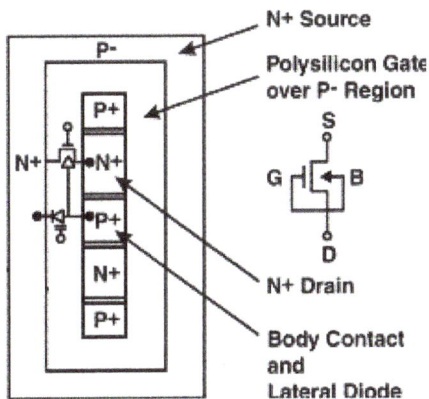

Fig. 2. DTMOS internal structure in a CMOS Transistor [8].

DTMOS Technique has been proposed with the goal of decreasing V_{th} and off-state leakage current. However, it can be used when transistor is in the saturation region. As seen in Fig. 2, in this technique the gate and body terminals are connected together and the input signal is applied to these terminals. If $V_{BS} > 0$, V_{th} decreases according to (1).

$$V_{th} = V_{th0} + \lambda(\sqrt{|2\phi_F - V_{BS}|} - \sqrt{|2\phi_F|})$$ (1)

in which V_{BS} is source-body voltage of the transistor, V_{th0} is the threshold voltage when $V_{BS} = 0$, λ is the body effect coefficient, ϕ_F is the Fermi potential (that has a value between 0.3 to 0.4 V) [7]. In Fig. 3, the input signal is applied to the gate and body terminals of M_1 and M_2 which are biased in DTMOS structure.

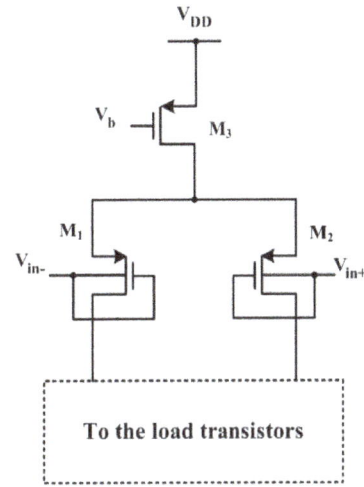

Fig. 3. DTMOS structure used in an Op-Amp input stage

Since lower voltage (compared to the case input signals are applied to the gates of transistors) is needed for driving transistors, we expect that input voltage range increases in the common mode and also the bias current decreases, resulting in saving power [8-9].

2.2. Positive Feedback Amplifier (PFA)

There are different techniques for increasing the output impedance and voltage gain such as cascading transistors, Gain Boosting and Positive Feedback Amplifier (PFA). In the first and second approaches, at least extra four transistors are needed and voltage swing is limited. Hence, these techniques cannot be employed in low voltage applications [10].

In the PFA technique, due to achieving negative conductance coefficient and also high output impedance, we can expect maximum output voltage swing and high DC voltage gain [11].

Fig. 4 shows the structure of a differential PFA. Equivalent circuit for this circuit, is presented in Fig. 5. In this schematic, feedback factor is g_{m2} and the closed-loop gain can calculated as following:

$$A_{CL} = (\frac{vo}{vi})_{CL} = g_{mi}A_{OL} \Big/ (1 - \beta A_{OL})$$ (2)

It is obvious that if $\beta A_{OL} \approx 1$, a high gain is achieved. This condition doesn't cause instability due to the negative feedback used in the Op-Amp [4].

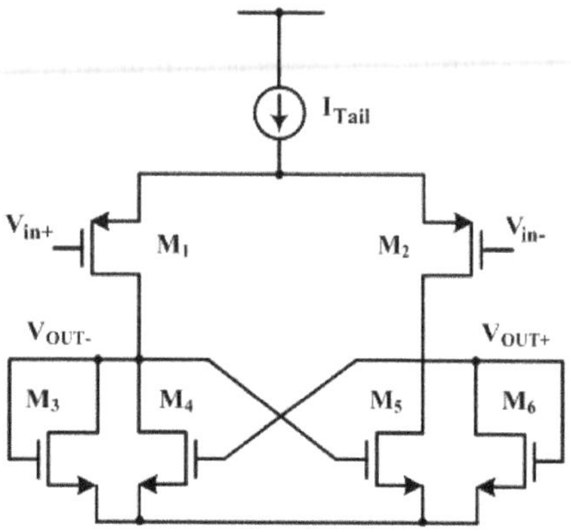

Fig. 4. *Schematic of a differential PFA*

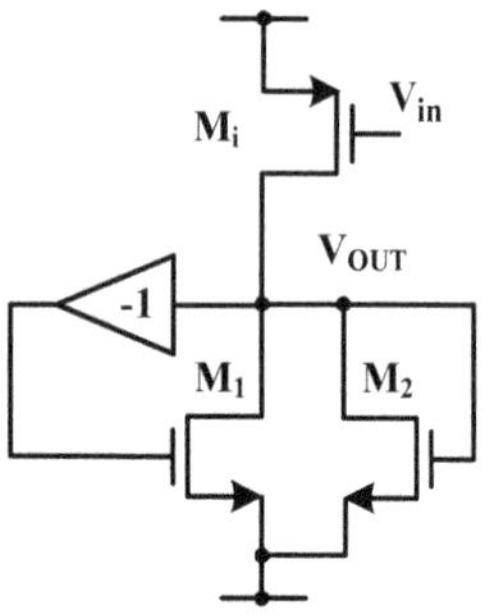

Fig. 5. *Half equivalent circuit of a differential PFA*

To achieve high gain, this factor can be used as following:

$$\beta A_{OL} = g_{m1} + g_{o1} + g_{o2} + g_{oi} \qquad (3)$$

It is worth mentioning that using this technique, high gain can be achieved without using cascading technique or cascading amplifier stages. Also, there is no need to the common mode feedback (CMFB) due to combination of M_4 and M_5.

2.3. Flipped Voltage Follower (FVF)

A simple Flipped Voltage Follower (FVF) configuration is seen in Fig. 6. This structure can be used as a voltage shifter or buffer. In this structure, the output current depends on the current passing through the main input transistor (M_{FVF}). Hence, by approaching to $V_{SG(MFVF)}$, the gain becomes 1 [4].

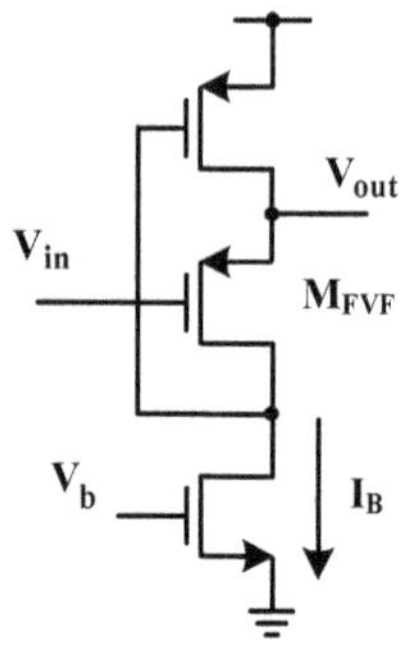

Fig. 6. *Simple FVF structure*

The most important feature of this structure is its low output impedance (20 to 100 Ω) that makes it suitable for use in negative feedback structures [5]. This structure can be employed as the input stage in an Op-Amp that is shown in Fig. 7. In this circuit, current passing through M_2 increases when $V_i > V_{i+}$. In this case, the gate voltage of M_2 decreases due to reducing V_{i+}, while its source voltage increases from another path. If this structure used in the input stage, it can increase slow rate of the amplifier.

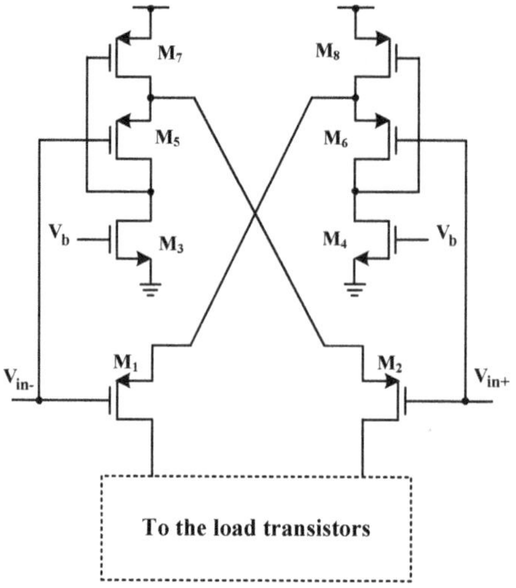

Fig. 7. *FVF structure of an Op-Amp input stage*

2.4. Super Class-AB OTA

Fig. 8 shows a Class-AB OTA. In this circuit, LCMFB is used for current stabilizing in main transistors. In order to increase bandwidth of the circuit and slow rate, two resistors are used in this structure [2]. Combination of M_4 and M_5 cause these transistors to operate in the ohmic region. Slow rate nan calculated by SR = I_{Bias}/C_L. For increasing the solw rate, we need to increase I_{Bias} which leads to more power dissipation. If this current is choosen low, the static power decreases. Size of R_1 and R_2 should be selected such that suitable phase margin is achieved. Using this technique, Autors in [4] can improve the slow rate 3 times compared to a conventional OTA.

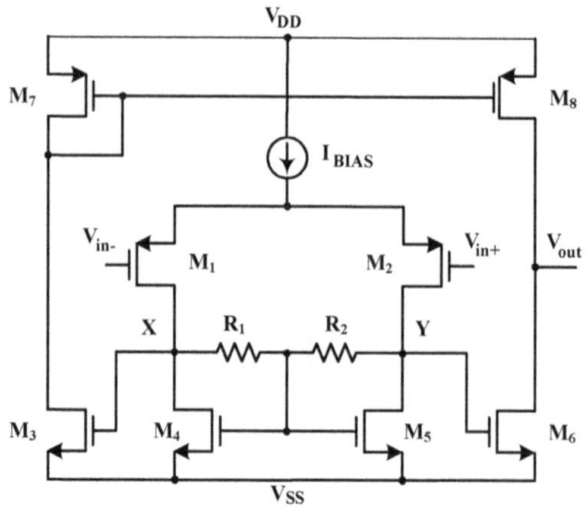

Fig. 8. *Class-AB OTA circuit*

When common mode (CM) input is applied in this circuit, current passing through resistors becomes zero and voltage of M_3 nad M_6 becomes equal. In this condition, I_{Bias} is devided between M_1 and M_2 equal. If differntial input is applied to the circuit, the currents of resistors and hence voltage of X and Y nodes become different.

When $V_i=V_{(i+)}-V_{(i-)}$ is applied and $I_d=I_1-I_2$ is generated, a current is produced throughout resistors. Maximum difference between X and Y nodes can be calculated by:

$$\Delta V_{GS}{}^{MAX} = RI_{Bias}/2 \qquad (4)$$

Calculated voltage difference in the above equation produces maximum voltage swing in current of the output stage. DC open-loop gain can be writhen as:

$$A_{OpenLoop} = g_{m2}R_Y g_{m6}R_{out} \qquad (5)$$

in which R_Y is impedance of the Y node and R_{out} is impedance of the output node. Also, R_Y can be expressed by:

$$RY = R\|r_{o5}\|r_{o2} \qquad (6)$$

Furthermore, resistor R can determine the maximum output current that is:

$$I_{Out}{}^{Max} = \beta(V_{DS4,5} + \Delta V_{GS}{}^{MAX})^2 \qquad (7)$$

It is worth mentioning that in the differential mode, gates of M_4 and M_5 are grounded and parasitic capacitances don't affect X and Y nodes. In Fig. 8, generated differential current is proportional to $V_{id}{}^2$[11-12]. Also the output current is proportional to $I_d{}^2$ and hence, it is proportional to $V_{id}{}^2$. Therefore the proposed circuit is a super class-AB OTA [13].

In this paper, two useful techniques of PFA and DTMOS are investigated separately and then are applied simultaneously on the implemented super class-AB OTA in the 0.18 μm CMOS technology. In order to make comparison fair, figure of merit (FOM) is calculated as follows:

$$FOM = \frac{A_{OpenLoop}.GBW}{P_d}\cdot\frac{1}{V_{Supply}} \qquad (8)$$

in which GWB is production of gain and bandwidth and $A_{OpenLoop}$ is open-loop gain[14].

3. Circuit Implementation

For simulating the proposed structure, first a differential class-AB OTA, -in which FVF technique is used- is implemented in the 0.18 μm CMOS technology using Hspice software. Then DTMOS technique is applied to the first stage. In this case, operating voltage is limited to ±0.5 V. Bias currents are selected 2 μA. Then without using DTMOS technique, M_4 and M_5 transistors along with R_1 and R_2 are replaced by PFA and simulation is carried out again. In this case supply voltage is ±1 V. The proposed technique in this paper is combination of two stated techniques. Using the proposed structure supply voltage can be limited to ±0.5 V and many circuit parameters are improved. Fig. 9 shows circuit of the proposed structure.

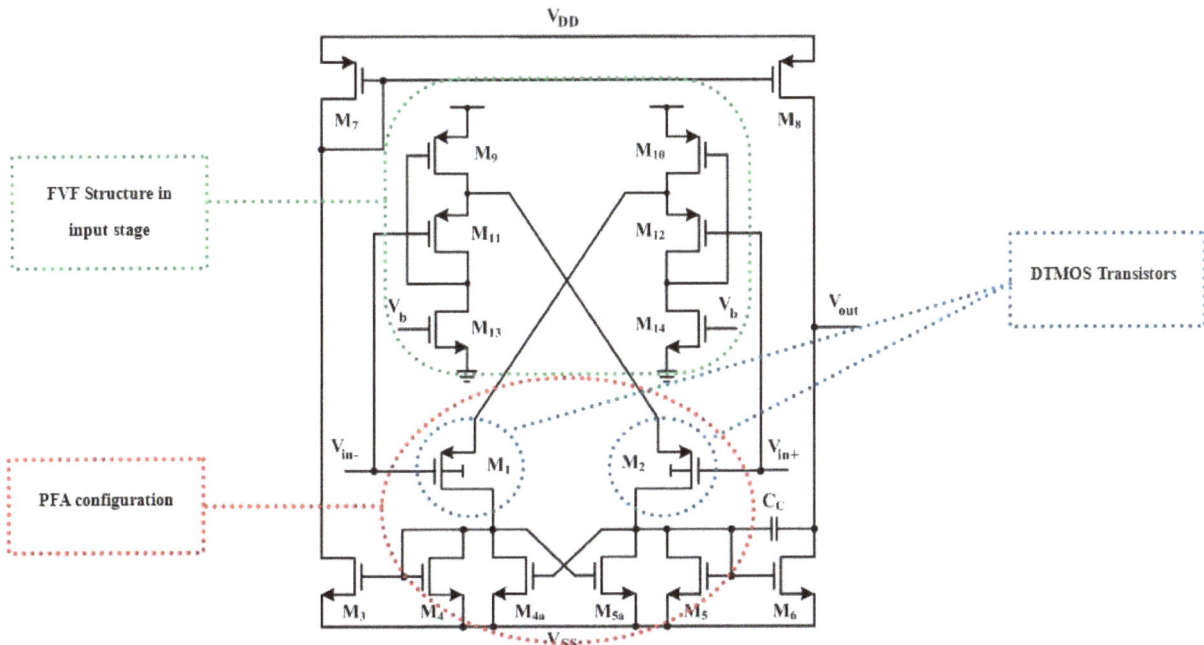

Fig. 9. *Schematic of the proposed circuit. Structure of the super class-AB OTA in which DTMOS and PFA techniques are used.*

4. Simulation Results

Circuits explained in the previouse section are simulated

and their frequency response are demonstrated in Fig. 10. The voltage gain and bandwidth of the first circuit are 52.6 dB and 103.51 MHz.

When PFA technique is used for implementing the amplifier, the power consumption reduced more than 50%, the voltage gain improves more than 47%, and CMRR enhances to 86.5 dD. The open-loop gain and bandwidth of the proposed circuit are 77.4 dB and 59.12 MHz, respectively in which the supply voltage is ±0.5 V. Table 1 shows simulation results of DTMOS, PFA and the proposed structures and also techniques proposed in [4] and [5] (in which three bias float circuits (A), (B), and (C) are used). As observed in Table 1, the proposed circuit has the best CMRR, least power and the best FOM.

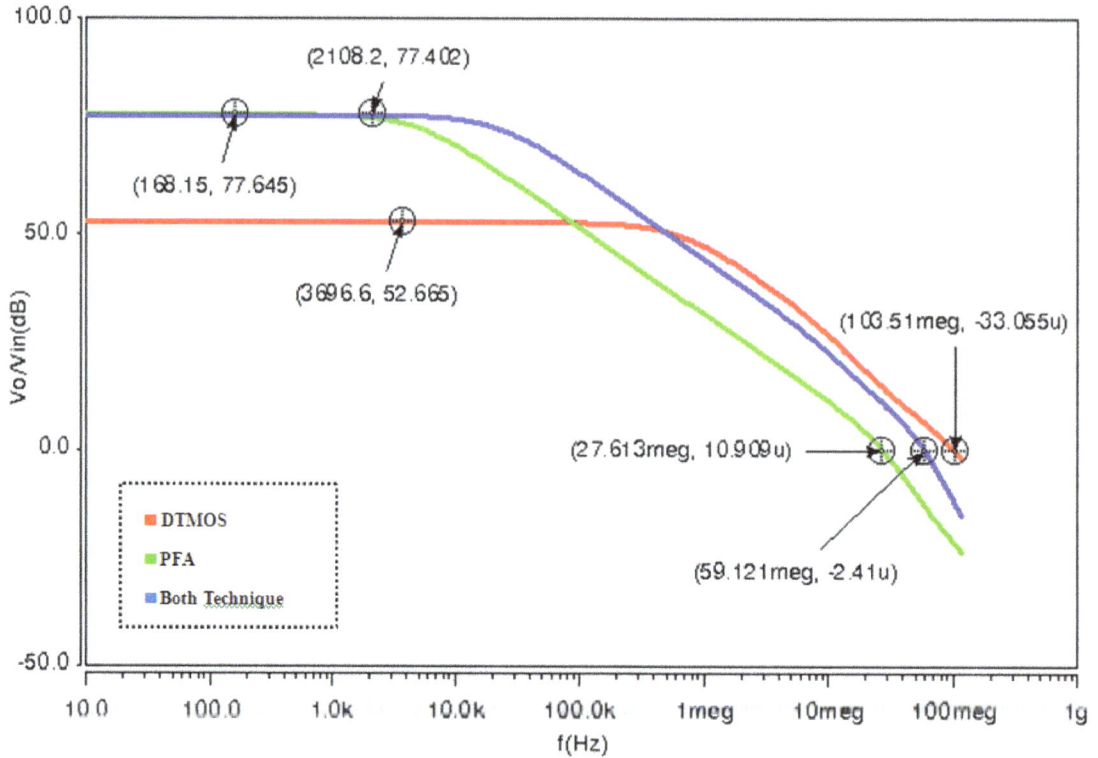

Fig. 10. *Voltage gain cuves of DTMOS, PFA and the proposed technique.*

Table 1. *Comparison of different techniques parameters.*

Parameter	[4]	[5] (A)	[5] (B)	[5] (C)	Simulated [5]-(A) circuit in this paper	This Work		
						DTMOS	PFA	Proposed
Technology (μm)	0.18	0.5	0.5	0.5	0.18	0.18	0.18	0.18
V_{Supply} (V)	1.8	±1	±1	±1	±1	±0.5	±1	±0.5
I_{Bias} (μA)	--	10	10	10	10	2	2	2
GBW (MHz)	69	0.725	0.41	0.47	22.1	103.51	27.6	59.12
Gain (dB)	83.7	43	37.5	37.5	50	52.6	77.64	77.4
Phase Margin (°)	87	90	90	90	90	89	85	88
CMRR (dB)	--	68	70	69	52.1	63.2	86.5	92.1
Output Swing (V)	1.6	--	--	--	±0.85	±0.45	±0.85	±0.4
Load Capacitor (pF)	2	80	80	80	5	2	2	2
Output Noise (μV$_{rms}$)	--	230	230	252	6.5	8	12	9.5
Power (μW)	450	120	120	140	280	248	112	97
PSRR(+ , -) (dB)	--	55-58	50-53	57-46	86-81	98-80	80-84	98-86
Slew Rate +/- (V/μS)	226	100	92	42	203	148.4	450	105
	---	-78	-76	-80	-86.5	-50	-177	-44.5
FOM*	7.12	0.13	0.064	0.063	1.975	21.95	9.56	47.17

* FOM : Figure of Merit

5. Discussion and Conclusion

In this paper, two practical techniques of DTMOS and PFA are investigated separately and are applied simultaneously on a Class-AB Amplifier in the 180 nm CMOS technology. In the first proposed technique, Simulation results show that operating voltage can be limited to ±0.5 V in which the voltage gain and bandwidth are 52.6 dB and 103.51 MHz, respectively. In the second proposed technique, the power consumption is reduced more than 50%, the open-loop gain is enhanced 47% and CMRR improves to 86.5 dB.

By applying combination of two techniques for designing

the amplifier, CMRR increases to 92.1 dB and the power consumption reduces to 97 μW with the bandwidth of 59.12 MHz.

Simulation result provided in the previous section proves that the proposed circuit improves many parameters of the amplifier. Decreasing operating voltage, power consumption and enhancing parameters including voltage gain, bandwidth and CMRR make the proposed structure a good choice for implementing high gain amplifiers with minimum voltage supply and also low-power portable and wireless applications.

References

[1] P. K. Sinha, A. Vikram and Dr. K. S. Yadav, "Design Of Two Stage CMOS Op-Amp With Low Power And High Slew Rate", International Journal of Engineering Research & Technology, Vol. 1 Issue 8, ISSN: 2278-0181, 2012.

[2] L. W. Zhong, Zh. Guang, Y. Ya, W. Sihong and C. Pan, "Progress in nanogenerators for portable electronics", Materials today from Elsevier Journals, Vol. 15, No. 12, pp. 532-543, 2012.

[3] A. Nechibvute, A. Chawanda and P. Luhanga, "Piezoelectric Energy Harvesting Devices: An Alternative Energy Source forWireless Sensors", Hindawi Publishing Corporation, Smart Materials Research, Article ID 853481, 13 pages, DOI :10.1155/2012/853481, 2012.

[4] I. Sarbishaci, T. K. Toosi, E. Z. Tabasy and R. Lotfi, "A High-Gain High-Speed Low-Power Class-AB Operational Amplifier", 48th IEEE International Midwest Symposium on Circuits and Systems, Vol. 1, pp. 271 - 274 , 2005.

[5] J. López-Martín, S. Baswa, J. Ramirez-Angulo and R. G. Carvajal, "Low-Voltage Super Class AB CMOS OTA Cells With Very High Slew Rate and Power Efficiency", IEEE Journal Of Solid-State Circuits, Vol. 40, No. 5, pp. 1068-1077, 2005.

[6] S. Baswa, M. Bikumandla, J. Ramírez-Angulo, A. J. López-Martín, R. G. Carvajal and G. Ducoudray-Acevedo, "Low-Voltage Low-Power Super Class-AB CMOS Op-Amp with Rail-to-Rail Input/Output Swing", 5th IEEE International Caracas Conference on Devices, Circuits and Systems, Vol. 1, pp. 83-86, 2004.

[7] Lasanen and Kimmo, "Integrated analogue CMOS circuits and structures for heart rate detectors and other low-voltage, low-power applications", University of Oulu, Oulu, Finland, 2011.

[8] S. Izadpanah-Tous, M. Behroozi and V. Asadpoor, "Design of 0.4 V operational amplifier using low-power techniques", Majlesi Journal of Telecommunication Devices, Vol. 2, No. 1, pp. 145-149, 2012.

[9] L. Qiang, T. Kuo Hwi Roy, H. Teo Tee and S. Rajinder, "A 1-V 36-μW Low-Noise Adaptive Interface IC for Portable Biomedical Applications", IEEE European Solid-State Device Research Conference, 11-13 September 2007.

[10] K. Bult and G. Geelen, "A Fast-Settling CMOS Op Amp for SC Circuits with 90-dB DC Gain", IEEE J. Solid-state Circuits, Vol. 25, No. 6, pp. 1379-1384, 1990.

[11] S. Voldman, D. Hui, D. Young, R. Williams, D. Dreps, J. Howard, M. Sherony, P. Assaderaghi and G. Shahidi, "Silicon-on-insulator dynamic threshold ESD networks and active clamp circuitry", Electrical Overstress/Electrostatic Discharge Symposium, pp. 29-40, 2000.

[12] J. Ramirez-Angulo, R. G. Carvajal, A. Torralba, J. Galan, A. P. Vega-Leal and J. Tombs, "The Flipped Voltage Follower: A useful cell for low-voltage low-power circuit design", in Proc. ISCAS 2002, Scottsdale, AZ, pp. II 615-618, 2002.

[13] M. M. Amourah and R. L. Geiger, "A High Gain Strategy With Positive-Feedback Gain Enhancement Technique", IEEE International Symposium on Circuits and Systems, Vol. 1, pp. 631- 634, 2001.

[14] E. Kargaran, M. Sawan, Kh. Mafinezhad and H. Nabovati, "Design of 0.4V, 386nW OTA Using DTMOS Technique for Biomedical Applications", 55th IEEE International Midwest Symposium on Circuits and Systems, pp. 270–273, 2012.

34

Cost effective method to locate the vulnerable nodes of circuits against the electrical fast transients

Behnam Gholamrezazadeh Family, Vahid Hamiyaty Vaghef, Maryam Shabro

Communication Department, Niroo Research Institute, Tehran, Iran

Email address:

bfamily@nri.ac.ir (B.G. Family), vvaghef@nri.ac.ir (V. H. Vaghef), mshabro@nri.ac.ir (M. Shabro)

Abstract: Electrical Fast Transients (EFT) pulses may cause a large number of circuits to fail. Switching power supplies, inductors, contact relays, and high voltage switches electrically or electromagnetically strike the data, address and control lines of processors, memory elements, or even analog parts and leads to soft or permanent errors. In general, compliant test is accomplished to address the susceptibility of circuits to EFT pulses. However, extremely high cost of these tests encounters the compliant test with difficulty. As a result, in this paper a low-cost EFT simulator circuit is proposed to locate the vulnerable parts of the circuit. The approach is easily applicable to any point of any circuit. The design is performed such that the proposed circuit does not damage the Equipment Under Test (EUT). Experimental results show that the proposed approach effectively detects the vulnerable circuits and practically has been used at the design phase of the DTPS-8C device.

Keywords: Electromagnetic Compatibility, EFT/B, Switching, Protection Devices

1. Introduction

Interaction between several electronic equipments significantly increases the vulnerability of current electric equipments. Thus, electromagnetic compatibility (EMC) is an important requirement to achieve the reliable operation.

EMC is defined as the proper function of equipment in its electromagnetic environment without introducing intolerable disturbances to neighbor equipment. On the other hand, immunity is defined as the ability of a device, equipment or system to work without performance degradation at the presence of an electromagnetic disturbance. Electrical Fast Transient/ Burst (EFT/B) is a major test to measure the immunity of equipments, as described in IEC 61000-4-4 (2012).

For industrial devices, taking all guidelines in hardware and software components into account is an essential step. EMI reduction and EMC improvement will be achieved using protection devices such as shields, grounding, filters, isolators, chocks, ferrite beads, varistors, transient voltage suppressors (TVS) and proper PCB layout design[2]. Typical EFT/B failures are due to an inadequate or unavailable interface for ground reference. Therefore, designers should try to use appropriate ground reference and protection at the incoming or outgoing ports. In addition, to achieve less susceptible systems to EFT pulses, the software layer should be taken into account. Software problems include the data loss, system malfunction, retrieving an incorrect data stored in memory, or executing an incorrect portion of the memory. Software should design to detect and correct errors before a device failure. Reading the critical inputs several times, verifying the data contents, counters, and pointers, using serial protocols instead of parallel ones, watchdog timers, etc are useful approaches to avoid software malfunction.

In general, resolving the problems caused by the presence of several kV pulses is a sophisticated process. Theoretical considerations of immunity against disturbances and radiation are effective, but insufficient. Despite an electronic device's seemingly complete design, sensitive parts of the circuit against EFT/B may still exist and performing standard tests on the entire system may not exactly identify the sensitive parts of the circuit (for example the reset pin of a microcontroller [3]). These sensitive parts may lead the system to fail an immunity test. However, Inspecting the susceptible points is difficult in general, and it requires several trial and error tests which are very time consuming and costly. Designers need low-cost and non-destructive tests to successfully transition from their model designs to finished industrial designs. Thus, the use of simulations during the system and board design phases is recommended for extracting the critical points.

This paper presents a simple and inexpensive way for detecting the critical parts of a circuit with respect to EFT/B pulses, as a fundamental evaluation of the system. The proposed method is low-cost while it benefits from non-destructivity and easy application to internal nodes of the EUT. Thus, it reduces the cost of reliable design significantly.

The remainder of the paper organizes as follows. The electrical fast transient pulses are described in section 2. The EFT pulse generator is proposed in section 3 and the conclusion is presented in section 4.

2. Electrical Fast Transients/ Burst

The EFT/B test aims to simulate the disturbances created by a 'showering arc' at the contacts of an ordinary AC mains switch or relay contacts as it opens. This test evaluates the immunity of electrical and electronic equipment when subjected to EFT/B on supply, signal, control and earth ports. This phenomenon is modeled as Fig. 1. The characteristic of burst generator, test setup, test verification, and test procedure are described in IEC 61000-4-4 standard. It has fast rise time, low energy, high voltage, and consists of a single unidirectional impulse repeated at a 5kHz(100kHz) in bursts lasting 15(0.75) milliseconds each, with three bursts per second. Different voltage test levels are defined in standard as Table 1. The test severity level and frequency are selected according the equipment installation environment.

3. Design and Implementation of a Cost Effective EFT Pulse Generator

The simplified circuit diagram of the EFT/B generator is given in Fig. 2. In this circuit, a high voltage source charges a Cc capacitor and the charging and discharging of capacitors is controlled by a Switch. The desired pattern is produced, as shown in Fig. 1.

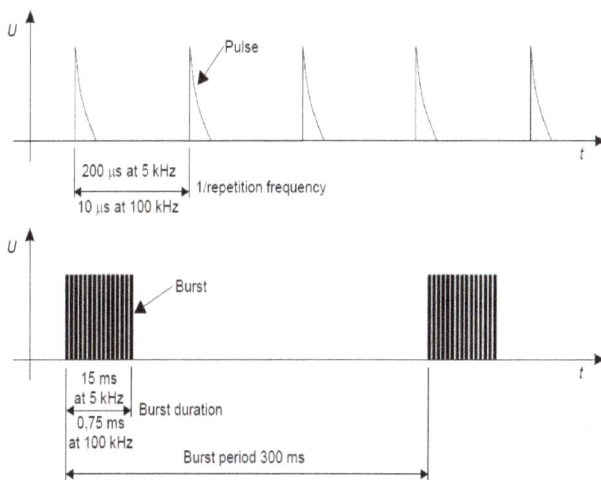

Figure 1. Representation of an EFT/B according to IEC 61000-4-4 [1].

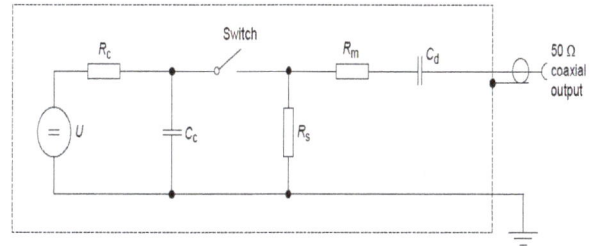

Figure 2. Simplified circuit diagram of an EFT/B according to IEC 61000-4-4.

Table 1. Test voltage and repetition frequency of the impulses according to IEC 61000-4-4 [1].

Level	Power ports, earth port (PE)		Signal and control ports	
	Voltage peak kV	Repetition frequency kHz	Voltage peak kV	Repetition frequency kHz
1	0.5	5 or 100	0.25	5 or 100
2	1	5 or 100	0.5	5 or 100
3	2	5 or 100	1	5 or 100
4	4	5 or 100	2	5 or 100
X	special	special	special	special

The use of 5kHz repetition frequency is traditional; however, 100kHz is closer to reality. Product committees should determine which frequencies are relevant for specific products or product types.

"X" can be any level, above, below or in between the others. The level shall be is in specified in the dedicated equipment specification.

This study explains the design and implementation of an EFT pulse generator in order to determine the critical nodes of an electronic circuit. The block diagram of the proposed circuit is shown in Fig. 3.

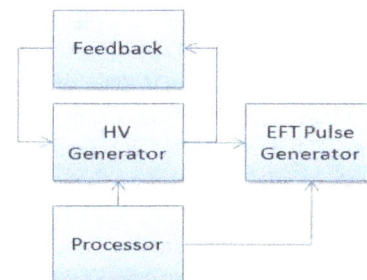

Figure 3. Block diagram of the designed EFT/B generator

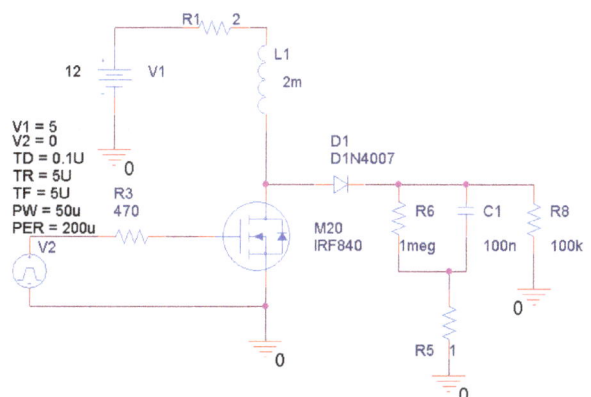

Figure 4. High voltage generator circuit.

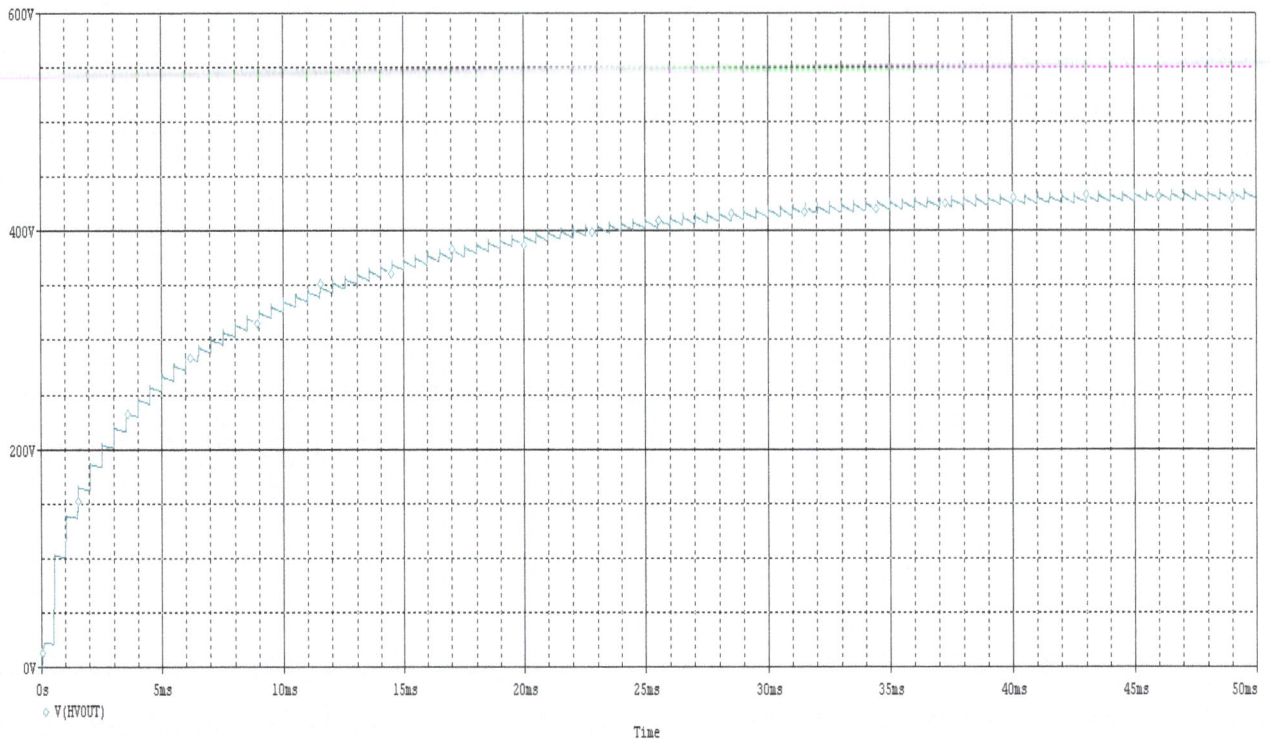

Figure 5. *Output voltage of Fig 4.*

As mentioned before, a high voltage circuit is required to produce EFT pulses [4]. Since these pulses have high voltage with low energy, the low-cost circuit in Fig. 4 is proposed to meet the requirements. The output voltage of this circuit is simulated by HSPICE software as shown in Fig. 5. This graph shows the voltage across the C1 capacitor. This circuit produces voltages higher than 400 V in less than 50ms. The main advantage of this circuit is its ability to generate high voltages using a voltage of 12 V. This makes the circuit portable. The power consumption of the circuit is also significantly low. The portability of this circuit allows users to test different parts of a large system easily.

To generate higher voltages, MOSFET transistors with higher breakdown voltages can be used. For example: FS10SM (with 800V breakdown) or IXZR08N120B (with 1200V breakdown).

To control the output voltage variations due to the load, the output voltage is sampled and accordingly varies the input frequency or duty cycle of the square wave. The efficiency of this circuit is low; however, using the feedback circuit shown in Fig. 6 the output voltage is controlled with higher efficiency.

Using the feedback circuit, the output voltage varies according to the input pulse shape which prevents excessive output voltage. The output voltage level is set by R14, R15 and the Zener diode D2. Several voltage levels can be selected using this circuit.

A simple circuit shown in Fig. 7 is used to generate the EFT pulses. The transistor pulsed on and off simultaneously with input pulses and discharges of high voltage stored in the C2 capacitor in the minimum possible time. The output of the

circuit is shown in Fig. 8. This output can be applied to the EUT using another capacitor.

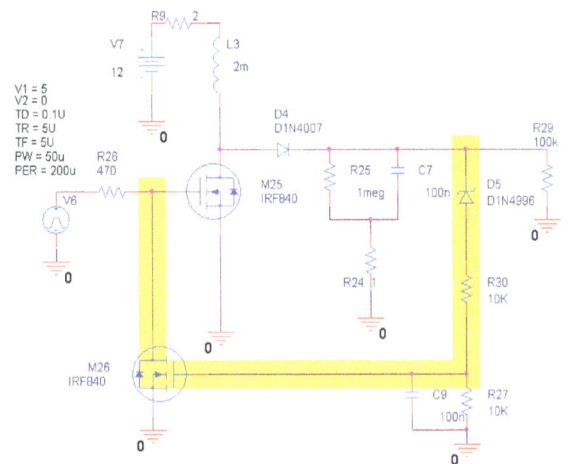

Figure 6. *Feedback for voltage control.*

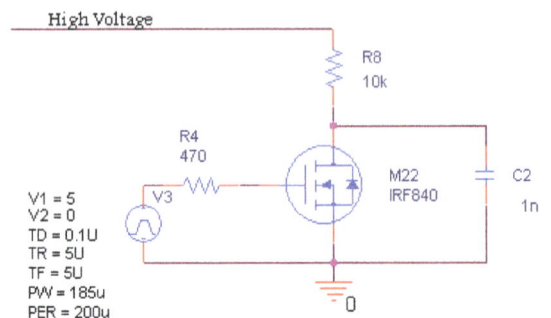

Figure 7. *EFT pulse generator circuit.*

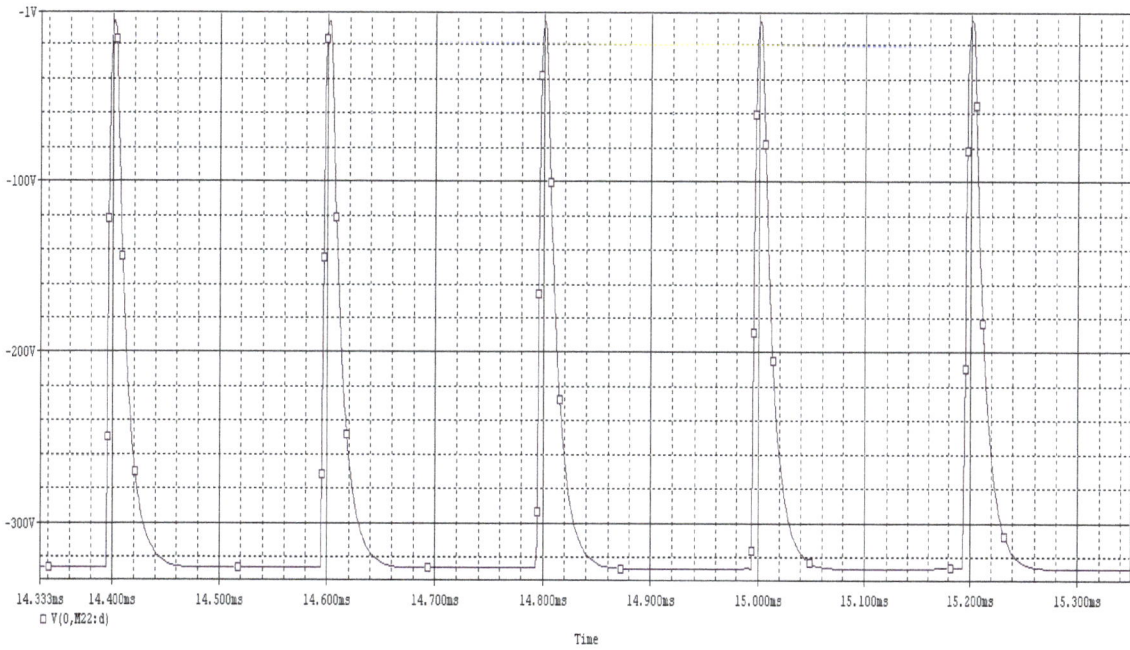

Figure 8. *EFT pulses of Fig 7.*

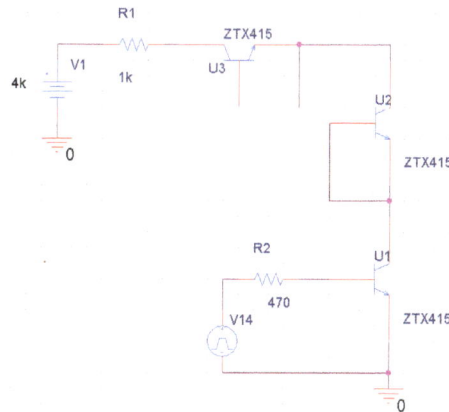

Figure 9. *Series of avalanche transistors.*

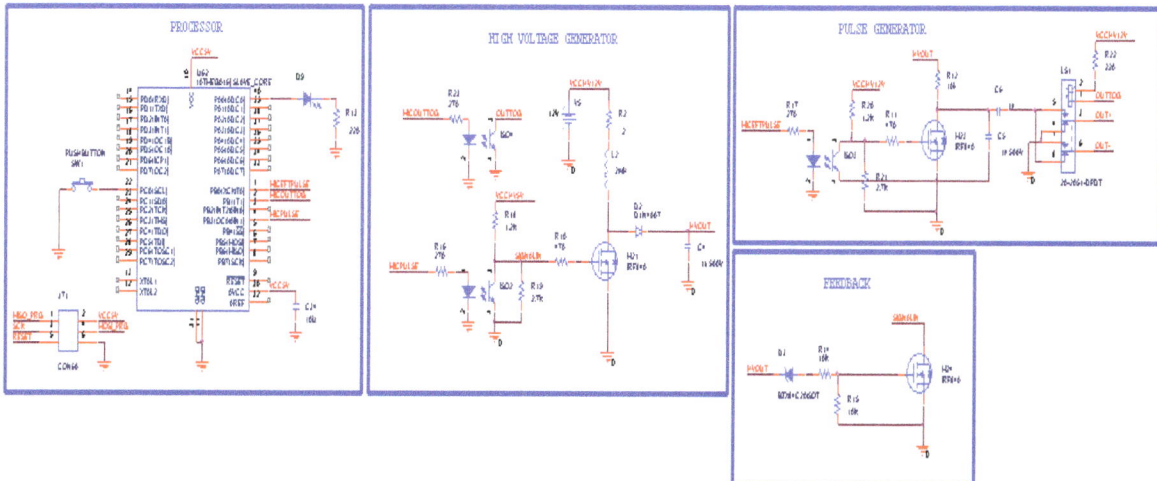

Figure 10. *Schematic of designed EFT simulator circuit.*

Figure 11. Designed EFT generator board.

The output pulses generated by this circuit are not fully compatible to the pulses described in the standard IEC61000-4-4 because of the capacitor's internal series resistance and the transistor's on delay time. But they can reveal the sensitive parts of the circuit under testing to standard EFT pulses.

It is also possible to use avalanche transistors connected in series and operated close to their avalanche breakdown as shown in Fig. 9 for switching very high voltage pulses [5]. For example, a ZTX415 transistor is suitable for this purpose.

In order to generate the input pulse for switching a MOSFET transistor, a microcontroller is used. The microcontroller output pin must be isolated from a high voltage generator; otherwise the distortion produced by this circuit prevents the proper operation of the microcontroller. Figures 10 and 11 show the schematic and layout of the designed circuit and the EFT pulses generated by this simulator are shown in Fig. 12 and Fig. 13. The impulses are repeated at 5 kHz in bursts which each one lasts 15ms and the period of bursts is 300ms.

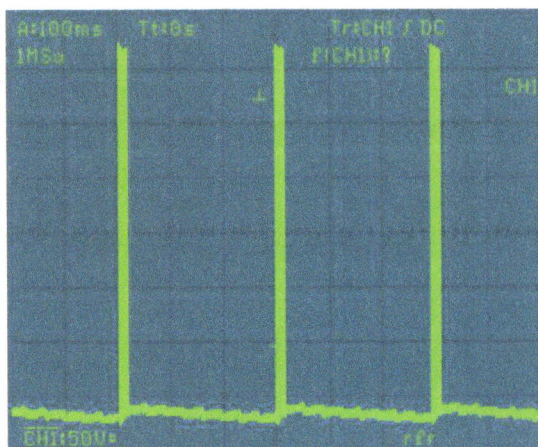

Figure 12. EFT pulses generated by simulator.

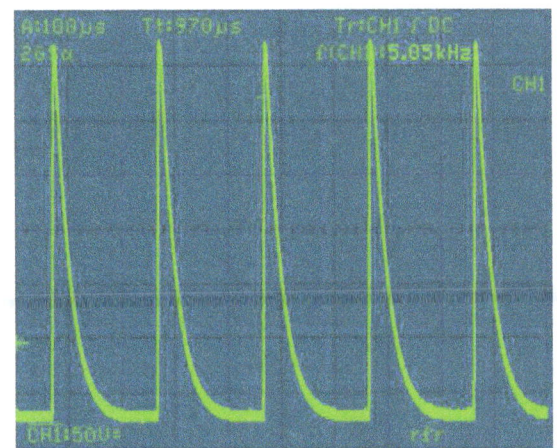

Figure 13. Waveform of a single burst.

According to IEC 61000-4-4 the EFT pulse is applied with a coupling network to the power ports and with a coupling clamp to input/output cables of the EUT. Due to the use of protective components in these ports, the amount of EFT/B pulses energy will decrease. Therefore, if the output of the proposed method with lower voltage than the standard EFT/B pulses is applied to the internal nodes of the circuit under testing, it approximately simulates the effect of a standard EFT/B pulse. Therefore, the sensitive parts of the circuit under test are easily identified using this method. Due to the lower energy of the proposed circuit, it is directly applicable to all components on the circuit during its normal operation while the circuit reaction is studying.

4. Discussion and Conclusion

A novel and low-cost circuit to find the highly vulnerable nodes of a circuit is proposed. The circuit is simple and efficiently locates these nodes. In addition to cost reduction, the proposed approach easily is used in design phase and approving the compliant tests. Experimental results of the proposed method to find the vulnerable nodes of a digital teleprotection system against EFT pulses, confirms its effectiveness. Furthermore, the proposed approach is non-destructive and easily applicable to internal node of the EUT.

Acknowledgements

This work was supported by the Niroo Research Institute (NRI), Iran and PKG under Grant No. JCMPN02.

References

[1] IEC61000-4-4, "Electromagnetic compatibility (EMC), Part 4-4: Testing & measurement techniques – Electrical fast transient/ Burst immunity test", edition 3.0, 2012.

[2] J. R. Barnes, "Designing Electronic Systems for ESD Immunity", Conformity magazine, Feb. 2003. Online at: http://www.dbicorporation.com/esd-art3.pdf.

[3] S. Hyung, J. Young, "Analysis of Coupling Mechanism and Solution for EFT Noise on Semiconductor Device Level", Proceedings of the International Conference on Electromagnetic Interference and Compatibility, New Delhi, India, 6-8 Dec. 1999.

[4] E. Rogers, "Understanding Boost Power Stages in Switch mode Power Supplies", Application Report, Mixed Signal Products, TI Literature Number SLVA061, 1999. Online at: http://www.ti.com/lit/an/slva061/slva061.pdf.

[5] A. Tamuri, "Nanoseconds Switching for High Voltage Circuit Using Avalanche Transistors", Applied physics research, Vol. 1, No. 2, 2009. Online at: http://www.ccsenet.org/journal/index.php/apr/article/viewFile/3287/3601.

[6] B. Gh. Family, V. H. Vaghef, and M. Shabro, "A Method for Determining the Critical Parts of Electronic Circuits into EFT/B", 7thSASTech international conference, Bandar-Abbas, Iran, 2013.

Permissions

All chapters in this book were first published in JEEE by Science Publishing Group; hereby published with permission under the Creative Commons Attribution License or equivalent. Every chapter published in this book has been scrutinized by our experts. Their significance has been extensively debated. The topics covered herein carry significant findings which will fuel the growth of the discipline. They may even be implemented as practical applications or may be referred to as a beginning point for another development.

The contributors of this book come from diverse backgrounds, making this book a truly international effort. This book will bring forth new frontiers with its revolutionizing research information and detailed analysis of the nascent developments around the world.

We would like to thank all the contributing authors for lending their expertise to make the book truly unique. They have played a crucial role in the development of this book. Without their invaluable contributions this book wouldn't have been possible. They have made vital efforts to compile up to date information on the varied aspects of this subject to make this book a valuable addition to the collection of many professionals and students.

This book was conceptualized with the vision of imparting up-to-date information and advanced data in this field. To ensure the same, a matchless editorial board was set up. Every individual on the board went through rigorous rounds of assessment to prove their worth. After which they invested a large part of their time researching and compiling the most relevant data for our readers.

The editorial board has been involved in producing this book since its inception. They have spent rigorous hours researching and exploring the diverse topics which have resulted in the successful publishing of this book. They have passed on their knowledge of decades through this book. To expedite this challenging task, the publisher supported the team at every step. A small team of assistant editors was also appointed to further simplify the editing procedure and attain best results for the readers.

Apart from the editorial board, the designing team has also invested a significant amount of their time in understanding the subject and creating the most relevant covers. They scrutinized every image to scout for the most suitable representation of the subject and create an appropriate cover for the book.

The publishing team has been an ardent support to the editorial, designing and production team. Their endless efforts to recruit the best for this project, has resulted in the accomplishment of this book. They are a veteran in the field of academics and their pool of knowledge is as vast as their experience in printing. Their expertise and guidance has proved useful at every step. Their uncompromising quality standards have made this book an exceptional effort. Their encouragement from time to time has been an inspiration for everyone.

The publisher and the editorial board hope that this book will prove to be a valuable piece of knowledge for researchers, students, practitioners and scholars across the globe.

List of Contributors

Lung-Jen Wang and Chia-Tzu Shu
Dept. of Computer Science and Information Engineering, National Pingtung University, Pingtung, Taiwan, R. O. C.

Zhengxi Wei and Jinming Liang
School of Computer Science, Sichuan University of Science & Engineering, Zigong Sichuan 643000, PR China

Song Yaqi
Department of Computer Science, North China Electric Power University, Baoding, China

Hamada Esmaiel and Danchi Jiang
School of Engineering, University of Tasmania, Hobart, Australia

Vahid Hamiyaty Vaghef, Maryam Shabro and Behnam Gholamrezazadeh Family
Communication Department, Niroo Research Institute, Tehran, Iran

Lee Cheuk Wing and Zhong Jin
Department of Electrical and Electronic Engineering, The University of Hong Kong, Hong Kong, China

Zeinab Shamaee and Mohsen Mivehchy
Department of Engineering, University of Isfahan, Isfahan, Iran

Farzam Saeednia
Department of Electrical Engineering, Kazerun Branch, Islamic Azad University, Kazerun, Iran

Shapour Khorshidi
Air-Sea Science and Technology Academic Complex, Shiraz, Iran

Mohssen Masoumi
Department of Electrical Engineering, Jahrom Branch, Islamic Azad University, Jahrom, Iran

Esmail Kaffashi and Soheyla Amirian
Information Technology and Computer Engineering Department, Amirkabir University of Technology (Tehran Polytechnic),Tehran, Iran

Sahar Hematian Bojnourdy and Soheila Zahraee
Industrial department, University of Applied Science, SID Bojnourd Center, Bojnourd, Iran

H. Lale Zeynelgil
Department of Electrcal Eng., Electric & Electronics Faculty, Istanbul Technical University, Maslak Istanbul, Turkey

Esmail Kaffashi, Ahmad Madadi Mousavi, Forough Khademsadegh and Soheila Amirian
Information Technology and Computer Engineering Department, Amirkabir University of Technology (Tehran Polytechnic), Tehran, Iran

Hamid Rezaei Rahvard
Pardazeshgaran Gam Aval, Mashhad, Iran

Sahar Hemmatian Bojnordi
Industrial department, University of Applied Science, SID Bojnourd Center, Bojnourd, Iran

Navid Khalili Dizaji
Department of Mechatronics Engineering, Tabriz Branch, Islamic Azad University, Tabriz, Iran

Aidin Sakhvati
Department of Electrical Engineering, Tabriz Branch, Islamic Azad University, Tabriz, Iran

Seyed Hossein Hosseini
Department of Electrical & Computer Engineering, University of Tabriz, Tabriz, Iran

Seyyd Mahmood Mosavi
Khorasan Regional Electric Company, Mashhad, Iran Department of Electrical Engineering, College of Ebne Yamin, Sabzevar, Iran

Sakine Shirvaliloo
Department of Medical Physics, Iran University of Medical Sciences, Tehran, Iran

Young Researchers Club, Mahabad, Iran
Hale Kangarloo
Faculty of Science, IAV, Branch Urmia, Urmia, Iran

Nikhil Marriwala
Electronics & Communication Engineering Department, University Institute of Engineering and Technology, Kurukshetra University, Kurukshetra, India

Om Prakash Sahu
Electronics & Communication Engineering Department, National Institute of Technology, Kurukshetra, India

Anil Vohra
Electronics & Science Department, Kurukshetra University, Kurukshetra, India

Ahmed Mohammad Nahhas
ECED, CEL, Umm-Al-Qura University, Makkah, Saudi Arabia

Mohamed Ali Belaïd
ECED, CEL, Umm-Al-Qura University, Makkah, Saudi Arabia
SAGE-ENISo, University of Sousse, Tunisia

Momamed Masmoudi
GPM-UMR CNRS, University of Rouen, Saint Etienne du Rouvray, France

Chen Jie, Zhou Li, Hu Libin, Li Chenying and Cao Jingying
State Grid Jiangsu Electric Power Research Institute, Nanjing, China

Li Hongze
State Grid Jiangsu Electric Power Company, Nanjing, China

Nourdine Bounasla and Kamel Eddine Hemsas
Laboratory of Automatic, Department of Electrical Engineering, Ferhat Abbas University Setif-1, Sétif 19000, Algeria

Hacene Mellah
Department of Electrical Engineering, Hassiba Benbouali University Chlef, Chlef 02000, Algeria

Valentin Andrei, Horia Cucu and Lucian Petrică
Speech and Dialogue Research Laboratory, University "Politehnica" of Bucharest, Bucharest, Romania

Zahra Alizadeh
Sepehr Industrial Group Corporation, Ahwas, Iran

Fereshteh Kheirabadi
Islamic Azad University, Science and Research Branch of Khorasan Razavi, Neyshabour, Iran

Seyyed Reza Talebiyan
Imam Reza International University, Computer and Electronics School, Mashhad, Iran

Navid Maleki
Department of Electrical Engineering, Saveh Branch, Islamic Azad University, Saveh, Iran

Mohammad Reza Alizadeh Pahlavani and Iman Soltani
Faculty of Electrical Engineering, Malek-Ashtar University of Technology (MUT), Tehran, Iran

Laleh Haddadi and Seid Babak Mozafari
Department of Power Engineering, Science and Research Branch, Islamic Azad University of Tehran, Tehran, Iran

Abolfazl Pirayesh Neghab
Department of Power Engineering, Shahid Beheshti University, Tehran, Iran

Morteza Ghasem Salamroodi
Power Distribution Company of West of Mazandaran, Noshahr, Iran

Mohammad Seifi, Aliakbar Kargaran Erdechi and Ahmad Hajipour
Department of Electrical and Computer Engineering, Hakim Sabzevari University, Sabzevar, Iran

Syed Mohamad Reza Haji Mirzaie and Amin Nikbakht
Electrical Engineering Department, Khavaran Institute of Higher Education, Mashhad, Iran

Hodeiseh Gordan
Department of Electrical Engineering, Sobhan Institute of Higher Education, Neyshabur, Iran

Jalil Shirazi
Department of Electrical Engineering, Islamic Azad University, Gonabad Branch, Gonabad, Iran

Akhilesh Sharma
NERIST, Nirjuli, Arunachal Pradesh, India

Neeraj Kumar
DIET, Rishikesh, Uttarakhand, India

Gunjan Gupta
Invertis University, Bareilly, U.P., India

Mozhde Elahi
Department of Control Engineering, Gonabad University, Gonabad, Iran

Mahsa Gharaee
Department of Electronic Engineering, Gonabad University, Gonabad, Iran

Mona Sameri
Electrical and Communication Engineering department, Islamic Azad University - South Tehran Branch, Tehran, Iran

Farokh Hojat Kashani
Department of Electrical Engineering, Iran University of Science & Technology, Tehran, Iran

Maryam Shabro, Behnam Gholamrezazadeh Family and Vahid Hammiaty Vaghef
Communication Department, Niroo Research Institute, Tehran, Iran

Fatemeh Zahedi and Zahra Zahedi
Electrical and Computer Engineering, Shiraz University, Shiraz, Iran

Li Juan, Liang Naifeng, Zhao Puzhi, Zhou Erbiao and Li Xiaoguang
State Grid Xinjiang Electric Power Compangy, Xinjiang, China

Hassan Mohamed EL-Kamchouchi and Ahmed Samir EL-Torgoman
Electrical Engineering, Alexandria University, Alexandria, Egypt

Hossein Movahedi-Aliabad
Department of Electrical and Electronics, Quchan Technical Institute, Technical and Vocational University, Quchan, Iran

Akram Norouzi, Sepideh Soltanmoradi and Manijeh Shahi
Department of Electrical Engineering, Bojnourd Branch, Islamic Azad University, Bojnourd, Iran

Mahshid Nasserian
Department of Electrical Engineering, Ferdowsi University of Mashhad, Mashhad, Iran

Behnam Gholamrezazadeh Family, Vahid Hamiyaty Vaghef and Maryam Shabro
Communication Department, Niroo Research Institute, Tehran, Iran

Index